Biology *for the* Informed Citizen

Biology *for the* Informed Citizen

Donna M. Bozzone
Saint Michael's College

Douglas S. Green
Saint Michael's College

New York   Oxford
OXFORD UNIVERSITY PRESS

Oxford University Press is a department of the University of Oxford.
It furthers the University's objective of excellence in research,
scholarship, and education by publishing worldwide.

Oxford   New York
Auckland   Cape Town   Dar es Salaam   Hong Kong   Karachi
Kuala Lumpur   Madrid   Melbourne   Mexico City   Nairobi
New Delhi   Shanghai   Taipei   Toronto

With offices in
Argentina   Austria   Brazil   Chile   Czech Republic   France   Greece
Guatemala   Hungary   Italy   Japan   Poland   Portugal   Singapore
South Korea   Switzerland   Thailand   Turkey   Ukraine   Vietnam

For titles covered by Section 112 of the US Higher Education Opportunity
Act, please visit www.oup.com/us/he for the latest information about
pricing and alternate formats.

Published in the United States of America by
Oxford University Press
198 Madison Avenue, New York, NY 10016
http://www.oup.com

CIP data is on file at the Library of Congress
ISBN: 978-0-19-538198-6

Printing number: 9 8 7 6 5 4 3 2 1

Printed in the United States of America
on acid-free paper

For all of our teachers, especially Allison and Samantha

Contents in Brief

Contents

UNIT 1. THE SCIENTIFIC STUDY OF LIFE

UNIT 2. REPRODUCTION, INHERITANCE, AND EVOLUTION

CHAPTER 5 Cancer 127
How Can It Be Prevented, Diagnosed, and Treated?

case study   Xeroderma pigmentosum 128

CHAPTER 6. Reproduction 155
What Kind of Baby Is It?

case study   The Fastest Woman on Earth 156

CHAPTER 7 Plants, Agriculture, and Genetic Engineering 185
Can We Create Better Plants and Animals?

case study   Golden Rice 186

CHAPTER 8 Health Care and the Human Genome 217
How Will We Use Our New Medical and Genetic Skills?

case study   Carrie Buck and the American Eugenics Movement 218

CHAPTER 9 Evolution 245
How Do Species Arise and Adapt?

case study Lactose Intolerance and the Geographic Variation of Human Traits 246

CHAPTER 10 The Evolution of Disease 283
Why Do We Get Sick?

case study Deadly Malaria 284

UNIT 3. INTERACTING WITH NATURE

CHAPTER 13 Human Population Growth 373
How Many People Can a Single Planet Hold?

▤ *case study*   A Story About Bacteria 374

Preface

We wrote *Biology for the Informed Citizen* because we love biology and are convinced that everyone should have a basic understanding of biology to function as a fully engaged, contributing member of society. As we researched topics, designed our approach, and wrote, we were thinking of you—students majoring in fields other than biology. You possess perspectives, interests, dispositions, and expectations that differ somewhat from those of most students majoring in biology, and that is what makes it so much fun and so rewarding to teach you.

One of the challenges in trying to foster an understanding and appreciation of the importance of biology is that our educational system and society tend to compartmentalize science rather than seeing it as a central aspect of modern life. In this era of deep specialization, we are even more in need of conversation, communication, and understanding among specialists and one another. In reality, the integration of knowledge—not simply within biology but also among sciences, social sciences, humanities, and the arts in general—is essential for confronting and finding solutions to the challenges we all face. You have the potential to play an important role in meeting these challenges and helping to find solutions precisely because your particular interests allow you to see biology from different perspectives. And the biology you will learn will enrich your understanding of and strengthen the connections among the things you already know.

Biology for the Informed Citizen presents biology in the context of important cultural and social issues you are likely to encounter now and in the future. In writing this book, we chose to address biology in a way that will help you learn what you need to know about biology to make informed decisions in your life; become effective, engaged citizens; and understand, at least in principle, the new opportunities and challenges modern biology provides. Although you may be interested in studying biology for its own sake, we recognize that you may be most interested in the consequences of biology: what it says about your health, disease, and the environment, for example.

Although our motivation for writing this book was to teach you, along with the guidance of your course instructor, the major concepts of biology, evolution, and the process of science so that you can apply your knowledge as informed consumers and users of scientific information, we also benefited in some unexpected ways. We became more informed scientists, teachers, and parents. We hope that you have as much pleasure reading and learning as we did in creating this book for you.

Sadly, Doug Green, my husband and coauthor of this book, passed away before its completion. However, I am delighted to see the project come to fruition.

Donna Bozzone
Saint Michael's College
Colchester, Vermont

Approach: Cases, Concepts, and Consequences

Our goal for *Biology for the Informed Citizen* was to write a book that, more than any other non-science-majors biology book, helps students **connect the concepts of biology to the consequences of biology**—the consequences that students can and should see in every facet of their lives, if only they are trained to identify them. This text teaches the concepts of biology, evolution, and the process of science so that students can apply their knowledge as informed consumers and users of scientific information.

In order to help students become biologically and scientifically literate, we wove two major themes into every chapter: the process of science and the theory of evolution. Our rationale is that if students are going to learn and then apply what they have learned, they need to know **not only "what we know," but also "how we know what we know."** Therefore, each chapter includes stories of real scientists who had interests and curiosities—not altogether different from some of the students reading this text. We hope these stories will motivate students to think critically in their daily lives. Throughout this book we also emphasize the theory of evolution—the most central of all biological concepts—to help students see the big picture underlying the magnificent diversity and awe-inspiring mechanisms of the living world.

Features

Because students come to their biology courses, and this text, with a rich set of interests, we included features to help students make connections between their present knowledge and the biology they are learning.

Case Studies

Each chapter opens with a rich case study that highlights an issue or challenge with biological significance and focuses on the consequences of biology. These cases motivate the material in each chapter and demonstrate ways in which an understanding of biology can be used to make informed decisions about important issues. Examples of cases we introduce include "Sickle Cell Disease, Malaria, and Human Evolution" (Chapter 4); "The 'Infidelity Gene'" (Chapter 1), and *Xeroderma pigmentosum* (Chapter 5), which address how genes influence our health and personal relationships. In the remaining sections of the chapter, we weave in the biology needed for a fuller understanding of the issue or challenge. As a result, students learn specific biological concepts in a context that shows why they are important and enables students to make connections between biology and other fields of study.

case study

Sickle Cell Disease, Malaria, and Human Evolution

FIGURE 4.1 Malaria is caused by the parasite *Plasmodium falciparum*, which is carried by the *Anopheles* mosquito. When one of these mosquitoes bites an individual, the parasitic cells take up residence in the blood, destroying red blood cells, the cells that carry oxygen through the body.

malaria :: a disease caused by a parasite carried by the *Anopheles* mosquito; it is often fatal

protozoan :: a simple, single-celled organism

red blood cell :: a cell that carries oxygen through the blood

Tony Allison grew up in Kenya His father, a farmer, had relocated the family from England in 1919. As a boy, Tony went on long excursions with professional naturalists to observe and help collect birds for the Natural History Museum in London. He also visited the archeological excavation site of Louis Leakey, the preeminent anthropologist of his time, and became intrigued with human evolution and the relationships among the various tribes he saw in Kenya.

During one of the Allisons' holidays on the beaches of Malindi in Kenya, Tony contracted malaria; he was only 10 years old. **Malaria** is a terrible disease and often fatal. It is caused by a **protozoan**, a simple single-celled organism, and carried by a specific type of mosquito. When one of these mosquitoes bites a person, it injects this protozoan, *P. falciparum*, into the individual (FIGURE 4.1). These parasitic cells take up residence in the blood and destroy **red blood cells**, the cells that carry oxygen through the blood. There is currently no vaccine against malaria. Tony's experience with malaria led him to switch gears: rather than becoming a naturalist, he decided to become a physician.

After earning his undergraduate degree in South Africa, Tony moved to England to finish his medical training at Oxford University. Although he enjoyed medical school, he had not lost his keen interest in human evolution. Tony was convinced that there had to be a way to measure human evolutionary relationships more precisely than anthropologists did by looking only at bones.

In 1949, when Oxford University sent a group of scientists to Kenya to survey and study plants and animals all over the country, Tony jumped at

Process of Science

Biology in particular, and science in general, represent one way of asking questions and evaluating the answers; it is not the only way. Still, the specific manner in which scientists go about learning about the natural world is both powerful and successful. And as a way of thinking, it is practical for many questions, not just scientific ones. Thus, we highlight the process of science in the ***How Do We Know?*** essays featured in each chapter. These essays help students move beyond memorizing facts to get them thinking critically about how we know what we know.

All too often, people do not appreciate the human dimension of biology. Everything ever discovered or solved is literally the result of a person or many people thinking about that the question being pursued was the *most* interesting and important thing to be found. They simply *had* to work on this problem—it was like an itch that had to be scratched. ***Scientist Spotlight*** essays in each chapter provide biographical information and historical context about the real individuals whose scientific discoveries have made tremendous impacts on all of our lives.

Real-World Applications

Biology does not exist as a disconnected field of study. In fact, to understand biology well, one needs to be conversant with the ways that biology connects to the larger culture. The inverse is also true: to understand our culture fully, one needs to be familiar with biology. More specifically, biological research, ideas, and knowledge intersect with global issues, ethics, and social responsibility. ***Life Application*** essays in each chapter present real-world examples illustrating how biological knowledge has been used to help individuals and society at large make informed decisions on a range of issues.

Advances in scientific research directly affect us in our day-to-day lives. A great deal of what is known in biology and continues to be studied depends upon the development and implementation of specific methods and techniques. ***Technology Connection*** essays in each chapter provide students with information on specific methods or techniques that biologists use to answer questions. These essays show how the tools of scientific research are being used to shape the world in which we live.

How Do We Know?

Pedigree Analysis

In 1866 the physician Paul Broca wondered whether the breast cancer from which his wife suffered was hereditary. He observed a cluster of cases in his wife's own family: 10 out of 24 women had breast cancer. Broca mapped a family history, or pedigree, to see if he could figure out the pattern of inheritance. The idea that some traits "run in the family" is a very old one. Broca saw that they also run in patterns.

Today, pedigree analysis helps genetic counselors determine the risk of developing or passing on traits, such as disease. First, a counselor interviews the individual about the occurrence of the disease in other family members. The counselor then constructs a chart and applies the rules of inheritance.

Scientist Spotlight

Rosalind Franklin (1920–1958)

James Watson and Francis Crick are most closely associated with discovering the structure of DNA. One strand of the double helix is even called Watson, and the other is Crick. However, the work of other scientists was essential to their discovery. Maurice Wilkins shared the 1962 Nobel Prize in Physiology or Medicine with Watson and Crick. Another scientist, Rosalind Franklin, received little recognition.

Born in London to a wealthy family, Rosalind Franklin was interested in getting answers to questions, even as a child. She was fortunate to attend one of the few schools in London that taught chemistry and physics to girls, and she decided at age 15 to become a scientist. Franklin passed the entrance exam for Cambridge University in 1938, but her father refused to pay. He was against *any* woman attending university. Fortunately, Franklin's aunt stepped in and offered to fund her niece's education. Franklin's mother also sup-

An expert in X-ray crystallography, Rosalind Franklin (A) produced a beautiful image of crystallized DNA (B) that was essential for figuring out the structure of this molecule.

She soon discovered that crystals of DNA take two forms. When these forms mix together, it is impossible to get

---

Life Application

The Effectiveness of Genetic Screening

Scientists and physicians have the ability to screen for carriers of many genetic diseases. Yet the outcomes of screening programs are quite variable. If we measure success as a reduction in the number of infants born with a particular disorder, screening for Tay Sachs disease has been spectacularly effective, whereas screening for sickle cell disease is disappointing so far. The difference in how well the screening programs have worked shows that it is important to educate patients. It is important, too, for healthcare professionals to be culturally sensitive and aware.

Tay Sachs disease afflicts children who are born with two copies of a mutant allele. The normal allele encodes a protein called *hexosaminidase A*. This protein breaks down a type of fatty molecule called GM2, which is present in the brain and nervous system. A child with Tay Sachs dis-

In contrast, the screening program for sickle cell disease reads like a script for how *not* to do something properly. Testing of African Americans began in the early 1970s—but without consent, without community outreach, and without adequate public education. When people learned that they had the sickle cell trait, they were rarely told they were *not* going to be sick. Nor did they learn the chances that their children would develop the disease. As if this were not bad enough, individuals who tested positive were discriminated against and unable to get life or health insurance. Until 1981, the U.S. Air Force Academy refused entrance to any applicant who carried the sickle cell trait. Even after these discriminatory practices were stopped, it was too late to build the trust necessary for this program to succeed. Instead, the program's legacy was an enhanced suspicion that medical genetics has racist intentions. ::

Technology Connection

Electrophoresis

"Who done it?" On television and in the movies, DNA tests often hold the answer. You know the scenario. DNA is collected from the saliva on a cigarette butt carelessly left behind at the crime scene. If the police have a suspect and there is enough evidence, a court order can require the accused to provide a DNA sample. But how do we know whether the samples match?

One of the first steps is to chop up the DNA into manageable fragments. Chemicals called enzymes can cut a DNA molecule at specific nucleotide sequences. For example, the enzyme EcoR1 cuts the molecule between A and G everywhere the sequence GATTC appears. Because each of us has unique DNA sequences, the enzymes will produce DNA sequences of different lengths. By comparing the DNA fragments from different samples, scientists can determine whether the DNA came from the same person.

might have taken a piece of paper towel or filter paper and dotted it with food coloring. When you dip the edge of the paper in water, it absorbs fluid, which travels up the paper past the colored dots, and the colors separate. The green dot, for example, reveals that it is actually a mixture of yellow and blue. Like chromatography, electrophoresis separates mixtures of substances by the movement of molecules. Unlike chromatography, it relies on their movement through an electrical field.

With electrophoresis, a mixture of DNA fragments is placed in a well cut into the gel, a solid matrix similar in consistency to gelatin. The gel, submerged in a conductive liquid, is subjected to an electric field. Because DNA is negatively charged when in solution, it will migrate toward the positively charged end of the field. Small fragments migrate the fastest,

Pedagogy

Every chapter in *Biology for the Informed Citizen* includes carefully crafted tools to help students learn and reinforce biological concepts.

Chapter Learning Objectives at the start of each chapter (based on Bloom's taxonomy) correspond to the main headings and provide a framework for the key concepts to help students focus on what is most important.

Questions-Based Chapter Titles and Section Headings model the spirit of inquiry at the heart of the scientific process.

Simple and Clear Illustrations in each chapter help students visualize important concepts. The art program uses a consistent format to help guide students through complex processes. For example, Chapter 2 introduces the steps of the scientific method, and the figures that highlight scientific experiments throughout the book help to reinforce these steps—blue: observations and facts; purple: hypotheses; pink: predictions; light green: hypothesis testing; and dark green: evaluation and/or results. Brief ***figure captions*** provide a running summary of the chapter, reinforcing main points discussed in the text.

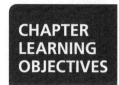

CHAPTER LEARNING OBJECTIVES

After reading this chapter, you should be able to answer the following questions:

4.1 What Is Sickle Cell Disease?
Specify the cause and consequences of sickle cell disease and the relationship between sickle cell disease and malaria.

4.2 Could Molecular Medicine Prevent Sickle Cell Disease?
Explain how Linus Pauling's research into the age of molecular medicine.

4.3 Where Is Our Genetic Information Stored?
Describe the study that disproved the concept of pangenesis.

4.4 How Did Mendel Discover the Rules of Inheritance?
Outline how Mendel conducted his experiments related to inheritance and the resulting set of rules.

4.5 How Much Do Mendel's Rules Explain?
Identify the three reasons that Mendel's rules fail to explain inheritance completely.

FIGURE 4.11 In the 1880s, August Weismann performed an experimental test of pangenesis using mice. Weismann concluded that the information to construct the tails of the mice did not reside in the tail itself.

FIGURE 2.7 Every event or outcome in nature has a cause or source. A scientist can learn about causes by observing the effects that occur; note that in this figure all of the causes on the left contribute to the effects on the right.

A ***Marginal Glossary*** defines key terms in the margins of the pages on which the terms appear, so students can easily find definitions and explanations when preparing for exams.

allele :: a different form of a gene

Each chapter concludes with a ***Biology in Perspective*** section that places the chapter concepts in larger context.

Bulleted ***Chapter Summaries*** at the end of each chapter are organized around the chapter learning objectives and highlight and reinforce the main concepts.

Chapter Summary

4.1 What Is Sickle Cell Disease?
Specify the cause and consequences of sickle cell disease and the relationship between sickle cell disease and malaria.
- Sickle cell disease is a severe inherited genetic disorder.
- In sickle cell disease, red blood cells that are normally disc shaped and flexible become rigid and pointy shaped.
- Sickled cells get stuck in blood vessels, which can lead to blood clots and other serious physical consequences, including kidney failure, paralysis, and heart failure.
- The inheritance of the gene associated with sickle cell disease also confers resistance against malaria.

4.2 Could Molecular Medicine Prevent Sickle Cell Disease?
Explain how Linus Pauling's research into sickle cell disease ushered in the age of molecular medicine.
- Linus Pauling and his colleagues examined two types of hemoglobin (HbA and HbS).

4.4 How Did Mendel Discover the Rules of Inheritance?
Outline how Mendel conducted his experiments related to inheritance and the resulting set of rules.
- Mendel chose his experimental organism carefully: garden peas, which are simple to cultivate and whose mating can be controlled manually.
- Mendel focused on one trait at a time and followed the cross for more than one generation, collecting quantitative data and keeping detailed records.
- Mendel's research revealed some basic rules of inheritance.
  o Genes can come in more than one form, or allele.
  o Alleles of a particular gene sort individually into gametes during meiosis.
  o Certain traits are dominant, while others are recessive.

4.5 How Much Do Mendel's Rules Explain?
Identify the three reasons that Mendel's rules fail to explain inheritance completely.

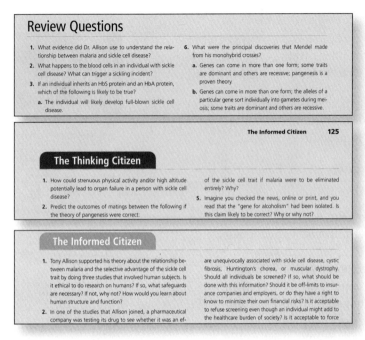

Basic multiple choice and short-answer **Review Questions** at the end of each chapter ask students to recall core information presented in the chapter. Answers to the multiple choice questions appear at the end of the book.

The Thinking Citizen advanced questions at the end of each chapter ask students to think critically and analytically about the main chapter concepts.

The Informed Citizen advanced questions at the end of each chapter ask students to apply biological concepts to relevant cultural and social issues.

Organization

Biology for the Informed Citizen covers the **foundational concepts** that comprise a standard non-science-majors biology course but does so on a "need-to-know" basis, placing biological topics within the context of important cultural and social issues, but **without excessive detail**. We thought carefully about which topics to include and which to omit, with the goal of providing the needed biological coverage in a framework that we hope students will enjoy reading! This book is organized into three units. (*Biology for the Informed Citizen* is also available *"with Physiology"*: including five chapters on homeostasis; circulation and respiration; the nervous system; infectious disease and the immune system; and nutrition, activity, and wellness.)

Unit 1: The Scientific Study of Life

In the first unit, we introduce the main themes of the text. In **Chapter 1**, we address the relevance of biology to informed citizenship, how biology affects our view of nature, ourselves and our society, and we also introduce the main features of life. Chapter 1 emphasizes the centrality of evolution in biology, an emphasis we maintain throughout the text. **Chapter 2** addresses the importance of science to the study of biology and medicine. We discuss the scientific method in theory and by example, and make several points about the philosophy of science. Finally, we compare science with other ways of knowing and contrast science and pseudoscience.

Unit 2: Reproduction, Inheritance, and Evolution

In the second unit, we explore the molecular, cellular, and evolutionary basis of life. The first three chapters provide foundational information about cells, genes, and inheritance. In **Chapter 3**, we use human development, from fertilization to childbirth, to introduce our discussion of cell biology. **Chapter 4** explores how inheritance works and how the molecules DNA, RNA, and protein contribute to the production of the physical and functional characteristics of organisms, including those seen in sickle cell disease. In **Chapter 5**, we address cancer and use it to motivate our discussion of how cell division is regulated.

In the next three chapters we emphasize topics related to informed citizenship. These topics include: fertility and genetic screening, cloning and genetic engineering, and gene therapy and stem cells. **Chapter 6** examines reproduction. We think that understanding the biology of reproduction is essential for both responsible family planning and stewardship of our natural resources. Our discussion of genetically engineering plants in **Chapter 7** provides an opportunity to introduce plant structure and photosynthesis. **Chapter 8** addresses biological determinism and the extent to which genes explain complex human characteristics.

The unit concludes with a closer examination of evolution, addressing the evolutionary basis of medicine and health. In **Chapter 9**, we complete the story of what evolution is and how it works, pointing out interesting and puzzling aspects of the natural world, and showing how they can be accounted for by evolution operating over long periods of time. In **Chapter 10**, we examine why it is that we get sick, how diseases evolve, and what evolution can tell us about how diseases can be prevented and controlled.

Unit 3: Interacting with Nature

The final unit looks at ecology. In **Chapter 11**, we examine how individuals of different species interact with one another and with nature and how those interactions drive an ecosystem's function. **Chapter 12** discusses biodiversity and the impact humans have on biodiversity and the impact of biodiversity on us. Finally, in **Chapter 13**, we examine human population growth and its consequences. Advances in agriculture, medicine, engineering, and technology have supported healthy population growth for most of our history, but the continued growth of the human population is unsustainable. Because we can recognize the problem, however, we can also discover ways to solve it.

Learning Package

Oxford University Press offers instructors and students a comprehensive ancillary package, designed to help students become fully informed citizens and to assist instructors in meeting this objective. The following resources are available for qualified adopters of *Biology for the Informed Citizen*.

For Students

Companion Website at www.oup.com/us/bozzone
The FREE and OPEN companion website offers a number of study tools, including online quizzes (over 1000 questions) and a curated guide to relevant animations, videos, podcasts, and more.

Biology for the Informed Citizen Dashboard Online Homework
Dashboard delivers quality content, tools, and assessments to track student progress in an intuitive, web-based learning environment. Assessments created by Everett Weber, Brandi King, and Heather Passmore (Murray State University) are designed to accompany *Biology for the Informed Citizen* and automatically graded so instructors can easily check students' progress as

they complete their assignments. The color-coded gradebook illustrates at a glance where students are succeeding and where they can improve so on-the-fly instructors can adapt lectures to student needs. For students, this means quality content and instant feedback. Dashboard features a streamlined interface that connects instructors and students with the functions they perform most, and simplifies the learning experience by putting student progress first. All Dashboard content is engineered to work on mobile devices, including the iOS platform. Our goal is to create a platform that is simple, informative, and mobile. For more information about Dashboard please visit www.oup.com/dashboard.

Study Guide

Authored by Sharon Gilman (Coastal Carolina University), who is also the author of the test item file (see below for more info), the Study Guide provides students with brief summaries and step-by-step analyses of each chapter, additional review questions, and thoughtful advice and study tips. (ISBN 9780199958078)

For Instructors

In addition to Dashboard described above, qualified adopters of *Biology for the Informed Citizen* can access the following teaching tools for immediate download at the companion website (**www.oup.com/us/bozzone**):

Text Image Library

All text images are available in PowerPoint and jpeg formats. Instructors who adopt the text gain access to every illustration, photo, figure caption, and table from the text in high-resolution electronic format, and some multipart figures are optimized by being broken down into their constituent parts for clear projection in large lecture halls. The image library includes:

- **Art slides** in PowerPoint and jpeg formats with figures exactly as they appear in the text
- **Unlabeled art slides**, in which text labels are turned off
- **Lecture note slides** with outlines for each chapter that can be edited, which makes preparing lectures faster and easier than ever.

Instructor's Resource Manual with Video and Animation Guide

The Instructor's Resource Manual is a collection of materials designed to help instructors build and implement a course around *Biology for the Informed Citizen*. The manual includes several kinds of supplemental instructional aids ranging from a list of chapter learning outcomes, to full chapter outlines and summaries, to question prompts for in-class discussions and activities. At the heart of the manual is a *curated and annotated guide to high-quality and freely available animations, movie clips, videotaped lectures, podcasts, and presentations* of core concepts covered in the text, all vetted and collected in one place for convenient access. The Video and Animation Guide for each chapter includes a web link to a customized YouTube playlist, which includes several relevant videos that highlight, illustrate, and expand on the concepts covered in the text.

Test Item File

Written by Sharon Gilman (also author of the Study Guide), the test item file includes over 1200 multiple choice, true/false, and short-answer questions in editable Microsoft Word format. Questions are organized by chapter section number and learning objectives, and each item is identified according to Bloom's taxonomic categories of knowledge, comprehension, and application.

Computerized Test Bank

Using the test authoring and management tool Diploma, the computerized test bank that accompanies this text is designed for both novice and advanced users. Diploma enables instructors to create and edit questions, create randomized quizzes and tests with an easy-to-use drag-and-drop tool, publish quizzes and tests to online courses, and print quizzes and tests for paper-based assessments. Available on the Ancillary Resource Center (ARC).

Ancillary Resource Center (ARC)

The Ancillary Resource Center (ARC) is a convenient, instructor-focused single destination for resources to accompany your text. Accessed online through individual user accounts, the ARC provides instructors access to up-to-date ancillaries at any time while guaranteeing the security of grade-significant resources. In addition, it allows OUP to keep instructors informed when new content becomes available. The ARC for *Biology for the Informed Citizen* includes the Test Item File, the Computerized Test Bank, the Text Image Library, and the Instructor's Resource Manual with Video and Animation Guide. For more information about ARC please visit www.oup-arc.com.

Ebook

Available through CourseSmart.

Acknowledgments

There are many people who deserve heartfelt thanks for their support and help throughout the creation of this book. These generous individuals fall into four categories: family, friends, the wonderful people at Oxford University Press, and the many individuals who reviewed chapters and artwork throughout many stages of this project. In the family category, we are grateful for the encouragement, support, and love of our daughters, Samantha and Allison. Their interests, good questions, perspectives, and suggestions made this a better book. Also, they never lost patience—at least not noticeably—with their distracted parents. We would also like to thank Bill and Janet Bozzone for their moral support and good dinners too.

All of our friends in the biology department at Saint Michael's deserve a thank you for good discussions and for making it a lovely thing to come to work. Denise Martin, Declan McCabe, and Doug Facey went above and beyond the call of duty in ways that are known to them and difficult to describe. Along with those three, my friends and colleagues in our education department Valerie Bang-Jensen and Mary Beth Doyle kept me going. We are also appreciative of the support that Saint Michael's College provided (so much so that the College is included in the friend category), especially a sabbatical leave for Doug in fall 2009 and one for Donna in spring 2010.

The team at Oxford University Press was extraordinary. Jason Noe, senior editor, in his cheerful, tenacious way kept this project moving forward with the right combination of encouragement (prodding) and keeping hands off. He is terrific. John Haber, our initial development editor, was instrumental in helping us to say what we were trying to convey in a more lucid and pleasing way. When he left OUP, we felt a bit panicked, but fortunately our fears turned out to be utterly unfounded. Lisa Sussman, senior development editor, stepped in and she has been fantastic—a gifted editor and talented writer in her own right as well as indispensable for the development of the *Biology for the Informed Citizen* art program. We would like to thank the editorial assistants who helped us over the years of developing the text, including Melissa Rubes, Katie Naughton, Caitlin Kleinschmidt, and Andrew Heaton, along with executive assistant Ross Yelsey. We would also like to thank Patrick Lynch, editorial director; John Challice, vice president and publisher; Jason Kramer and David Jurman, marketing managers; Christine Naulty and Meghan Daris, marketing assistants; Frank Mortimer, director of marketing; Jolene Howard, market development; and Bill Marting, national sales manager. We would also like to thank the exceptional and dedicated production team who helped take this book from ideas and drafts to a final, published reality, including, Lisa Grzan, production manager; Barbara Mathieu, senior production editor; Michele Laseau, art director; and the team at Precision Graphics.

Reviewers, Class Testers, and Focus Group Participants

We are especially grateful to the extraordinary group of dedicated colleagues teaching non-science majors who provided thoughtful commentary as reviewers, class testers, and focus group participants as we developed the text's manuscript, illustrations, and supplements program. We started this journey with the goal of providing instructors with a set of tools that could help them reach out to their students. We wanted students to see how by becoming scientifically literate, they could improve their lives and those of their fellow citizens. Your comments and suggestions were invaluable in that effort and in helping us to refine the final version of the first edition.

Reviewers

Over the course of development, we extensively reviewed *Biology for the Informed Citizen* at 145 colleges and universities, with approximately 175 reviewers. We read each review and incorporated feedback wherever we could in order to develop this first edition so it would be the best option for you and your students. We extend our heartfelt appreciation to the following reviewers:

Manuscript Reviewers

Sylvester Allred	Northern Arizona University
Tara Devi S. Ashok	University of Massachusetts Boston
Yael Avissar	Rhode Island College
Ellen Baker	Santa Monica College
Andrew Baldwin	Mesa Community College
Roberta Batorsky	Middlesex County College
Emma Benenati	Northern Arizona University
Morgan Benowitz-Fredericks	Bucknell University
Brenda Bourns	Seattle University
Mark Buchheim	The University of Tulsa
Sara Carlson	The University of Akron
Aaron Cassill	The University of Texas at San Antonio
Deborah Cato	Wheaton College
Michelle Cawthorn	Georgia Southern University
Thomas T. Chen	Santa Monica College
Thomas F. Chubb	Villanova University
John L. Clark	University of Alabama
Michael F. Cohen	Sonoma State University
Claudia Cooperman	University of South Florida
James W. Cosgrove	Montgomery College
Helen Cronenberger	The University of Texas at San Antonio
Michael S. Dann	Pennsylvania State University
Paula Raelynn Deaton	Sam Houston State University
Buffany DeBoer	Wayne State College
Leif D. Deyrup	University of the Cumberlands
Hartmut Doebel	The George Washington University
Paul Farnsworth	University of New Mexico
Michele Finn	Monroe Community College
Susan W. Fisher	Ohio State University
Brandon Lee Foster	Wake Technical Community College
Lori Frear	Wake Technical Community College
Wendy Jean Garrison	University of Mississippi
Vaughn M. Gehle	Southwest Minnesota State University
Sharon L. Gilman	Coastal Carolina University
Mary Gobbett	University of Indianapolis
Brandon Groff	Eastern Michigan University
Laine Gurley	Harper College
Kristy Halverson	The University of Southern Mississippi
Janelle Hare	Morehead State University
Mesha Hunte-Brown	Drexel University
Allison Hunter	Worcester Polytechnic Institute
Evelyn F. Jackson	University of Mississippi
Arnold Karpoff	University of Louisville
Christopher J. Kirkhoff	McNeese State University
Peter Kourtev	Central Michigan University
Jeff Kovatch	Marshall University
Ellen Shepherd Lamb	The University of North Carolina at Greensboro
Ann S. Lumsden	Florida State University
Molly MacLean	University of Maine
Lisa Maranto	Prince George's Community College
Karen McCort	Eastern New Mexico University, Ruidoso Branch Community College

Diane L. Melroy	University of North Carolina Wilmington
Tim Metz	Campbell University
Scott M. Moody	Ohio University
Brenda Moore	Truman State University
Cynthia E. Morgan	Austin Peay State University
Mario Mota	University of Central Florida
Ann Murkowski	North Seattle Community College
Rajkumar Nathaniel	Nicholls State University
Fran Norflus	Clayton State University
Paul Eugene Olson	University of Central Oklahoma
David Oppenheimer	University of Florida
Wiline Pangle	Central Michigan University
Mark A. Paulissen	Northeastern State University
Ashley Ramer	The University of Akron
Terri S. Richardson	California State University, Northridge
Darryl Ritter	Northwest Florida State College
Pamela Sandstrom	University of Nevada, Reno
Georgianna Saunders	Missouri State University
Malcolm Schug	The University of North Carolina at Greensboro
Caryn Self-Sullivan	Georgia Southern University
Alice Sessions	Austin Community College
Justin Shaffer	North Carolina Agricultural and Technical State University
Edward A. Shalett	Mesa Community College
Jack Shurley	Idaho State University
Ayodotun Sodipe	Texas Southern University
Bethany B. Stone	University of Missouri
Mark Thogerson	Grand Valley State University
Jeffrey Thomas	Queens University of Charlotte
William Unsell	University of Central Oklahoma
Sheela Vemu	Northern Illinois University
Janet Vigna	Grand Valley State University
Timothy S. Wakefield	John Brown University
Kristen L.W. Walton	Missouri Western State University
Everett Weber	Murray State University
Aimee K. Wurst	Lincoln University

Illustration Reviewers

Sylvester Allred	Northern Arizona University
Megan Anduri	California State University, Fullerton
Yael Avissar	Rhode Island College
Emma Benenati	Northern Arizona University
Gayle Birchfield	Austin Peay State University
Lisa G. Bryant	Arkansas State University Beebe
Mark Buchheim	The University of Tulsa
Sara Carlson	The University of Akron
Aaron Cassill	The University of Texas at San Antonio
Deborah Cato	Wheaton College
Hartmut Doebel	The George Washington University
Lori Frear	Wake Technical Community College
Wendy Jean Garrison	University of Mississippi
Vaughn M. Gehle	Southwest Minnesota State University
Renaud Geslain	College of Charleston
Brandon Groff	Eastern Michigan University

Arnold Karpoff   University of Louisville
Janet E. Kübler   California State University, Northridge
Diane L. Melroy   University of North Carolina Wilmington
Krista Peppers   University of Central Arkansas
Malcolm Schug   The University of North Carolina at Greensboro
Justin Shaffer   North Carolina Agricultural and Technical
                              State University
Jack Shurley   Idaho State University
Ayodotun Sodipe   Texas Southern University
Bethany B. Stone   University of Missouri
Jeffrey Thomas   Queens University of Charlotte
Kristen L.W. Walton   Missouri Western State University

Supplements Program Reviewers

Kim Atwood   Cumberland University
Yael Avissar   Rhode Island College
Roberta Batorsky   Middlesex County College
Erin Baumgartner   Western Oregon University
Sara Carlson   The University of Akron
Deborah Cato   Wheaton College
Buffany DeBoer   Wayne State College
Hartmut Doebel   The George Washington University
Lori Frear   Wake Technical Community College
Seth Jones   University of Kentucky
Peter Kourtev   Central Michigan University
Brenda Leady   University of Toledo
Rajkumar Nathaniel   Nicholls State University
Justin Shaffer   North Carolina Agricultural and Technical
                              State University
Jeffrey Thomas   Queens University of Charlotte
Jennifer Wiatrowski   Pasco-Hernando Community College
Dwina W. Willis   Freed-Hardeman University

Market Development Reviewers

We would also like to thank the extensive list of faculty at over eighty schools who participated in our first-edition market development review program for sharing their insights and suggestions.

Fernando Agudelo-Silva   College of Marin
Justin Anderson   Radford University
Kim Atwood   Cumberland University
Melissa Barlett   Mohawk Valley Community College
Katrinka Bartush   University of North Texas
Tonya Bates   University of North Carolina at Charlotte
Jane Beers   John Brown University
Tiffany Bensen   University of Mississippi
Donna Bivans   Pitt Community College
Michelle Boone   Miami University
Steven Brumbaugh   Green River Community College
Rebecca Bryan   Metropolitan State University of Denver
Lisa Bryant   Arkansas State University Beebe
Stylianos Chatzimanolis   University of Tennessee at Chattanooga
Peter Chen   College of DuPage
Catherine Chia   University of Nebraska Lincoln

Claudia Cooperman	University of South Florida
Jacquelyn Duke	Baylor University
Kari Eamma	Tarrant County College Northeast
Deborah Gelman	Pace University
Sandra Gibbons	Moraine Valley Community College
Paul Gier	Huntingdon College
Mary Gobbett	University of Indianapolis
Andrew Goliszek	North Carolina A&T State University
Laine Gurley	Harper College
Judy Haber	California State University Fresno
James Harper	Sam Houston State University
Olivia Harriott	Fairfield University
Timothy Henkel	Valdosta State University
Kelley Hodges	Gulf Coast State College
Jill Holliday	University of Florida
Sue Hum-Musser	Western Illinois University
Carl Johansson	Fresno City College
Scott Johnson	Wake Technical Community College
Staci Johnson	Southern Wesleyan University
Claudia Jolls	East Carolina University
Barbara Juncosa	Citrus College
Brenda Leady	University of Toledo
Sarah Leupen	University Of Maryland Baltimore City
David Luther	George Mason University
Paul Luyster	Tarrant County College
James Malcolm	University of Redlands
Christiane Meyer Healey	University of Massachusetts Amherst
Liza Mohanty	Olive-Harvey College
Jamie Moon	University of North Florida
Rajkumar Nathaniel	Nicholls State
Dana Newton	College of The Albemarle
Trent Nguyen	Tarrant County College
Fran Norflus	Clayton State University
Brandi Norman	University of North Carolina at Pembroke
Katrina Olsen	University Of Wisconsin Oshkosh
Mark Pilgrim	Lander University
Nicola Plowes	Mesa Community College
Mary Poffenroth	San Jose State University
Gerald Posner	Broward College
Vanessa Quinn	Purdue University North Central
Erin Rempala	San Diego City College
Angel Rodriguez	Broward College
Rob Ruliffson	Minneapolis Community and Technical College
Lynn Rumfelt	Gordon State College
Ann Rushing	Baylor University
Michael Rutledge	Middle Tennessee State University
Anna Schmidt	University of Wisconsin Platteville
Erik Scully	Towson University
Jyotsna Sharma	University of Texas at San Antonio
Heidi Sleister	Drake University
Marc Smith	Sinclair Community College
Ayodotun Sodipe	Texas Southern University
Kathryn Spilios	Boston University
Sonja Stampfler	Kellogg Community College
Louise Steele	Kent State University

Zuzana Swigonova	University of Pittsburgh
Pamela Thinesen	Century College
Rani Vajravelu	University of Central Florida
Sheela Vemu	Northern Illinois University
Paul Verrell	Washington State University
Stephen Wagener	Western Connecticut State University
Lisa Weasel	Portland State University
Jennifer Wiatrowski	Pasco-Hernando Community College
Dwina Willis	Freed-Hardeman University
Kimberly Zahn	Thomas Nelson Community College
Martin Zahn	Thomas Nelson Community College
Ted Zerucha	Appalachian State University

Class Testers

Over the course of development, we extensively class-tested *Biology for the Informed Citizen* at eighteen schools with eighteen instructors and 400 student participants. We thank all of them sincerely for their involvement, comments, and enthusiasm, which helped shape the final version of the first edition.

Michael Anders	Tarrant County College
Gayle Birchfield	Austin Peay State University
William Blaker	Furman University
Karen Bledsoe	Western Oregon University
Brenda Bourns	Seattle University
Yavuz Cakir	Benedict College
Sarah Cooper	Arcadia University
James Courtright	Marquette University
Kenneth Filchak	University of Notre Dame
Lori Frear	Wake Technical Community College
Andrew Goliszek	North Carolina A&T State University
MaryKate Holden	Greensboro College
Staci Johnson	Southern Wesleyan University
Jennifer Landin	North Carolina State University
Jonas Okeagu	Fayetteville State University
Brian Shmaefsky	Lone Star College Kingwood
Ayodotun Sodipe	Texas Southern University
Aaron Sullivan	Houghton College

Focus Group Participants

We would also like to thank those individuals who participated in our focus group activities at the NABT (National Association of Biology Teachers) meetings:

Dallas, Texas, 2012:

Erin Baumgartner	Western Oregon University
Claudia Womack Cash	Tarrant County College Northeast Campus
Sehoya Cotner	University of Minnesota
Kathy Gallucci	Elon University
Kristy Halverson	University of Southern Mississippi
Mickey Laney-Jarvis	Rogue Community College
Sunita Rangarajan	Collin College
David Tanner	University of North Texas at Dallas
Lisa Turnbull	Lane Community College

Atlanta, Georgia, 2013:

Pieter deHart	Virginia Military Institute
Maria Fernandez	Georgia Gwinnett College
Kathy Gallucci	Elon University
Tammy Goulet	University of Mississippi
Kristy Halverson	University of Southern Mississippi
Joseph Hawkins	College of Southern Idaho
Jennifer Kneafsey	Tulsa Community College
Sharon Lee-Bond	Northampton Community College
Melissa Masse	Tulsa Community College
Fran Norflus	Clayton State University
Samantha Parks	Georgia State University
Kim Purcell	Lander University
Eileen Roark	Central Connecticut State University
Theus Rogers	Georgia Gwinnett College
Danilo Sanchez	San Joaquin Delta College
Denise Shipley	Mountain View College
Brian Shmaefsky	Lone Star College Kingwood
Patricia Smith	Valencia College
Michelle Taliaferro	Auburn University at Montgomery
Sue Thomson	Auburn University at Montgomery
Shalini Upadhyaya	J. Sargeant Reynolds Community College
Daniel Ward	Waubonsee Communtiy College

Biology *for the* Informed Citizen

CHAPTER LEARNING OBJECTIVES

After reading this chapter, you should be able to answer the following questions:

1.1 How Does Biology Affect Your Life?
Explain how understanding biology can help you make informed decisions as a citizen.

1.2 What Are the Features of Life?
Use the four principal features of life to discriminate living things from nonliving things.

1.3 How Do Organisms Function?
Describe the relationship among atoms, macromolecules, and cells, and explain how enzymes maintain cell function.

1.4 How Do Organisms Reproduce?
Describe the relationship among DNA, genes, and alleles, and compare how genes are inherited in sexual and asexual reproduction.

1.5 How Does Life Evolve?
Create a scenario describing how a population of organisms might adapt to a new environment.

1.6 What Patterns of Diversity Are Found in Nature?
Explain the connection among common ancestors, the diversity of environments on Earth, and the scientific classification system.

The natural world is full of wonders, mysteries, and surprises.

CHAPTER 1

The Nature of Biology and Evolution

Why Does Biology Matter to You?

Becoming an Informed Citizen . . .

Biological research advances at an unprecedented rate. Breakthroughs change our society and our lives, providing us with many opportunities. However, they also raise moral, ethical, and personal challenges. To understand and evaluate these challenges, you must understand the science behind them.

utterflies are beautiful—but have you ever wondered why they are so brightly colored? Eating an apple or a cookie can give you the jolt you need to get going. How does your body use the energy stored in the foods that you eat so that you can function? People say you look a lot like your mother but not so much like your father. Why? The blind mole rat never sees a thing. Not only is it blind, it lives its entire life underground. But why does it still have eyes—and why are they completely covered with skin? And why are there over 1 million species of beetles but only four kinds of great apes, the group to which we belong?

The natural world is full of puzzles, mysteries, and surprises. Nature has inspired poets, artists, scientists, and explorers. Biology is your key to understanding nature. This book seeks to satisfy your curiosity about biology and the natural world.

But curiosity is only one reason for studying biology. Biological research advances at an unprecedented rate. Breakthroughs change our lives and our society. They provide us with many opportunities, but they also raise moral, ethical, and personal challenges. If you want to understand and evaluate these challenges, you will need to understand the science behind them. This book introduces you to concepts that will help you understand how biology plays an increasingly important role in your life.

To illustrate this role, consider John and Mary Smith: a married couple who suddenly realized they needed to know more biology.

case study

The "Infidelity Gene"

John and Mary Smith met in college and married soon after graduation. They were lucky enough to find good jobs in the same area, bought a condo, and settled into a comfortable life.

Eventually John and Mary decided to start a family. They heard about DNA testing: a new family planning procedure using DNA samples to screen for possible genetic differences that could affect their child's health and development. This sounded like a wise precaution, and they ordered the tests. When the results came back, John and Mary were relieved to learn that they each had a clean bill of genetic health as far as their future children were concerned. There was one personal glitch, however: the tests revealed that John had the "infidelity gene" (**FIGURE 1.1**).

Mary had never heard of an infidelity gene. In fact, Mary wasn't entirely sure what a gene was, and she had no idea how—or whether—a gene could cause her husband to be unfaithful. Mary consulted her doctor and searched online to find out more about this gene, but what she read was confusing. The following is a sample of her findings.

The term "infidelity gene" is the popular name for a gene called *AVPR1a*. (The names of human genes are always capitalized and italicized.) *AVPR1a* controls the production of a protein known as the *arginine vasopressin receptor*. This protein is found on the membranes of nerve cells in the brain and is responsible for transporting the hormone *arginine vasopressin* from the

FIGURE 1.1 The term "infidelity gene" is the popular name for an allele of the *AVPR1a* gene, which controls production of a protein responsible for transporting the hormone arginine vasopressin from the bloodstream to the nerve cell. This particular allele produces less protein and so transports less hormone, which some scientists believe results in less interest in a mate.

bloodstream to the nerve cell. Several versions of the *AVPR1a* gene are found in humans. Different versions of the same gene are called **alleles**. John's particular allele of *AVPR1a* made less protein than other alleles, which means that less of the hormone made it to John's brain cells. Some scientists conclude that less hormone means less interest in a mate, which could lead to infidelity.

Mary understood very little of what she learned about *AVPR1a*. How do genes control protein production? What are hormones? Why do hormones need receptors? What do hormones do inside your brain? The one thing Mary did understand, however, was that scientific studies showed that males with this particular allele of *AVPR1a* are more likely to commit adultery. Was their marriage in trouble?

allele :: different version of the same gene

1.1 How Does Biology Affect Your Life?

For most people, Mary's story will raise more questions than answers—just as it did for her. Can genes really control your behavior this way? (For further discussion on this topic, see *Technology Connection: Identifying the "Infidelity Gene."*) If you plan a family, should you get screened for possible genetic problems? What will you do if you learn of any? Do other genes affect our behavior

Technology Connection

Identifying the "Infidelity Gene"

Our case study is hypothetical, but the infidelity gene is real—though whether it affects humans is an open question.

The *AVPR1a* gene controls the transport of arginine vasopressin, a hormone, from the blood to the brain. Sexual activity and other social interactions between individuals release this hormone in the brain, which helps to make you feel good. Scientists believe this hormone helps to strengthen the pair bond between two individuals. If this is true, then individuals with alleles of *AVPR1a* that transport less of the hormone to brain cells may experience a weaker pair bond. Such individuals may therefore be more likely to seek sexual gratification outside the pair bond.

The original scientific work on *AVPR1a* was done on several species of *voles*, a mouselike organism, not on humans. The prairie vole is *monogamous*: one male and one female form a pair bond and mate only with each other. The meadow vole and the montane vole are *promiscuous*: both sexes mate several times with different partners, and no pair bond forms. When researchers examined the *AVPR1a* gene in these species, they discovered that the monogamous species and the promiscuous species had different versions (called alleles) of the gene. The popular press picked up on this result and labeled the allele found in the promiscuous species the "infidelity gene."

Researchers are now beginning to study humans to see whether versions of *AVPR1a* alleles have the same effect in us as they do in voles. In 2008, researchers studied 1000 couples in Sweden who had been together at least five years. Couples who reported being in a less intimate or less stable relationship were also more likely to have an allele of *AVPR1a* that transports less of the hormone to the brain cells. In fact, this allele is similar to the one found in promiscuous species of voles.

A second study examined game-playing behavior instead of potential infidelity. Scientists invited 200 subjects to play certain games. In this case, the scientists found that the

Prairie voles are monogamous; one male and one female form a pair bond and mate only with each other.

Meadow voles are promiscuous; both sexes mate several times with different partners and no pair bond forms.

The original scientific work on *AVPR1a* was done on several species of *voles*, a mouselike organism.

"infidelity" allele was associated with increased levels of selfish, merciless game playing. This result has led the popular media to give the allele a second name, the "ruthless gene." Neither name is a scientific term, and, to add to the confusion, both should be identified as an allele of a gene, not the gene itself.

A quick Internet search will yield thousands of hits for the infidelity gene or the ruthless gene, and even the prestigious scientific publication *Nature* has succumbed to calling these alleles by their popular names. And there really is a company that will test your DNA (or that of your significant other) to determine whether you have the allele—only US$99 as of this writing!

Despite the popular attention, these studies are at best preliminary. A relatively small number of people have been tested, and the studies have not been repeated. As we will see in Chapter 2, the scientific method relies on studies that can be repeated in order to see how widely applicable the results are. We will also see that an *association* between an allele and a behavior, even if it clearly exists, does not prove that the allele *causes* the behavior. Whether a test for this particular allele of the *AVPR1a* gene provides useful information is still an open question. ::

and health? Why do we have the genes we do—and why would anyone have genes that appear to do more harm than good?

Advances in biology have given us new abilities and a lot more information about our lives. But these advances require us to learn more if we want to use them effectively. One of the main goals of this book is to help you understand biology better and enable you to make better decisions about how to use your new biological skills. In later chapters, we will discuss in greater detail genes, hormones, receptors, and other questions raised here. For now, try putting yourself in John and Mary's place. What do you know, what do you need to know, and what actions would you take?

During your life, you'll face many questions about your personal health: diet, exercise, injuries, diseases and ways to treat them, and eventually (and sadly) questions about aging, cancer, cardiovascular problems, and end-of-life choices. Scientists and medical researchers will continue to discover new information that will affect the answers to all these questions. Of course, no one can keep up with the discoveries in all these areas. But luckily you won't have to go it alone. For all these decisions, you'll be working with doctors and counselors who can advise you about your choices. Even so, understanding the biology behind the choices will help you make decisions that are right for you.

Biology affects more than your personal life, however. Every year science and biology issues raise questions that affect your public life, life in your community, in your country, and on the planet. The Internet provides a wealth of information that is just a click away, a resource undreamed of a decade or two ago. You can consult newspapers, television broadcasts, blogs, and many other sources. But ultimately you need to evaluate what you learn and make responsible decisions that contribute positively to your life and the lives of others.

Suppose, for example, that you serve on a jury, and DNA evidence plays a large role in the case. Your understanding of the biology of DNA could affect whether you send a person to prison for life. Suppose you sit on the school board for your local district, and a citizen's group petitions to have intelligent design taught alongside evolution in science class. Your decision will influence the education of hundreds or even thousands of children (**FIGURE 1.2**). Is there a right to life, or a right to die? Should we support stem cell research or genetic

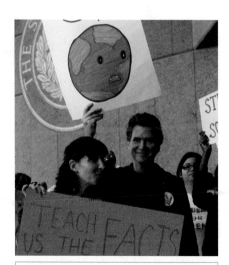

Science writers display signs in advance of a hearing in 2008 by the Texas State Board of Education. The Board was preparing to hear testimony on school standards that would encourage teaching alternative explanations for evolution.

The State of State Science Standards 2012, Thomas Fordham Institute:
http://www.edexcellencemedia.net/publications/2012/2012-State-of-State-Science-Standards/2012-State-of-State-Science-Standards-FINAL.pdf

FIGURE 1.2 A 2012 analysis of science standards in education gave a majority of states mediocre to awful grades. One of the criteria used to determine the grades was whether or not state standards undermined evolution.

FIGURE 1.3 The diversity of life is astounding. Organisms come in an amazing array of shapes, sizes, and complexity.

engineering, gay marriage, or DNA databanks? Should we allow the hunting of top predators, such as wolves, or allow farmers to plant genetically modified crops?

Your understanding of biology can influence how you exercise your most fundamental right and responsibility as a citizen: your vote. You might not vote on these issues directly, but you certainly will vote for legislators who will take positions on them and who will control funding for different avenues of scientific research. Good citizens take an active role in making decisions that affect their community, and making good decisions requires that you be informed about the issues.

The issues you will confront are complex and multifaceted—involving politics, economics, ethics, and other disciplines. As biological research advances, more and more of these complex issues will involve biology as well. We have written this book to help you understand the biology that underlies these issues. We hope that your understanding will lead you to exercise good citizenship by making informed decisions about the biological issues that affect you and your community.

The social consequences of biology are important, but biology is fascinating on its own as well. Our planet is home to an astonishing collection of living organisms (**FIGURE 1.3**). Life comes in every size, shape, and color, and the ways that organisms go about making their living are equally diverse. In fact, diversity is one of the main features of life. In spite of this diversity, living organisms have many features in common. Understanding the diversity and common features of life is in fact the subject of the theory of evolution, as we will see in this chapter, and it is the first step to understanding biology.

 # What Are the Features of Life?

Most complicated concepts are difficult to define, and life is no exception. Because life is so diverse, almost any definition will have some exceptions. A better approach is to *describe* life, not *define* it. We can describe life best by describing the principal features that all living things have in common (**FIGURE 1.4**) and the characteristics of life itself that arise as a result:

- *Metabolism:* All living organisms take in energy and matter (i.e., food) and use them in a controlled way to promote their responsiveness, survival, and reproduction.
- *Inheritance and reproduction:* All living organisms can reproduce, and each organism inherits biological information from its parents or parent.
- *Evolution:* All living organisms are related, and all species change over time. Through **evolution**, new species arise and populations adapt to their environments. Evolution is the central feature of life because *all* the other features are the result of evolutionary change.
- *Diversity:* All living things have evolved into distinct, unique forms, but this diversity contains patterns that confirm life's evolutionary past.

evolution :: theory that explains how all living organisms are related and how existing populations adapt to their environments and new species arise

This book examines each of life's main features in detail. In the remaining sections of this chapter, we will lay the groundwork by introducing each feature and explaining why it is important for understanding biology.

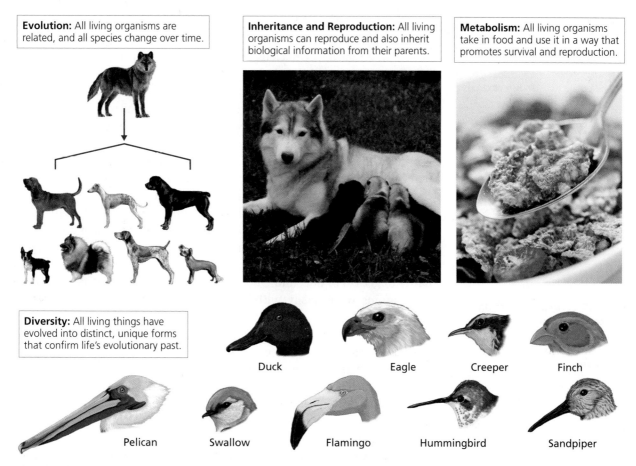

Evolution: All living organisms are related, and all species change over time.

Inheritance and Reproduction: All living organisms can reproduce and also inherit biological information from their parents.

Metabolism: All living organisms take in food and use it in a way that promotes survival and reproduction.

Diversity: All living things have evolved into distinct, unique forms that confirm life's evolutionary past.

Duck   Eagle   Creeper   Finch

Pelican   Swallow   Flamingo   Hummingbird   Sandpiper

FIGURE 1.4 All living organisms have four principal features in common: metabolism, inheritance and reproduction, evolution, and diversity.

1.3 How Do Organisms Function?

Although John and Mary may not have thought of it in this way, John's infidelity gene is part of his biochemistry: to understand it, we have to apply the basics of chemistry to biology. The gene is composed of DNA, which itself is made from other subunit molecules called nucleotides. These subunits are made from even smaller parts: atoms. One of life's wonders is that atoms can be arranged to make complicated structures that contribute to complex characteristics and behaviors—maybe even affecting whether a person is faithful to his or her partner. To understand how organisms are put together, function, and behave, it is essential to understand certain fundamental concepts of chemistry and biochemistry.

Atoms, Chemical Bonding, and Molecules

Every physical thing in existence is composed of **atoms**. There are more than 90 unique kinds of atoms, one for each element. An **element** is a substance composed of only one type of atom. Atoms themselves are composed of three types of subatomic particles: **protons**, **neutrons**, and **electrons**. All atoms have the same general structure: a **nucleus**, or central sphere made of protons and neutrons, and electrons that orbit around the nucleus (**FIGURE 1.5**). Protons are positively charged, electrons have a negative charge, and neutrons

atom :: the building block of every physical thing in existence; there is one type of atom for each element

element :: a substance composed of only one type of atom

proton :: positively charged subatomic particle inside the nucleus of an atom

neutron :: neutral subatomic particle inside the nucleus of an atom

electron :: negatively charged subatomic particle that orbits the nucleus of an atom

nucleus :: central sphere of an atom made of protons and neutrons, with electrons orbiting

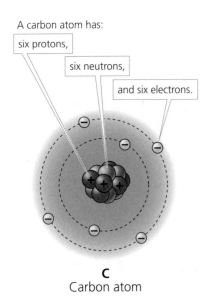

A carbon atom has:

six protons,

six neutrons,

and six electrons.

C
Carbon atom

FIGURE 1.5 Every physical thing in existence is composed of atoms. All atoms have the same general structure: electrons orbiting around a nucleus or central sphere made of protons and neutrons.

orbital :: the outermost shell of an atom that can accommodate a specific number of electrons

chemical bonding :: sharing of electrons in the outer shell of more than one atom

molecule :: formed through the chemical bonding of atoms

have no electrical charge. Each type of element has atoms of a unique size and with a specific number of protons. For example, a carbon atom has six protons and six neutrons in the nucleus and six electrons orbiting.

The chemistry of living things deals with a relatively small subset of elements—only around 25. And four elements account for close to 96% of all atoms in organisms: carbon, oxygen, hydrogen, and nitrogen. Yet this subset of the elements can make a variety of structures, including us.

The personality of an atom—whether it is reactive or inert—depends on the number and arrangement of electrons it has orbiting its nucleus. The outermost **orbital**, or shell, of an atom can accommodate a specific number of electrons. When the shell is filled or complete, the atom is nonreactive. No electrons need to be gained or lost to complete the outer shell. In contrast, if an atom needs just one electron to complete its outer shell, or if an atom has only one electron in its outer shell, it will be highly reactive. Atoms like sodium and chlorine are each one electron away from a completed shell: sodium needs to get rid of one, and chlorine needs to gain one (**FIGURE 1.6**). Both of these atoms react easily with one another to make sodium chloride, or common table salt.

Many atoms have partially filled outer shells. They can't get rid of enough electrons or gain enough to complete their outer shells in this way. Instead, they share electrons. The atoms get close enough together so that the electron orbitals of each overlap. The electrons in the outer shell of each atom orbit *both* atoms. As a result, each atom has a complete shell at least some of the time. This sharing of electrons is referred to as **chemical bonding** (**FIGURE 1.7**). Chemical bonding combines atoms to make **molecules**. For example, carbon and oxygen can chemically bond to each other to make carbon dioxide, the gas each of us exhales.

Macromolecules

Living cells (the fundamental units of life) and organisms are comprised of mostly water and carbon-based **macromolecules**, or large molecules made

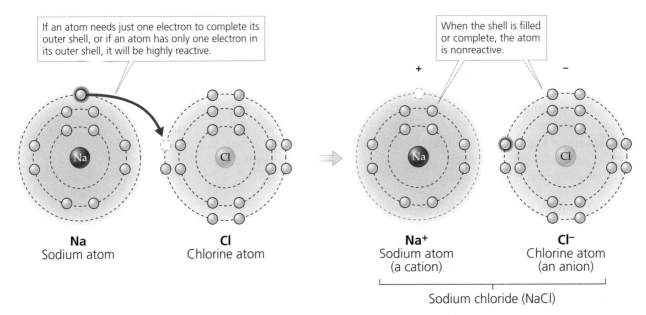

If an atom needs just one electron to complete its outer shell, or if an atom has only one electron in its outer shell, it will be highly reactive.

When the shell is filled or complete, the atom is nonreactive.

Na
Sodium atom

Cl
Chlorine atom

Na⁺
Sodium atom
(a cation)

Cl⁻
Chlorine atom
(an anion)

Sodium chloride (NaCl)

FIGURE 1.6 Atoms like sodium and chlorine are each one electron away from a completed shell, making them highly reactive. When they react with one another, they make sodium chloride, or common table salt.

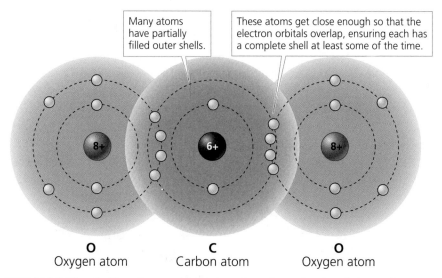

Many atoms have partially filled outer shells.

These atoms get close enough so that the electron orbitals overlap, ensuring each has a complete shell at least some of the time.

O
Oxygen atom

C
Carbon atom

O
Oxygen atom

FIGURE 1.7 Chemical bonding combines atoms to make molecules. Carbon and oxygen can chemically bond to each other to make carbon dioxide.

up of many atoms. These macromolecules are the nucleic acids, proteins, carbohydrates, and lipids (**FIGURE 1.8**). Each type of macromolecule is made of specific types of molecular subunits. For instance, amino acids are hooked together to make proteins. Because there are 20 different types of amino acids in cells that can be combined in any order, a bewildering array of proteins can be made. Let's consider each of these macromolecules, the subunits that comprise them, and examples of biological functions.

- *Nucleic acids:* As we saw in Mary and John's case, DNA, one type of nucleic acid, is the physical material of which the infidelity gene and other hereditary instructions are made. Nucleic acids are made from subunits called nucleotides.
- *Proteins:* Amino acids make up proteins that serve functional roles like regulating the chemical reactions in cells. For example, John's infidelity gene controls the production of the arginine receptor protein. Proteins also have structural functions: they make the many parts of cells and organisms.
- *Carbohydrates:* Sugars, such as glucose, are joined together to make carbohydrates like cellulose, the structural material of plant cell walls. Some carbohydrates, like starch, act as a food source.
- *Lipids:* Subunits called *fatty acids* and *glycerol* join together to make lipids, or fats. Besides being an excellent energy source and the molecule animals use to store excess food, fats insulate the body and serve as shock absorbers for internal organs. Lipids are also an essential structural component of cell membranes.

macromolecule :: large molecule made up of many atoms

Cells

Cells, the smallest units of life, are discrete worlds that maintain the conditions needed to support life (**FIGURE 1.9**). The boundary of the cell, the cell membrane, regulates precisely what can enter and leave. Because cells and organisms contain so much water, the chemical reactions related to life actually take place in a unique aqueous, or watery, environment. Anything inside

cell :: the smallest unit of life

Lipids are made up of subunits called fatty acids and glycerol; they are an excellent energy source.

Glycerol subunit   **Fatty acids**

Amino acids make up **proteins** that regulate the chemical reactions in cells; protein-rich foods have all the amino acids we need.

Serine, an amino acid

Sugars, such as glucose, join together to make **carbohydrates**, some of which act as a food source.

D-Glucose

Nucleic acids are made from subunits called nucleotides; DNA is one type of nucleic acid.

FIGURE 1.8 Living cells and organisms are comprised of mostly water and carbon-based macromolecules. These macromolecules are the nucleic acids, proteins, carbohydrates, and lipids.

Cell membrane of a single-celled amoeba.

The amoeba engulfs its food: a paramecium.

FIGURE 1.9 Cells are the units of all life. The cell membrane regulates precisely what may enter and leave, and most water-soluble molecules on either side cannot pass across the membrane unassisted.

Scientist Spotlight

Carol W. Greider (1961–)

Carol Greider was doing laundry in the early morning before getting her two young daughters up and ready for school when the phone rang unexpectedly. She picked it up. "Hello?" A voice at the other end had astonishing news: "Congratulations, you have been awarded the 2009 Nobel Prize in Physiology or Medicine!" Carol Greider decided to finish the laundry later. Not your stereotypical scientist.

Born in 1961 in California, Greider began doing research as an undergraduate at the University of California at Santa Barbara. She loved it and decided to become a professional scientist. Although a gifted student with excellent grades, research experience, and letters of recommendation, Greider had earned such bad GRE scores (as a result of dyslexia that had previously been unknown to her) that almost no graduate schools would consider her for admission. Fortunately, the University of California at Berkeley interviewed and accepted her. In fact, it was in the lab of Elizabeth Blackburn at UC Berkeley that Greider began the project that led ultimately to a Nobel Prize for them both.

Greider and Blackburn's big breakthrough was the discovery of a molecule called *telomerase*. Telomerase is like a cell's fountain of youth. It helps prevent some aging effects and allows a cell to divide over and over again. This is particularly important in multicellular organisms. In a healthy body, cells divide to replace others that are damaged or worn out. As the organism ages, its cells become less effective at dividing and replacing other cells, which accounts for some of the problems aging produces.

Telomerase works by placing a protective cap, or telomere, on the ends of chromosomes inside the cell. This cap protects chromosomes during the cell's life and during reproduction. The chromosomes contain the cell's genetic information in the

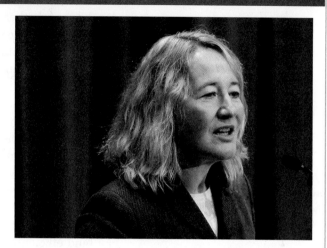

Carol Greider won the 2009 Nobel Prize in Physiology or Medicine as a result of her work with the molecule telomerase.

form of DNA. If the chromosomes become damaged, the cell may die or become unable to reproduce.

In the research Greider has done since the discovery of telomerase, she has shown that as cells age, telomeres shorten. Once telomeres get too short, they lose their protective power. The chromosomes are damaged and the cell dies. Greider has also discovered that cancer cells have overactive telomerase and telomeres that are too long. As a result, cancer cells are immortal. Evidently, telomere length regulates the life span of a cell, and telomerase is the molecule responsible for that length.

Greider's research has propelled her to turn these lab results into strategies for treating patients. She is driven by her intense curiosity and the challenge of, in her words, "diving into the unknown." ::

the cell that can dissolve in water, such as salts and many small molecules, will therefore exist in a solution state. In contrast, molecules such as fats, which do not dissolve in water, will remain separate.

Cell membranes enclose cells so successfully in part because most water-soluble molecules on either side cannot pass across the membrane unassisted. If that were to happen, the cells would die. (To learn more about the relationship between cell aging and cancer, see *Scientist Spotlight: Carol W. Greider*).

As we will discuss in later chapters, there are thus two chemical worlds in cells and organisms, a water-soluble one and a fat-soluble one. And that has important consequences for a range of functions—including cell communication, fertilization, digestion, water balance, and even thinking.

Chemical Reactions and Enzymes

Two broad categories of chemical reactions take place in cells and organisms: molecules are taken apart to produce smaller components, or molecules are joined together to make larger structures. Whether a reaction occurs at all depends on whether there is energy available. Some reactions require a large input of energy because the products of the reactions are structurally more complex than the starting materials. In contrast, some reactions occur without a net input of energy.

Reactions inside organisms depend upon balancing energy input and output. Biosynthetic reactions need a net input of energy. For instance, DNA is more structurally complex than the nucleotides that form it. It costs energy to make this macromolecule. Breakdown reactions, like the digestion of starch from a potato, yield energy ultimately but the reaction needs a kick start; enzymes serve this purpose.

Most **enzymes** are proteins that make it easier for chemical reactions to occur. They lower the degree to which "start-up" energy is needed. Enzymes do this in a variety of ways such as orienting reacting molecules so they can bind efficiently or by actually bending molecules so they are easier to break into parts (**FIGURE 1.10**). Enzymes play an especially important role in regulating all chemical reactions in cells and organisms, including those responsible for digesting food, growing new cells, or replicating DNA, for example.

enzyme :: protein that makes it easier for chemical reactions to occur

Energy Extraction and Use

The responsiveness, survival, and reproduction of all organisms depend on **metabolism**—the chemical reactions in cells that sustain life. The ultimate source of energy that supports life is the sun. Plants are able to capture light energy from the sun and convert it into chemical bond energy in sugars and related molecules. Organisms extract energy from the food they eat. The chemical bond energy present in food molecules is converted to a form that can be used for any of a variety of functions, including protein production, walking, or mating. The reactions of metabolism are carefully regulated by the actions of specific enzymes.

metabolism :: all of the chemical reactions in cells that sustain life

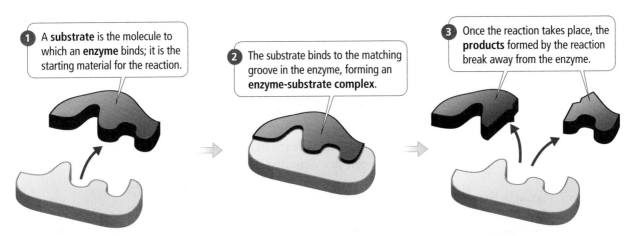

1 A **substrate** is the molecule to which an **enzyme** binds; it is the starting material for the reaction.

2 The substrate binds to the matching groove in the enzyme, forming an **enzyme-substrate complex**.

3 Once the reaction takes place, the **products** formed by the reaction break away from the enzyme.

FIGURE 1.10 Enzymes are proteins that play an important role in regulating all chemical reactions in cells and organisms. They help reacting molecules to bind efficiently or to bend, so they are easier to break into parts, as shown in this example.

1.4 How Do Organisms Reproduce?

In our case study, John learned that he had the infidelity gene—or, more accurately, the infidelity allele. We've discussed some of the consequences of having this allele. But we haven't discussed *why* John had it in the first place. Part of the answer is that John inherited the allele from his father or mother. This fact might have helped John and Mary evaluate their concerns about the allele. If John's parents had a happy marriage—in spite of having the infidelity allele—the prospects for John and Mary look better. At least they could conclude that the allele doesn't necessarily *cause* infidelity. What else can we learn about life by studying inheritance and reproduction?

Inheritance

Inheritance is the way life perpetuates itself from one generation to the next. Organisms like us, for example, inherit from two sources: the mother and father. Each parent has the same set of genes (if we ignore sex differences). But each parent has a different combination of alleles for these genes. This means that you inherit the same set of genes as all humans, but your particular combination of alleles, some from the father and some from the mother, is unique (unless you have an identical twin or other identical multiple-birth sibling). You probably resemble your parents and your siblings, but each individual varies genetically from others (**FIGURE 1.11**).

Inheritance involves passing *information* from one generation to the next. For us, only a microscopic egg and sperm cell are required. But these tiny entities contain all the genetic information needed to make a human (**FIGURE 1.12**). These are the instructions a new organism needs to grow,

FIGURE 1.11 Every human being inherits the same set of genes as all other humans, but the particular combination of alleles is unique (unless you have an identical twin or other identical multiple-birth sibling). For example, individuals may have different combinations of alleles for hair color, resulting in different shades.

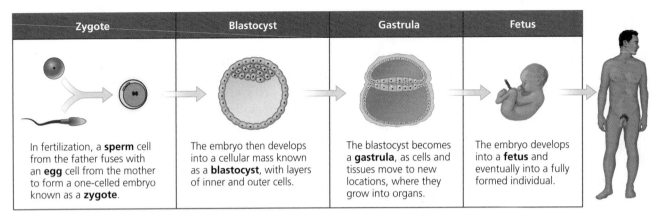

Zygote	Blastocyst	Gastrula	Fetus
In fertilization, a **sperm** cell from the father fuses with an **egg** cell from the mother to form a one-celled embryo known as a **zygote**.	The embryo then develops into a cellular mass known as a **blastocyst**, with layers of inner and outer cells.	The blastocyst becomes a **gastrula**, as cells and tissues move to new locations, where they grow into organs.	The embryo develops into a **fetus** and eventually into a fully formed individual.

FIGURE 1.12 Only a microscopic egg and sperm cell are required to pass on the genetic information needed to make a human. This information includes the instructions a new organism needs to grow, survive, and eventually reproduce.

survive, and eventually reproduce. An individual ("within" a generation) uses inherited information to direct the basic processes of living. This information, also passed on to offspring ("between" generations), allows them to do the same.

Reproduction

Inheritance works the way it does because of the way organisms reproduce. The steps in reproduction make new individuals and ensure that each inherits the full complement of genetic information. We know now that genetic information is stored in DNA. We also know that the chemical properties of DNA allow it to be used both to store information and to pass it on to the next generation. In nature this happens in two ways.

The simplest form of reproduction occurs in individual cells, like the cells in your body or in organisms like bacteria that have only a single cell. First the DNA is copied, so that two complete sets exist. Then the cell divides, and one copy goes into each daughter cell. These daughter cells are (almost) genetically identical to the parent. This type of reproduction, referred to as **asexual reproduction**, involves a single parent—so no sex (**FIGURE 1.13**).

The DNA copying mechanism is very accurate—but not perfect. Although the chances are much less than one in a million, sometimes errors occur when DNA is copied. These mistakes are called **mutations**. Because of mutations, any cell may have slight genetic differences from its parent. When a mutation occurs in a gene, it makes a new allele of that gene. That's why alleles exist, like the allele of *AVPR1a* that may be related to infidelity. While there is a *very* slight chance that John's copy of *AVPR1a* was a mutation, and that he didn't inherit it from his parents, this is unlikely.

A few multicellular organisms—flatworms, sea stars, and plants, for example—can reproduce asexually when they break into pieces. But most multicellular organisms, even those that reproduce asexually, reproduce sexually. **Sexual reproduction** (**FIGURE 1.14**) occurs when an egg cell from the mother fuses with a sperm cell from the father to form a new individual. Sexually formed individuals don't just have a copy of their parents' DNA: they have a unique combination of alleles. The actual traits a new individual shows depend on how the alleles in this particular set interact.

But why reproduce at all? Unfortunately, bodies wear out, the chances of fatal accidents accumulate, and over long time spans, habitats change too

asexual reproduction :: reproduction that involves a single parent and no sex

mutation :: a physical change in the DNA

sexual reproduction :: reproduction that occurs when an egg cell from a female fuses with a sperm cell from a male to form a new individual

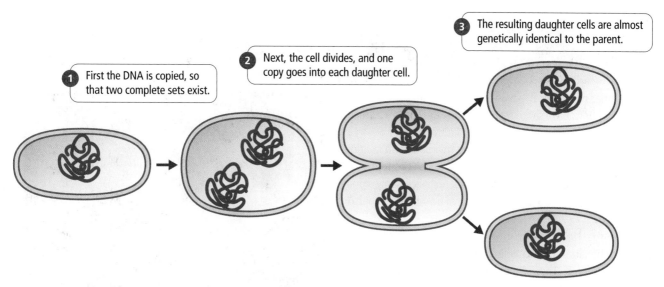

1. First the DNA is copied, so that two complete sets exist.

2. Next, the cell divides, and one copy goes into each daughter cell.

3. The resulting daughter cells are almost genetically identical to the parent.

FIGURE 1.13 Asexual reproduction involves a single parent—no sex. It occurs in the cells in your body and in single-celled organisms like bacteria.

much for a single individual to tolerate. Ironically, a universal feature of life is death (see *Life Application: Determining When Life Has Ended,* p. 18).

The fact that all individuals die makes reproduction and inheritance a very significant part of life. From an evolutionary perspective, reproduction is the single most important event in life. Two facts about these processes are particularly noteworthy:

• Living organisms are part of an unbroken chain of ancestors and descendants, extending back into time. Reproduction divides this

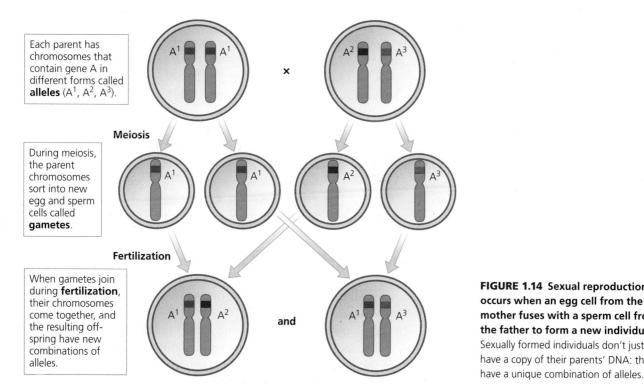

Each parent has chromosomes that contain gene A in different forms called **alleles** (A^1, A^2, A^3).

Meiosis

During meiosis, the parent chromosomes sort into new egg and sperm cells called **gametes**.

Fertilization

When gametes join during **fertilization**, their chromosomes come together, and the resulting off-spring have new combinations of alleles.

FIGURE 1.14 Sexual reproduction occurs when an egg cell from the mother fuses with a sperm cell from the father to form a new individual. Sexually formed individuals don't just have a copy of their parents' DNA: they have a unique combination of alleles.

Determining When Life Has Ended

The amazing medical advances achieved in the 20th and 21st centuries have saved many lives. Yet they have also made it more challenging to establish a definitive time of death. In the not-too-distant past, if a person's heart stopped beating, he or she was considered dead. With our abilities to provide medical life support, a still heart does not necessarily signify death at all. How do physicians determine whether a person has died?

In the 1960s, a group of physicians at Harvard Medical School began efforts to define clear criteria for neurological, or brain, death. They refined and formalized guidelines, and by 1980 the Uniform Determination of Death Act (UDDA) was drafted as a model of legislation for individual states to pass. Supported by the President's Commission for the Study of Ethical Problems in Medicine and Biomedical and Behavioral Research, the UDDA states: "An individual who has sustained either (1) irreversible cessation of circulatory and respiratory functions, or (2) irreversible cessation of all functions of the entire brain, including the brain stem, is dead."

Based on state guidelines, hospitals have strict protocols for determining brain death. First, only appropriately trained physicians who have met certain professional standards may designate brain death. Clinical evaluations precede this determination to verify that three essential features are present: coma, absence of brainstem reflexes, and apnea. A coma is a deep state of unconsciousness from which a person cannot be aroused. The brainstem is the most posterior part of the brain and is responsible for controlling certain critical but not neurologically complex functions. A person with a dead brainstem has no cough reflex, no gag reflex, no corneal reflex, no normal eye movement, and no pupil response—even when a physician tries to elicit these behaviors. Apnea means that a person has stopped breathing independently.

In the event that a patient appears to be brain dead, physicians perform a complex battery of tests twice, with several hours separating the two tests, to make sure that the

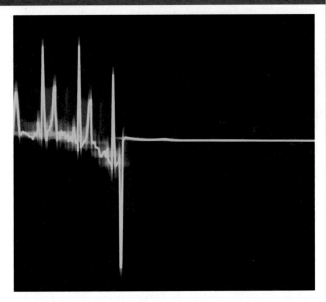

In the event that a patient appears to be brain dead, physicians perform a complex battery of tests twice. In some circumstances, the brain death diagnosis will be verified by an electroencephalogram (EEG).

individual has an irreversible loss of brain function. Precautions are also taken to rule out the possibility of some potentially reversible cause for a lack of brain function (a drug overdose, for example). In some circumstances, the diagnosis of brain death will be verified by a brain scan, to see whether there is any blood flow, or by an electroencephalogram (EEG)—a method of detecting electrical activity in the brain or determining whether it has "flatlined."

If a person meets the criteria for brain death, and if he or she is an organ donor, the certification of brain death generally requires the signatures of two physicians. Neither can be directly involved in performing the transplant operation. This extra step serves to allay concerns that a determination of death may have been influenced by a need for healthy organs for transplantation. ::

chain into generations, and inheritance guarantees some degree of genetic continuity. Ever since life got started, new life has come only from existing life, not from nonliving sources (see *How Do We Know? Spontaneous Generation*).

- The combination of mutation and sexual shuffling of alleles guarantees that individuals in one generation differ from each other and from the previous generation. This genetic variation is the basis for life's most central feature: evolution.

Spontaneous Generation

If you have ever had the unpleasant experience of coming across a dead animal in the woods or on the side of the road, you probably noticed that the carcass was covered with maggots. Perhaps you even knew that these creatures were simply fly *larvae*, the term for young insects. Most of us would find this observation unsettling or even repulsive. Yet no one would conclude that the dead animal had "turned into" maggots. People in ancient times, however, believed that maggots arose "spontaneously" and directly from an animal's remains. We know today that the only way new individuals arise is from their parents, through reproduction. But how was this result established scientifically?

The first person to tackle whether life could arise spontaneously was the Italian natural philosopher Francesco Redi. In 1668, Redi developed the hypothesis that maggots appear in rotting meat not as a result of "spontaneous generation," but because maggots develop from eggs laid by flies. To test this hypothesis, Redi set up a simple demonstration: he placed meat in glass jars. One set of jars was left open. A second set was sealed shut. The third set was covered with gauze. The results were conclusive. Maggots appeared only on the meat that was in the open jars. Flies entered the jars, landed on the meat, and laid eggs. No maggots were observed in the meat in the sealed jars or the jars covered by gauze.

Although Redi's experiment put to rest the idea that animals could arise spontaneously, many still held on to the notion that microorganisms such as bacteria could do so because they were such small and simple forms of life. It wasn't until almost 200 years after Redi's work that Louis Pasteur, a French microbiologist and chemist, performed the critical experiment that demonstrated that *all* forms of life come only from the reproduction of other living organisms. Pasteur was motivated by a contest sponsored by the French Academy of Sciences (FAS) in 1859. The FAS offered a prize for the best experiment that would either prove or disprove spontaneous generation.

Pasteur cultured bacteria in S-shaped glass flasks. He poured culture broth into these "swan-neck" flasks and then boiled the broth to sterilize it; no bacteria remained. The flasks were left open to the air, but because of the S-shaped neck, dust particles could not fall into the flask. As a consequence, even though air could enter, no bacteria did. The broth remained sterile. When Pasteur broke off the neck, so that the broth was directly under the opening of the flask, a teeming population of bacteria materialized within a day. Although Pasteur was not the only scientist to address whether microorganisms could generate spontaneously, he was certainly the person who did so most decisively. It should come as no surprise that Pasteur won the contest by a unanimous vote! ::

1 Pasteur boiled culture broth, so it was sterile and no bacteria remained.

2 Pasteur placed the cultured bacteria in S-shaped glass flasks.

3 Pasteur left the flasks open to the air.

The broth remained sterile.

4 Pasteur broke off the S-shaped neck of the flask, so that the broth was directly under the opening of the flask.

A teeming population of bacteria materialized within a day.

Louis Pasteur, a French microbiologist and chemist, performed the critical experiment that demonstrated that *all* forms of life come only from the reproduction of other living organisms.

1.5 How Does Life Evolve?

Let's return to the question of why John had the infidelity allele. Noting that John inherited it from one of his parents simply pushes the question back a generation—why did one of *them* have it?

Humans have about 25,000 genes, and many (perhaps most) of them have several alleles. Some alleles are very common. For example, alleles for blue eyes and blond hair are common in Scandinavia, while alleles for black hair and dark eyes are common in Africa. The chances of having a particular allele depend on how common it is in a population—that is, how many people in the group have it. In any human population, a certain number of people will have the infidelity allele of *AVPR1a*. Unfortunately for John, one or both of his parents fell into this category.

Over time the frequency of some alleles increases, while the frequency of other alleles decreases. If the frequency of an allele goes up, more people will have it, and if it decreases, fewer people will have it. Any change in allele frequency, for any reason, reflects *a genetic change in the population* (**FIGURE 1.15**). This is the modern biological definition of evolution. Later in this book, in Chapter 9, we will look thoroughly at how Charles Darwin made his discovery and the evidence for evolution. And, again and again, especially in Chapters 3–8, we will see how genes are part of our inheritance— and what humans are doing with that knowledge. Let's start, though, with an overview of Darwin's theory, to see why evolution serves as a unifying theme of biology.

Darwin's Theory

When Darwin proposed his theory of evolution in 1859, he did not know about DNA, genes, or alleles. But he did know about inheritance, and he discovered the significance of the fact that individuals vary. Here are the facts Darwin worked with:

- Individuals in a population vary from one another.
- Variations are inherited, at least in part.
- In every population, many or most of the offspring die before reaching maturity.

What conclusion would you draw from these easily verified facts? Darwin asked this question: *which* offspring would die? Darwin's discovery was

White arctic foxes can blend in better with their winter environment, so over time, alleles for darker-colored foxes decreased, while alleles for white-colored foxes increased.

FIGURE 1.15 Over time, the frequency of some alleles increases, while the frequency of other alleles decreases. Any change in allele frequency, for any reason, reflects a genetic change in the population.

simple and yet profound. Some of the variations, Darwin reasoned, would help certain offspring survive better than others. Offspring that were a little bigger, or a little faster, or that were able to hide or find food better—these offspring had a better chance of surviving than others that lacked these variations.

The survivors were more likely to reproduce, and when they did, their off-spring would inherit these variations. Variations that promoted survival would become more common, while variations that hindered survival would become less common. The next generation would not only be different from the previous one, it would be different in a specific way. It would have more individuals with variations that helped them survive the challenges of the particular habitat in which they lived.

Darwin referred to this process as "descent with modification" (**FIGURE 1.16**). He realized that a species could change over time, which was a revolutionary idea in the 1800s although other scientists were reaching the same conclusion. Darwin used the term **natural selection** to describe the *mechanism* by which these changes occurred—the ways that conditions in the natural environment "selected" which individuals would survive and reproduce.

natural selection :: the term Darwin used to describe the mechanism by which species change over time; conditions in the natural world select which individuals will survive and reproduce

Extending Darwin's Theory Through Time

Very often in science, ideas that seem simple can have a major impact on our thinking. If we open the door to the possibility that species can change, where does this concept lead?

One of the first realizations was that as species change, they become better adapted to their environment. In Darwin's time, geologists were beginning

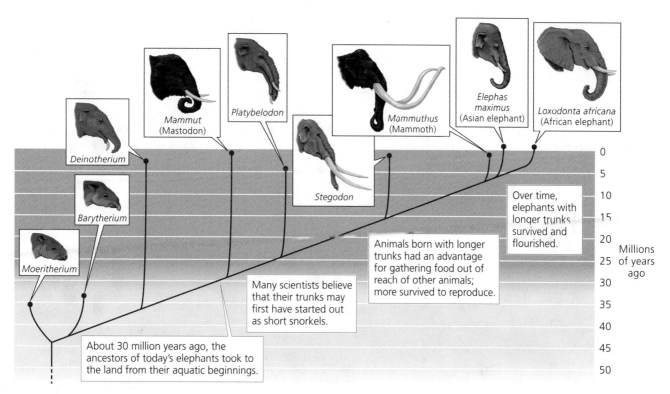

FIGURE 1.16 "Descent with modification" was the name Darwin gave to the idea that each generation would vary from the previous one in a specific way. Variations that promoted survival would become more common, while variations that hindered survival would become less common.

to show that conditions on Earth changed over time—almost as surprising as species changing. Clearly these processes would have to go hand in hand. If the environment changed, species would either change to keep up or become extinct.

A second realization was that if species changed enough, they might even change into a different species. A single species could have populations living in many different and widespread areas. If conditions in these areas changed independently of one another, as the different populations adapted to their local conditions they would become increasingly different. Given sufficient time, Darwin thought, they would separate into two distinct species where originally there had been only one (**FIGURE 1.17**). The idea that a new species could be created by natural means was earthshaking and difficult for many to accept.

But the most far-reaching conclusion of Darwin's theory emerges if we run this process *backward* in time. Instead of species changing, becoming different, and splitting into more species, we see species becoming more similar and merging together. Viewed from this perspective, two species are related because they descended from a common ancestor—just as you and your cousins are related through a common grandparent.

Because all forms of life are part of a continuous chain of reproduction, the logical conclusion of Darwin's theory is that *all* species are related, all part of an ever-expanding tree of life. In the century and a half since Darwin, biologists have learned a great deal about evolution. They have extended and modified Darwin's ideas, and found some new ones. But Darwin's original ideas about evolution remain largely in place today.

Evolution Is the Unifying Theme of Biology

The goal of science is to discover the reasons for the events and situations that occur in the natural world. Evolution implies that all forms of life—including ours—are related to some degree; this is a fact that can be difficult to see at

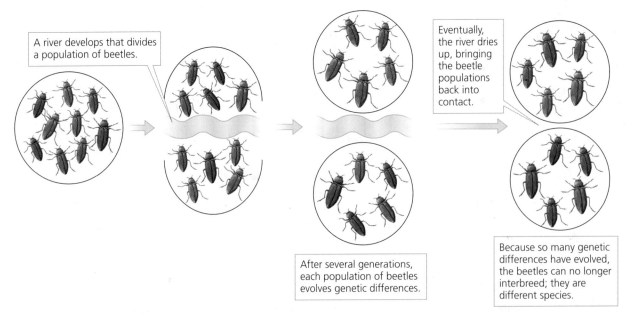

A river develops that divides a population of beetles.

After several generations, each population of beetles evolves genetic differences.

Eventually, the river dries up, bringing the beetle populations back into contact.

Because so many genetic differences have evolved, the beetles can no longer interbreed; they are different species.

FIGURE 1.17 Darwin came to the realization that if species changed enough, they might even change into a different species.
The idea that a new species could be created by natural means was earthshaking and difficult for many to accept.

first. But this idea explains so many facts of nature that biologists see evolution as the central theme that ties together all of biology.

Our own animal group provides a good example. We belong to the *vertebrates*: animals with internal skeletons made of bones (mostly) and with well-defined heads that contain our brain and sensory organs. Vertebrates are a diverse group (**FIGURE 1.18**). Whales, dinosaurs, and elephants are among the largest animals that ever lived. The dwarf goby fish, one of the smallest vertebrates, is less than a centimeter long. And a flea toad fits comfortably on your thumbnail. Some vertebrates are fierce predators; others eat plants. Cheetahs can run at over 70 miles per hour, and falcons dive at over 200 miles per hour. On the other hand, tree sloths typically move less than a foot per minute, and they may not move at all for long periods of time. Vertebrates live in the deepest oceans, on the highest mountains, and in the desert, the rainforest, prairies, and marshes—just about everywhere from the equator to the poles.

Since vertebrates show such differences, you might expect their bodies to be different too. But all vertebrates have the same body plan: a skull attached to a vertebral column, which supports ribs and ends in a tail. Vertebrates all have similar bone structure (**FIGURE 1.19**). Bones vary in size and shape and are sometimes lost, but for the most part the set of bones is the same. Vertebrates have many of the same muscles that move these bones. All vertebrates have the same organs in roughly the same places, make similar systems of blood vessels, and have the same organization in their nervous systems.

If you examine them in detail, you learn that all vertebrates are variations on a theme rather than new designs for a particular way of life. We see this pattern over and over in nature. And this is precisely the pattern that evolution

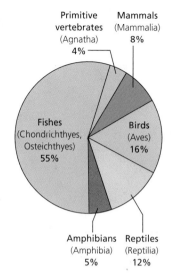

FIGURE 1.18 Vertebrates are a diverse group comprised of mammals, birds, reptiles, amphibians, and fish. They reside just about everywhere, and they are also some of the largest and smallest and the fastest and slowest animals that ever lived.

FIGURE 1.19 Vertebrates all have similar bone structure. Bones vary in size and shape and are sometimes lost, but for the most part, the set of bones is the same; the femur is highlighted in each illustration.

predicts. Vertebrates inherited their similarities from common ancestors and evolved differences as they adapted to environments. It is difficult to explain this pattern of diversity and similarity any other way. The eminent scientist Theodosius Dobzhansky put it succinctly in a 1973 paper: "Nothing in biology makes sense except in the light of evolution."

1.6 What Patterns of Diversity Are Found in Nature?

diversity :: the variety of life forms in the natural world

Scientists have described about 1.3 million species. But we have a long way to go: by some estimates, there may be 30 million living species and another 50 *billion* that have gone extinct. Life is diverse, and **diversity** is one of the most obvious facts of life. All you have to do is look around you and count the number of life forms you see to appreciate natural diversity. In Chapter 12, we will look at the challenge of preserving the great diversity of life for future generations.

Life's diversity presents biologists with two tasks. First, we must explain why this diversity exists. Second, we need to organize this diversity in a meaningful way, one that reflects its underlying causes.

Life Is Diverse

New species arise when older species adapt to new and different living conditions. The diversity of life goes hand in hand with the diversity of living conditions our planet has to offer (**FIGURE 1.20**). Every region on Earth presents a distinct set of physical conditions, like temperature, moisture, and amount of sunlight. Within each region, there are numerous ways to get food and shelter, and many ways to respond to other species. Over time, all these conditions change, as continents drift, ice ages come and go, climate varies, and species go extinct. The Earth is a dynamic place, and new species arise that can take advantage of the opportunities it provides.

Evolutionary Diversification Leads to Degrees of Relatedness

Evolution results in diversity, but it results in a particular kind of diversity that has a distinct pattern. Let's consider what this pattern is and why it occurs.

It's helpful to start with a more familiar process—your own family tree. You could trace your ancestry back to some distant great-great-grandparent and fill out your complete family tree. You would list the first ancestor, all of his or her children, all of their children, and so on until you reach your place on the family tree. Your parents and siblings would be on the tree, as would your cousins, uncles, and aunts, and ever more-distant relatives. You would be related to everyone on the tree because you all descended from a common ancestor. However, you would not be *equally* related to everyone. Your siblings and parents are particularly close relatives, and you may resemble them closely. On the other hand, you may have little in common with more distant relatives in the tree.

The same type of result occurs when species split and split again, giving rise to numerous descendant species (**FIGURE 1.21**). Each pair of species has a common ancestor. But if that common ancestor lived long ago, the species

FIGURE 1.20 The diversity of life goes hand in hand with the diversity of living conditions on the planet. Over time, these conditions change, leading some species to go extinct, while others arise that can take advantage of opportunities.

may not be closely related. There may be a long chain of species leading back to the common ancestor, and there will have been a lot of time for the species to diverge and become different.

On the other hand, two species may have a common ancestor that lived very recently—more like a mother species than a great-great-great-grandmother species. These species may have much in common, such as similar genetics, appearance, and behaviors. In fact, you might observe a species in the process of splitting, in which the various subgroups are so similar it's hard to tell them apart.

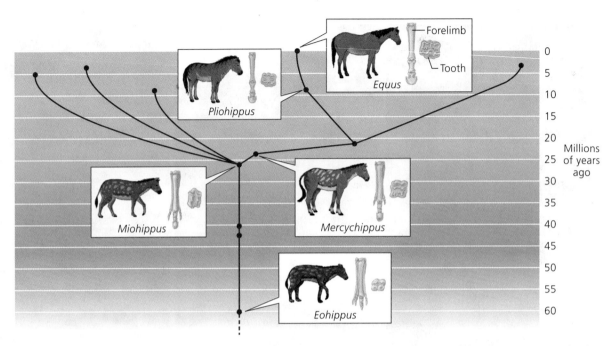

FIGURE 1.21 Each pair of species has a common ancestor. If that common ancestor lived long ago, however, there will have been a lot of time for the species to diverge and become different.

Our understanding of evolution allows us to make a strong prediction about diversity in nature. If new species arise when older species split, we should see patterns of relatedness: groups of species that are closely related (like a family), that are part of larger but less closely related groups (like an extended family), and that are part of still larger and less closely related groups (perhaps like a village founded long ago by only a few couples). This type of pattern is called a hierarchy. If life's diversity can be organized into a hierarchy of relatedness, this is good evidence that evolution produces new species in the way Darwin envisioned: older species split in two as they adapt to new and different conditions.

Organizing Hierarchies in the Diversity of Life

Species lie at one end of life's hierarchy. Members of a species are genetically so similar that they can interbreed successfully. Let's consider dogs as an example. Dogs vary dramatically, the result of our own "artificial" selection. But even a Great Dane and a Chihuahua can produce offspring (granted, they may need a little help). All dogs are in the same species.

It's easy to see that dogs are part of a larger group that includes wolves and coyotes. A still larger grouping includes foxes. Dogs are carnivores, which puts them in an even larger group that includes cats and seals. All of these species are mammals, and all mammals are vertebrates. Finally, all vertebrates are animals. Natural diversity consists of a hierarchical organization of groups within groups within groups.

People have undoubtedly observed this hierarchy of similarity and diversity for ages. The hierarchy was formally organized in the early 1700s by Carolus Linnaeus. He defined a hierarchy with seven levels, some of which may be familiar to you:

kingdom
  phylum (plural "phyla")
    class
      order
        family
          genus (plural "genera")
            species (both singular and plural)

A kingdom consists of one or more phyla, a phylum contains one or more classes, and so forth down to species. Each level contains fewer and more closely related groups than the one above it. Linnaeus used his hierarchy to *classify* every species known at the time. **Classification** means assigning every species to a genus, each genus to a family, continuing on to kingdom (**FIGURE 1.22**).

One consequence of Linnaeus' work is that each species has a unique **scientific name**—its particular classification from kingdom to species. These names for the specific levels are in Latin and designed to be descriptive of the grouping. The Latin word for dog is *canis*, for example, and *Canidae* is the name of the dog family. It's cumbersome to use the complete classification to name a species, so by convention a species' scientific name includes its genus and species names only. Both are italicized since the words are Latin; the genus is capitalized, and the species name is not.

The most important consequence of Linnaeus's work, though he did not realize it, is that a species' classification is its family, or, more specifically, its

classification :: the assignment of every species to a genus, family, order, class, phylum, and kingdom

scientific name :: the unique classification of each species from kingdom to species

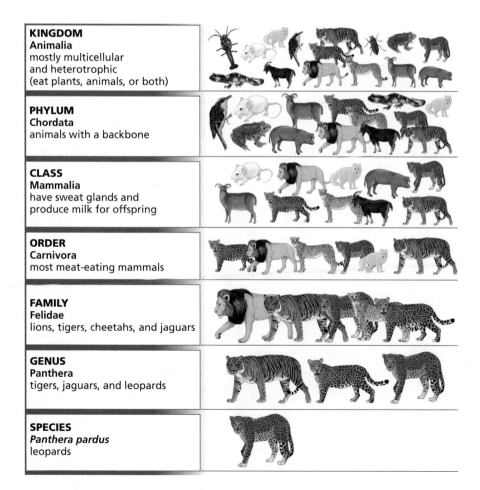

KINGDOM
Animalia
mostly multicellular and heterotrophic (eat plants, animals, or both)

PHYLUM
Chordata
animals with a backbone

CLASS
Mammalia
have sweat glands and produce milk for offspring

ORDER
Carnivora
most meat-eating mammals

FAMILY
Felidae
lions, tigers, cheetahs, and jaguars

GENUS
Panthera
tigers, jaguars, and leopards

SPECIES
Panthera pardus
leopards

FIGURE 1.22 Carolus Linnaeus formally organized the hierarchy of life in the early 1700s. He defined a hierarchy with seven levels: kingdom, phylum, class, order, family, genus, and species.

evolutionary tree. Tracing a species back through its evolutionary tree allows you to see what species it is related to, and how close the relationship is, just as you could do with relatives in your own family tree. Knowing these relationships helps you understand the biology of species better. For example, you know that if a species is in class *Mammalia*, it gives birth to live young, feeds them milk, has hair, and maintains a constant body temperature. The simple fact that species can be classified and organized this way is strong evidence for evolution and the origin of species.

Biology in Perspective

Why does biology matter to you? Knowing more about biology will help you make better personal decisions regarding your health and that of your family. It might even save you from the kind of worry and stress our case study couple, John and Mary, experienced. Understanding biology can also make you a better citizen. Citizenship requires making informed decisions on issues that affect your community, country, and planet. While these issues are complex and multidisciplinary, biology will play an increasingly important role as scientists make advances in biological and medical research.

Biology is the study of life. Life on Earth is astonishingly diverse, but all living things have several features in common. All living things have a metabolism and use molecules and energy in a controlled way to support their life. All living things store genetic information and pass it on to their descendants when they reproduce. And although individuals don't evolve, over time species evolve, becoming better adapted to their habitats and occasionally producing new species. The diversity of species and the adaptations they possess are all products of evolution, making evolution the central unifying theme of biology.

Learning about biology provides you with a window to the natural world around you. Adding a scientific approach gives your learning a new dimension and helps you develop a more detailed and accurate view of nature. The natural world is complex and fascinating, and it is satisfying to learn about how it works. The study of nature can provide you with a lifetime of rewards.

Chapter Summary

1.1 How Does Biology Affect Your Life?

Explain how understanding biology can help you make informed decisions as a citizen.

- Understanding biology can help you make informed decisions about health care, aging, cancer, disease, voting, and legislation, among many other things.
- New technologies offer opportunities but may also present risks.

1.2 What Are the Features of Life?

Use the four principal features of life to discriminate living things from nonliving things.

- Living things exhibit four principal features: metabolism, inheritance and reproduction, evolution, and diversity.

1.3 How Do Organisms Function?

Describe the relationship among atoms, macromolecules, and cells, and explain how enzymes maintain cell function.

- All substances are composed of atoms; each atom has a spherical nucleus, composed of neutrons (no charge) and protons (positive charge), orbited by negatively charged electrons.
- Elements are pure chemical substances in which all of the atoms contain the same number of protons.
- Living things are mostly composed of four elements: carbon, oxygen, hydrogen, and nitrogen.
- The reactivity of an atom is determined by the number of electrons in its outer shell; when atoms share electrons, they form chemical bonds.
- Living cells are composed mostly of water and macromolecules; these macromolecules—nucleic acids, proteins, carbohydrates, and lipids—are the smallest discrete units that maintain the conditions required to support life.
- Most chemical reactions require energy.
- Metabolism is the chemical reactions in cells that sustain life.

1.4 How Do Organisms Reproduce?

Describe the relationship among DNA, genes, and alleles, and compare how genes are inherited in sexual and asexual reproduction.

- Organisms inherit information that codes for traits on DNA from their parents; alleles are different versions of a gene.
- Asexual reproduction involves a single parent; the offspring have the same DNA as the parent.
- Most multicellular organisms reproduce sexually by combining the DNA from two parents. Eggs from the mother and sperm from the father carry the genetic information required for a new organism, and the exchange of information results in a new combination of alleles.
- Mistakes made copying DNA are called mutations; mutations and sexual reproduction create the variation required for evolution.

1.5 How Does Life Evolve?

Create a scenario describing how a population of organisms might adapt to a new environment.

- The modern biological definition of evolution is a change in allele frequency in a population.
- Natural selection is a process by which population allele frequencies change to those that are better adapted to the environment.
- Because all organisms require a living ancestor, every species is related to each other through common ancestors.
- All forms of life on Earth are related in some way; this is a unifying theme of biology.

1.6 What Patterns of Diversity Are Found in Nature?

Explain the connection among common ancestors, the diversity of environments on Earth, and the scientific classification system.

- The Earth is dynamic in space and time. Through evolution, species adapt to their specific environments.
- Species that evolved in the recent past have a more recent common ancestor. As a result, they have more common features.
- Biologists use a pattern of relatedness, based on a hierarchy of relatedness similar to a family tree, to classify organisms.
- The most important aspect of the classification system is that the tree represents the evolutionary history of organisms.

Key Terms

allele 5
asexual reproduction 16
atom 9
cell 11
chemical bonding 10
classification 27
diversity 24
electron 9

element 9
enzyme 14
evolution 8
macromolecule 10
metabolism 14
molecule 10
mutation 16
natural selection 21

neutron 9
nucleus 9
orbital 10
proton 9
scientific name 27
sexual reproduction 16

Review Questions

1. What does the "infidelity gene" (*AVPR1a*) do in your body? How does this affect your brain?

2. What is an allele?
   a. A different gene
   b. A gene that encodes for hair color
   c. A gene that encodes for eye color
   d. A similar gene
   e. A different version of the same gene

3. List the four major features of life.
   a. Metabolism, inheritance and reproduction, evolution, and diversity
   b. Metabolism, evolution, diversity, and decomposition
   c. Metabolism, creationism, inheritance and reproduction, and diversity
   d. Inheritance and reproduction, evolution, diversity, and death
   e. Inheritance, evolution, diversity, and death

4. How do atoms bond to make molecules?

5. What are enzymes, and what is their function?
   a. Enzymes are proteins that make it more difficult for chemical reactions to occur.
   b. Enzymes are lipids that make it easier for chemical reactions to occur.
   c. Enzymes are proteins that make it easier for chemical reactions to occur.
   d. Enzymes are molecules that break down other molecules.
   e. Enzymes are carbohydrates that make it easier for chemical reactions to occur.

6. What is the difference between asexual and sexual reproduction?
   a. Asexual reproduction occurs only in the cells of the human body; sexual reproduction occurs between two parents of different organisms.
   b. Asexual reproduction involves a single parent; sexual reproduction occurs when an egg cell from the mother fuses with a sperm cell from the father to form a new individual.
   c. Asexual reproduction involves two parents; sexual reproduction involves one parent.
   d. Asexual reproduction involves a single parent; sexual reproduction occurs only when a human egg cell from the mother fuses with a human sperm cell from the father to form a new individual.
   e. Asexual reproduction always involves a single parent; sexual reproduction may or may not involve two parents in forming a new individual.

7. List two reasons why children vary genetically from their parents.

8. What did Darwin mean by "descent with modification"? Is descent with modification an observable fact of nature?

9. Briefly describe how natural selection works.

10. What are the seven levels in the Linnaean system of classification? How is this classification system used to define scientific names?

The Thinking Citizen

1. What are some of the ways that nature inspires you or has a positive impact on your life?

2. What are some specific examples of issues or questions you might encounter in your life for which knowledge of biology might be helpful?

3. When cells divide, they copy their DNA and pass one complete copy to each daughter cell. Would this process work to make egg or sperm cells? Explain why or why not.

4. Would natural selection work if the younger generation were genetically identical to the older generation? That is, are mutation and (for organisms that reproduce sexually) the combining of DNA from both parents required for natural selection to work?

The Informed Citizen

1. (a) Use the Internet to search for the "infidelity gene" (or "ruthless gene"). Select several sites and try to evaluate each for clarity, accuracy, and usefulness. (b) Search for a commercial lab that will test your DNA for the presence of the infidelity gene. This process is called direct-to-consumer (DTC) marketing. In this case, DTC marketing gives people genetic information without the benefit of a trained genetic counselor or other medical professional. Do you think most people will be able to understand the test results on their own? Should the government regulate DTC marketing of genetic screening procedures? Should it be eliminated in favor of medically supervised tests?

2. Do you think your genes influence your behavior? Do they control your behavior? If a person has the "infidelity gene" and the hormone imbalance it produces, is fidelity still a choice this person can make? Why or why not?

3. Suppose a public hearing is convened in your town to discuss, and eventually vote on, the bulleted scenarios. For each of the scenarios, answer these questions: (a) What information would you need to draw your conclusion or to form a viewpoint? (b) From where or whom would you seek this information? (c) What specific questions would you need to have answered to make an informed choice or decision? (d) How would you vote?

- A biotechnology company has plans to build a facility in your community that will provide genetically modified organisms to local farmers.

- A group of parents submit a petition to your local school board asking that intelligent design be taught in biology courses as an alternative explanation to natural selection.

- The local university seeks state funding to build a new research facility where scientists will study human genetics, research animal cloning, and perform stem cell research.

- Your town proposes to add fluoride to the drinking water.

CHAPTER LEARNING OBJECTIVES

After reading this chapter, you should be able to answer the following questions:

2.1 How Would a Scientist Investigate Childbed Fever?
Outline the steps that Ignaz Semmelweis took to find the cause of childbed fever.

2.2 How Does Science Work?
Determine whether a given statement is an observation, hypothesis, or theory, and use the steps of the scientific method to design and evaluate a simple experiment.

2.3 What Assumptions Does Science Make About Nature?
Use the assumptions of science to determine whether a statement is scientific.

2.4 What Are the Principal Features of Science?
Explain the principal features of science.

2.5 How Does Science Differ from Other Ways of Knowing?
Distinguish science from other ways of knowing.

2.6 How Does Science Differ from Pseudoscience and Quackery?
Determine whether a claim is likely science, pseudoscience, or quackery.

Most scientists try to solve a particular problem, answer a specific question, or determine how a certain process works. The work of a scientist can increase our knowledge of the world, and it can help save lives and improve our quality of life.

CHAPTER 2

The Nature of Science

How Do We Know How the World Works?

Becoming an Informed Citizen …

There is a lot we do not understand about the world. What we do know comes from the fact that the world appears to operate in a regular, repeatable way. Science exploits the world's regularity in particular ways that allow us to better understand how it works.

Albert Einstein once noted that *"the most incomprehensible thing about the world is that it is comprehensible."* There is much that we do not understand about the world and the way it works. But it is surprising that we understand as much as we do. Our knowledge comes from the fact that the world appears to follow a set of rules. Science exploits the world's regularity in ways that enable us to discover these rules. As we learn more about the rules, we come to understand the world better. This is the principal aim of science.

Most scientists try to solve a particular problem, answer a specific question, or determine how a certain process works. Science includes a number of tools, procedures, and working assumptions that facilitate these endeavors. Sometimes a scientist is motivated simply by curiosity, but often there are important consequences to a scientist's work that impact medicine, industry, and society. A few scientists, such as Isaac Newton, Charles Darwin, and Albert Einstein, developed comprehensive theories that explain a great deal about how our world works. The work of a scientist can increase our knowledge of the world, and it can help save lives and improve our quality of life.

Consider how scientific reasoning helped a doctor discover a cause and treatment for a tragic disease, childbed fever.

case study

The Mysterious Case of Childbed Fever

The pregnant young woman slowly climbed the stairs of the most modern hospital in Austria, the *Allgemeines Krankenhaus*, or General Hospital, of Vienna (**FIGURE 2.1**). Her destination was the maternity ward. It was 1846, and she was in labor. At the top of the stairs, there was a doorway opening into

FIGURE 2.1 During the 1840s, Ignaz Semmelweis investigated the cause of childbed fever at the Allgemeines Krankenhaus (General Hospital) in Vienna.

FIGURE 2.2 Autopsies are thorough examinations conducted by doctors to determine the cause of death. This famous painting by Rembrandt, *The Anatomy Lesson of Dr. Tulp* (1632), features an autopsy of a male patient.

a reception area. On the right was a hallway leading to Division 1, where doctors and medical students delivered babies. The hallway on the left led to Division 2, where midwives and their students attended the laboring mothers. As the young woman was led to the right, she hesitated, remembering that other women had told her to avoid the doctors, since there were rumors that many of the women who went to the right never came back. She had heard stories about women in labor who, upon being sent to Division 1, would panic and beg to be discharged from the hospital and sent home.

The young woman delivered a baby boy, and at first it seemed to her that her future was a bright one. However, a few hours after her baby was born, she started to feel a pain in her lower abdomen. The pain was much worse when any of the sore areas were touched. She became nauseous, vomiting frequently, and her pain worsened. Suffering with a high fever and a racing pulse, her abdomen filled with gas, causing even greater pain. She continued to worsen in all regards, and just three days after her baby's birth she died.

Upon opening the body of the deceased woman during the autopsy (**FIGURE 2.2**), the doctors and the medical students who were there to observe encountered a terrible, foul stench in her abdomen. Inside, they observed that her uterus wall was swollen, and pus and putrid flesh were evident in the body cavity. The skin and flesh were filled with gaseous bubbles and a foul-smelling fluid. The doctors determined that the woman had died of childbed fever, a disease that claimed the lives of one out of every 10 women who delivered babies in Division 1 of the hospital.

Why was giving birth so dangerous in the 1840s, and what has changed? We will find out shortly, while learning more about how the world works, thanks to the science of biology. This chapter will introduce you to the nature of science.

How Would a Scientist Investigate Childbed Fever?

Ignaz Semmelweis was a Hungarian, born in 1818 in the city now called Budapest, the fifth of 10 children in a prosperous family. He earned his medical degree in obstetrics in 1844. For two years, he worked on the medical staff in Division 1 of the Allgemeines Krankenhaus before becoming head of the division in 1846. In that year alone, 459 women died of childbed fever in his division, and Semmelweis became obsessed with discovering the cause of the disease and a way to prevent it.

Looking for Clues

One of the first things Semmelweis did was to compile records of births and childbed-fever–related deaths at the hospital. His results for the years 1841–1846 are shown in **TABLE 2.1**.

These records provided Semmelweis with his first clues about the causes of childbed fever. He discovered that death rates were over 2.5 times higher in Division 1. During this period, doctors and their medical students (all male) treated the women in Division 1, while midwives and the midwife students (all female) treated the women in Division 2 (although occasionally doctors worked in Division 2 as well). In addition to these data from the hospital, Semmelweis had anecdotal information: that women who delivered babies outside the hospital, either at home or on their way to the hospital, hardly ever contracted childbed fever.

Possible Causes

Whatever the cause of childbed fever, it appeared to involve practices at the hospital itself rather than some epidemic among the public. Semmelweis quickly set about investigating the cause by narrowing down the possibilities.

Was it miasma? One of the first possibilities Semmelweis pursued was that childbed fever was caused by **miasma**: harmful, toxic vapor supposedly exhaled by sick people or exuded by garbage or sewers into the air around them. According to this widely held belief, breathing in miasma could cause people to become sick, which indeed happened regularly to people surrounded by other sick people or by garbage and sewers.

miasma :: harmful, toxic vapor supposedly exhaled by sick people or exuded by garbage or sewers into the air around them

Table 2.1 :: Maternal Births and Deaths by Division, 1841–1846

	DIVISION 1			DIVISION 2		
Year	Births	Deaths	Rate (%)	Births	Deaths	Rate (%)
1841	3036	237	7.8	2442	86	3.5
1842	3287	518	15.8	2659	202	7.6
1843	3060	274	9.0	2739	164	6.0
1844	3157	260	8.2	2956	68	2.3
1845	3492	241	6.9	3241	66	2.0
1846	4010	459	11.4	3754	105	2.8
TOTALS	20,042	1989	9.9	17,791	691	3.9

Sources: I. Semmelweis, *The Etiology, Concept, and Prophylaxis of Childbed Fever*, 1861; and adapted from D. Gillies, *Hempelian and Kuhnian Approaches in the Philosophy of Medicine: The Semmelweis Case*, n.d.

Semmelweis predicted that if miasma were the cause, he could reduce the death rate by increasing the ventilation and air circulation in the divisions. He took measures to keep windows open and found other ways to introduce fresh air. However, none of these measures reduced the incidence of childbed fever, which led Semmelweis to conclude that miasma was not the cause. Even if miasma had been implicated as a cause, Semmelweis realized that this would not explain the differences in death rate between the two divisions, since both were in the same building and even shared a common entryway.

Was it the birthing position? Semmelweis considered another possibility based on the method doctors used to deliver babies. In Division 1, doctors had the mother lie on her back during labor, while in Division 2, midwives had the mother lie on her side. If lying on one's back was somehow a cause of childbed fever, Semmelweis predicted that changing the birthing position would reduce the death rate. He instructed doctors to deliver babies while the mother laid on her side, the same practice used in Division 2. Unfortunately, changing the birthing position had no effect on the incidence of childbed fever.

Were the male medical students being too rough on the patients? One of the main differences between the two divisions was the presence of male medical students in Division 1 and the presence of female midwife students in Division 2. Thus, Semmelweis considered whether the male medical students were too rough when they examined and treated pregnant women. This idea led Semmelweis to reduce the number of medical students by half, from around 40 to 20, and to reduce the number of examinations that each patient underwent. Unfortunately, this practice also failed to reduce the death rate significantly.

What else could it be? As 1846 came to an end, Semmelweis became despondent. He had tested many different possibilities, but the results of his tests indicated that none of them caused childbed fever. In March 1847, Semmelweis tried to renew himself by taking a vacation to Venice. When he returned on March 20, he learned of the death of his friend and fellow doctor Jakob Kolletschka. Kolletschka's death was tragic, but it provided the final clue Semmelweis needed to determine the cause of childbed fever.

"Cadaverous Particles"

Jakob Kolletschka had been stuck in the finger by a medical student's scalpel during an autopsy on a woman who had died of childbed fever. Within a few days, Kolletschka died of a massive infection. When his body was autopsied, doctors found pus, putrefied tissue, and many of the other abnormalities typically found in the cadavers (dead bodies) of women who died from childbed fever. Semmelweis concluded that Kolletschka in fact died from childbed fever, even though he clearly was not pregnant. This led Semmelweis to the conclusion that the corpse of a victim of childbed fever contained "cadaverous particles" that caused the disease (**FIGURE 2.3**). Today, we would call these particles "germs," or more specifically **microorganisms** (like bacteria). If these particles entered the body of a healthy person, that person would catch the disease as well. In the case of Kolletschka, the cadaverous particles contaminated the scalpel during the woman's autopsy and entered his body when the scalpel cut his finger.

microorganism :: a living thing that cannot be seen without the use of a microscope, such as bacteria

Semmelweis's concept of cadaverous particles explained many aspects of the childbed fever mystery. Only doctors and medical students performed autopsies, most of them on women who died of childbed fever. Typically, doctors and their medical students performed autopsies in the morning and then went on to Division 1 to examine patients and deliver babies. Sterilization of equipment and latex gloves were not part of the medical repertoire in 1847.

FIGURE 2.3 Semmelweis discovered that childbed fever could be greatly reduced by washing hands and instruments thoroughly to eliminate "cadaverous particles" (what today we know as microorganisms). This 1966 painting by Robert Thom, *Semmelweis—Defender of Motherhood*, depicts medical professionals washing their hands.

Doctors and students washed their hands after performing autopsies, but often their hands still smelled from the corpse, suggesting that their hands could be contaminated with cadaverous particles. These particles could be transferred to patients during an examination or when delivering a baby, leading to infection of the mother. Since the midwives in Division 2 did not perform autopsies, patients in this area would be infected at a much lower rate.

This idea relating to cadaverous particles led to a simple prediction. If cadaverous particles on the hands of doctors and medical students caused childbed fever, then more rigorous hand cleaning would reduce the incidence of the disease. Semmelweis placed a bowl of diluted chlorine bleach at the door to Division 1. He required all medical personnel to wash their hands and instruments with it (**FIGURE 2.4**), using a brush to clean under fingernails, before entering the ward to examine women or deliver babies.

FIGURE 2.4 Semmelweis required all medical personnel to wash their hands before entering the ward to examine women or deliver babies. The process outlined here should take about the time it takes to sing "Happy Birthday" twice.

1 Apply soap and rub palms together.

2 Rub the backs of both hands.

3 Rub in between fingers.

4 Rub fingertips against palms.

5 Rub thumbs and wrists.

6 Scrub under fingernails.

Table 2.2 :: Maternal Births and Deaths by Division, 1847–1858

Year	DIVISION 1			DIVISION 2		
	Births	Deaths	Rate (%)	Births	Deaths	Rate (%)
1847	3490	176	5.0	3306	32	1.0
1848	3556	45	1.3	3319	43	1.3
1849	3858	103	2.7	3371	87	2.6
1850	3745	74	2.0	3261	54	1.7
1851	4194	75	1.8	3395	121	3.6
1852	4471	181	4.0	3360	192	5.7
1853	4221	94	2.2	3480	67	1.9
1854	4393	400	9.1	3396	210	6.2
1855	3659	198	5.4	2938	174	5.9
1856	3925	156	4.0	3070	125	4.1
1857	4220	124	2.9	3795	83	2.2
1858	4203	86	2.0	4179	60	1.4
TOTALS	**47,935**	**1712**	**3.6**	**40,870**	**1248**	**3.1**

Sources: I. Semmelweis, *The Etiology, Concept, and Prophylaxis of Childbed Fever*, 1861; and adapted from D. Gillies, *Hempelian and Kuhnian Approaches in the Philosophy of Medicine: The Semmelweis Case*, n.d.

This procedure yielded spectacular success. Within months, the mortality in Division 1 dropped to the level seen in Division 2, approximately 3%. Semmelweis kept childbed-fever–related death records for the next 12 years. The rates varied slightly over time but remained consistently low, with no major differences between the two divisions (**TABLE 2.2**).

With over 150 years of scientific and educational advances since Semmelweis's time, it is easy to understand why his approach was successful. His cadaverous particles were actually bacteria, which cause a lethal infection when they enter a person's bloodstream (*see Life Application: Childbed Fever*).

Life Application

Childbed Fever

Outbreaks of childbed fever occurred regularly in the cities of Europe and America for almost two centuries, between 1700 and 1900. During this time, the practice of medicine advanced dramatically, and many hospitals were built in the larger cities. But more hospitals and more doctors attending births just made the problem worse. There was considerable evidence linking childbed fever to hospitals. For example, in London between the years of 1831 and 1843, 10 mothers per 10,000 died of childbed fever with home delivery, while 600 per 10,000 died in hospital maternity wards. There was also evidence of the link between doctors and contagion. Many doctors across the globe pursued investigations similar to those of Semmelweis and reached similar conclusions. In the

1840s Boston physician Oliver Wendell Holmes Sr. published a paper arguing that doctors should wash their hands and sterilize their instruments to reduce the transmission of the disease.

Unfortunately, science doesn't always work out the way it should. The results reported by Semmelweis, Holmes, and others were attacked by a number of prominent physicians. Despite the evidence, these physicians did not accept the idea that childbed fever was contagious. In fact, they felt it was preposterous that well-trained doctors could be responsible for transmitting disease. When hospitals initiated hygienic practices, they saved the lives of many women. Yet it was not until the early 1900s that the majority of doctors accepted

(Continued)

Childbed Fever (*Continued*)

that childbed fever and other diseases are contagious and caused by germs. As the germ theory took hold, modern hygienic and sterilization practices became the norm.

Childbed fever is known to be caused by the bacterium *Streptococcus pyogenes*, shown here. This is the same bacterium that causes strep throat and other more serious diseases such as toxic shock and flesh-eating disease.

Today, childbed fever, more technically referred to as *puerperal sepsis*, is known to be caused by a bacterium: *Streptococcus pyogenes*. This is the same bacterium that causes strep throat, scarlet fever, rheumatic fever, and a variety of skin infections. *S. pyogenes* normally lives on the skin and inside the nasal passages. Typically, up to 20% of the population is infected with the bacterium, but as long as it lives outside of our bodies, it does no harm. Once it infects the bloodstream, however, the bacterium changes its genetic expression in ways that cause the diseases just mentioned.

Childbed fever still occurs occasionally, but with early detection it is easily treated with antibiotics, and death is rare. It is intriguing that the bacterium is still present, although outbreaks of the disease are gone. Some biologists hypothesize that the bacterium has evolved into a less virulent form, since for most parasites killing the host is a bad idea. ::

Chlorine bleach kills bacterial cells, and washing hands in bleach drastically reduces the risk of transmitting bacteria from doctor to patient. In 1847, however, none of this was known. Doctors at the time were well aware that microorganisms existed, but most did not believe that microorganisms could cause disease. As a result, many doctors did not adopt Semmelweis's practices, even though they clearly were successful in reducing death from childbed fever. It was not until about 25 years later that research by scientists like Louis Pasteur and Robert Koch would establish the link between microorganisms and disease, now known as the "germ theory" of disease (see *Scientist Spotlight: Robert Koch*).

2.2 How Does Science Work?

scientific method :: set of procedures scientists use in their investigations; includes four steps: observations and facts, hypotheses and predictions, testing, and evaluation and interpretation of results

Semmelweis found a treatment for childbed fever because he took a scientific approach to the problem. In this section, we examine the **scientific method**: the set of procedures scientists use in their investigations (**FIGURE 2.5**, p. 42). Depending on the particular question or problem being explored, scientists may not always follow all of these steps in the order described, but they will emphasize certain components of the scientific method, as appropriate.

Observations and Facts

observation :: what you can see, hear, smell, taste, or feel physically

fact :: something that has actual existence

A scientific investigation begins with **observations** (what you can see, hear, smell, taste, or feel physically) and **facts** (things you know to be true) about the natural world. These relate to some phenomenon, process, or pattern that we see in nature. Scientists are curious. They want to explain the observations and facts they see. They want to discover the cause, reason, or mechanism that explains these observations and facts. Semmelweis observed two facts at the outset of his investigation: (1) deaths due to childbed fever had higher incidence in the hospital than outside, and (2) within the hospital, the incidence

Robert Koch (1843–1910)

Robert Koch was born December 11, 1843, in the Upper Harz Mountains of Germany. Even as a boy, he was smart, determined, and persistent. He was also methodical. By the time Koch was five years old, he had actually taught himself to read by looking at newspapers. No one had helped him in this endeavor. These personal characteristics, already evident in childhood, were ultimately critical for the successes that Koch achieved as a scientist.

Robert Koch was instrumental in the development of the germ theory of disease.

Interested in nature and biology from childhood, Koch attended the University of Göttingen, earning an M.D. in 1866. He began private practice in 1867 and in 1870 volunteered for service in the Franco-Prussian War. After the war, in 1872, Koch served as the district medical officer in a rural area of Prussia that is now part of Poland. Although he was a skillful and effective physician, Koch's restless intellect was not satisfied with the practice of medicine alone. He set up a laboratory for research in his four-room house. It was here that he labored for the next four years, alone, with a microscope his wife gave him as a gift and other equipment he made himself. During this period, Koch had no access to libraries and no scientific colleagues with whom he could exchange ideas.

Koch first took an interest in *anthrax*, a deadly scourge for farm animals and farmers. Although the bacterium associated with anthrax had been discovered earlier, Koch wanted to show scientifically that this bacterium was indeed the specific *cause* of this disease. Koch injected mice, using slivers of wood that he made, with blood from the spleens of animals that had died from anthrax. For a comparison, or control, he also injected another group of mice with blood from spleens of healthy animals. Koch carefully observed both populations of mice, and to his delight, the results were unequivocal. The mice inoculated with the anthrax bacteria died of the disease, whereas the control mice remained healthy.

Koch continued his study of anthrax bacteria and developed pure cultures of them for microscopic study. He photographed the bacteria and the structures they formed. Anthrax bacteria make these structures, called spores, when growth conditions are poor. Koch showed that spores could germinate to create new bacterial cells. These observations solved a mystery associated with anthrax: how could animals get the disease even if they had not been exposed to other sick animals? The answer is that anthrax spores can remain dormant in the soil for many years until there is an opportunity to infect a new host. The ability of anthrax to form long-lasting, deadly spores is one reason for concern about its use as a biological weapon.

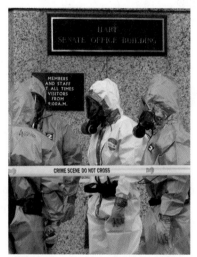

Koch's discoveries about the ability of anthrax to form long-lasting deadly spores led to concern about the use of anthrax in biological weapons. In 2001, an anthrax scare shut down the U.S. Capitol Complex, which houses the Senate and House of Representatives.

By 1880 Koch had established his own professional research lab, where he developed many new techniques for growing and observing bacteria. Koch also traveled all over the world studying the infectious origins of many diseases including tuberculosis, cholera, plague, malaria, and typhus. The recipient of many awards, Koch won the 1905 Nobel Prize in Physiology or Medicine for his work on the germ theory of disease. ::

Steps of Scientific Method

FIGURE 2.5 Semmelweis used the steps of the scientific method to investigate the cause of childbed fever.

was higher in Division 1 than in Division 2. His investigation focused on finding the cause of these two facts.

Hypotheses and Predictions

hypothesis :: possible cause or mechanism that could explain observations and facts

The second step in a scientific investigation is to develop a **hypothesis**. A hypothesis is a possible cause or mechanism that could explain the observations and facts. Semmelweis developed a number of hypotheses for the cause of childbed fever: miasma, birthing position, or rough treatment by male examiners. Scientists often develop several hypotheses during the course of their investigation if their initial hypothesis turns out to be incorrect.

prediction :: educated speculation about what an outcome will be

Every scientific hypothesis leads to a **prediction**, or an educated speculation about what an outcome will be. If you understand how something works, you can predict how it will change or work differently under different conditions. Semmelweis came up with a different prediction for each of his hypotheses. For example, if his hypothesis that miasma causes childbed fever were correct, the prediction would be that ventilation will reduce the incidence of the disease. All scientific predictions have this form: *if* the hypothesis is correct, *then* a certain set of conditions should yield a certain set of results.

Testing

test :: a procedure that sets up the conditions the predictions require

The third step is to **test** the hypothesis by designing a procedure to set up the conditions the predictions require. If you look back at the form predictions take, you can see how the test should be done. If the predicted results occur, your hypothesis may be correct. If they don't, your hypothesis is probably wrong. When Semmelweis tested his hypotheses, most of them turned out to be wrong. He added ventilation, he changed the birthing position, and he reduced examinations by male students—in each case, his hypothesis predicted that the incidence of childbed fever would go down. But it didn't. His hypotheses were not supported by the results. This is a common, and very frustrating, part of science. But Semmelweis's final hypothesis, that cadaverous particles transmit the disease, led to a prediction that came true when it was tested: washing hands and instruments to remove the particles reduced the number of cases. It took a while, but careful testing allowed Semmelweis to discover the cause of childbed fever, which led in turn to a way to prevent it.

There are several ways to test a hypothesis, but a **controlled experiment**—in which a scientist keeps all variables (possible factors that could affect the outcome of the test) the same except for the one under investigation—is one of the most common. Generally, there are many extraneous factors that can affect the outcome of a test. Suppose, for example, you developed a hypothesis that predicts a specific type of fertilizer will increase plant growth. If you grow your plants outside in a garden, factors like rainfall, clouds, and cold temperatures will affect your results and may overpower any effects the fertilizer might have. In an experiment, the scientist controls all the factors except the one under investigation. Plants could be grown in a greenhouse under constant temperature and in groups that all received equal amounts of light and moisture but varied in how much fertilizer they got. Because the amount of fertilizer is the only difference among the plant groups, the scientist can conclude that any corresponding differences in growth are due to the fertilizer.

controlled experiment :: a test or manipulation in which a scientist keeps all variables (possible factors that could affect the outcome of the test) the same except for the one under investigation

Evaluation and Interpretation of Results

The fourth and final step is to evaluate and interpret the test results. We have already discussed this somewhat. If the predicted results occur, this provides support for the hypothesis. If they don't, this is good evidence that the hypothesis is wrong. But given the results, what does the scientist do next?

In general, the first thing a scientist does is repeat the experiment. If the predicted results do not occur, it could be because some mistake was made performing the experiment. Repeating the experiment can determine whether this is the case. But if repeated failures convince the scientist that the hypothesis is incorrect, it's back to the drawing board. Like Semmelweis, the scientist comes up with a new hypothesis, a new set of predictions, and a newly designed test. There is no guarantee, but persistence often pays off with a successful test that supports the new hypothesis.

If the predicted results occur, the scientist wants to know whether they will occur consistently. If repeated experiments all yield the same results, the scientist has more confidence in the hypothesis.

Even if the test results consistently support the hypothesis, the scientist still is not done (as you might expect, science is never "done"). Most hypotheses lead to more than one set of predictions. To make the case stronger, a scientist will stick with the hypothesis but design other experiments to test these other predictions. Eventually, the scientist can accumulate a good deal of evidence, all of which may support the hypothesis. The accumulation of evidence helps the scientist to convince others that the hypothesis is true, perhaps in a published paper or a presentation at a conference.

But in science, no matter how much evidence you have to support a hypothesis, *the hypothesis is never proven with absolute certainty.* Proof like that happens only in mathematics, not in the natural world. Even if a large number of experiments support a hypothesis, you can't be certain that the next one will do so as well. And nature is so complex, with so many factors that affect one another, that even if you get your expected results, you can't be certain you have the correct cause. Some other factor you have not considered could be responsible for the results (see *How Do We Know? Hypothesis Testing and Scientific Proof*).

Despite these difficulties, it would be wrong to think that science can never lead to valid conclusions. Just take a moment to consider all the things that science has made possible: computers, improved medical care and increased longevity, and robotic probes that explore the surface of Mars, to name just a few. The fact that scientific ideas are not proven means that eventually they may be replaced with even more accurate ideas. But until then, the practical results of current scientific inquiry provide us with many benefits.

How Do We Know?

Hypothesis Testing and Scientific Proof

Science is based on a type of logic known as inductive reasoning—the search for general truths from specific observations. For example, suppose that the Republicans I know are conservatives. If I then generalize to say that all Republicans are conservatives, I have used inductive reasoning: *if* these Republicans are conservatives, *then* all Republicans are conservatives.

In this example, my observations (*if* . . .) are the *premises*, which are followed (*then* . . .) by the *conclusion*. If the premises are false, then the conclusion is false, too. It is possible that one of these Republicans is a libertarian or that he or she subscribes to some other ideology or set of beliefs. But even if all the premises are true, the conclusion can still be false. The conclusion makes a prediction, and further observations may prove that prediction wrong. After all, no matter how many conservative Republicans you observe, it takes only one liberal Republican that you did not observe to falsify the statement.

Even if a predicted result does occur, this does not prove that a hypothesis is true. There may be a number of different hypotheses that make the same prediction. For example, in the past, many people believed that the sun orbited the Earth. Let's make that our hypothesis: the sun orbits the Earth once a day, clockwise as seen from above the North Pole. Our hypothesis predicts that the sun will rise in the east and set in the west every day. And it does, but our hypothesis is still false. We neglected to consider another hypothesis that makes the same prediction: the Earth rotates about its axis once a day, counterclockwise, and this rotation is responsible for the rising and setting of the sun.

A

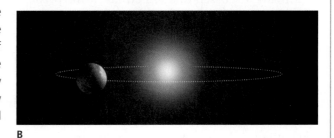

B

Even if a predicted result occurs, such as the rising and setting of the sun every day, this does not prove that a hypothesis is correct. Many people used to believe that the sun orbited the Earth (A), when in fact another hypothesis, the Earth rotates round its axis once a day (B), makes the same prediction.

Even though you can't prove inductive statements about nature, they are still useful and meaningful. Just like science, induction begins with a series of empirical observations. The world is a complex place, so you can never observe *everything* about it, or even a part of it. Since your observations are incomplete, any conclusions you draw from them could be incorrect. But the nice property of induction is that as you add more observations, you increase the chances that your conclusion is correct. In science, this means that repeated experiments and related work by other scientists provides more information about nature and brings us ever closer to understanding how nature truly works.

It is important to understand inductive reasoning if you are to understand the nature of scientific "proof." Even if a hypothesis (or theory) has support from prior research, it will still be false if it fails to make correct predictions in the future. A truly scientific hypothesis is falsifiable: it makes predictions that could fail repeated tests. It is wrong, however, to conclude that science cannot prove anything. If a hypothesis makes correct predictions under repeated testing, a scientist can conclude that the results support the hypothesis. We can then treat this hypothesis as proven until contradictory evidence arises. ::

Scientific Theories

You may be familiar with a number of scientists who are famous because they developed a scientific theory. Isaac Newton's theory of gravity, Charles Darwin's theory of evolution by natural selection, and Albert Einstein's theory of relativity are good examples of scientific theories, and these scientists are famous because of them. But what exactly is a scientific theory?

Theories are similar to hypotheses in some ways, but they provide a bigger picture of how some aspect of nature works, and they may weave together supporting evidence from several scientific fields. Both theories and hypotheses explain why certain events occur in the natural world. Both make predictions that can be tested. Hypotheses, however, focus on explaining smaller and more specific events. Theories are broader in scope and explain a wider variety of events. You can think of successful hypotheses as representing individual pieces in the puzzle of nature. Theories put these pieces together (**FIGURE 2.6**).

Darwin's theory of natural selection, which we encountered in Chapter 1, is a good example. There have been thousands of studies done to determine if natural selection can account for some specific trait or behavior in a given species: the peacock's tail or the lion's mane, or the behavior of a prairie dog that risks death by giving alarm calls instead of hiding when a predator approaches. Each of these studies tests a hypothesis about natural selection in a specific species. Darwin's theory of natural selection provides a general explanation of how adaptations in *all* organisms evolved. Every individual scientific study that tests some detail of natural selection has the potential to support Darwin's theory, as do results from a number of other scientific fields including geology, paleontology, and cosmology.

In science, a theory is as close to a proof as one can get. Like hypotheses, however, *theories are only supported by the evidence, not proven*. Given enough time, new theories can supersede earlier ones, though this can take a long time. Newton's theory of gravity lasted four centuries before Einstein's theory of relativity replaced it.

Scientific theories tie together results of numerous studies, often from several different disciplines. Theories also provide a framework for further investigations as scientists pursue new questions in new contexts to test them. Over time, a successful theory increases our understanding of the fundamental processes of nature and the rules by which nature operates. The idea that nature operates according to rules that we can discover and understand is relatively recent. As we see next, this idea is based on a number of assumptions that are critical to how science works.

theory :: an idea, supported by evidence, which provides a bigger picture than a hypothesis of how some aspect of nature works; it may weave together supporting evidence from several scientific fields

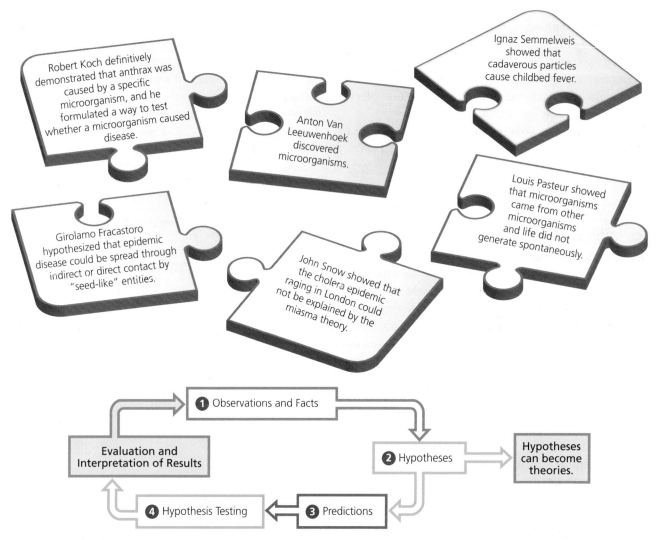

FIGURE 2.6 While hypotheses represent individual pieces in the puzzle of nature, theories put these pieces together. All of these hypotheses and observations made by different scientists contributed to the germ theory of disease, which states that infectious disease is caused by microorganisms.

2.3 What Assumptions Does Science Make About Nature?

Science is built on a number of assumptions about the nature of the world. These assumptions have come from years of research and experiments by scientists all over the globe. The assumptions are based on four related principles, or rules, that have been shown to hold true over hundreds of years of study: cause and effect, consistency, repeatability, and materialism.

Cause and Effect

**cause and effect :: ** every event or outcome in nature has a source; if a scientist sets up the correct conditions, the results can be predicted in advance

First and foremost, science assumes that the rules of **cause and effect** (**FIGURE 2.7**) hold for all natural phenomena. Every event or outcome in nature has a cause or source, and since cause and effect are related, a scientist can learn about causes by observing the effects that occur. As we saw previously, cause and effect is the key to testing a hypothesis. If a scientist sets up the correct causes (conditions), the effects (results) can be predicted in advance.

Cause: Carbon dioxide emissions

Cause: Methane emissions from animals

Cause: Crop fertilizers

In complex systems, cause and effect are often distant in time and space.

Effect: Polar ice caps are melting

Effect: Sea levels are rising

Effect: More violent storms are occurring

FIGURE 2.7 Every event or outcome in nature has a cause or source. A scientist can learn about causes by observing the effects that occur; note that in this figure all of the causes on the left contribute to the effects on the right.

Consistency and Repeatability

Science also assumes that events are consistent and repeatable. That is, if the same conditions (causes) are set up, the same results (effects) will occur. This assumption is important because it leads to the idea that there are general rules in nature, and that these rules will operate in a similar manner over and over again. It is also a reason why scientists repeat their experiments and try to repeat those of other scientists. Experimental results that can be regularly repeated become more strongly established and are more likely to be true. On the other hand, results that cannot be repeated by other scientists become suspect and may likely be the result of experimental error.

Experimental repetition has another consequence: it makes science *self-correcting*. Individual scientists can make mistakes in how they perform an experiment or interpret its results. Through the collective action of many scientists repeating these experiments, mistaken results may be weeded out.

Materialism

Another assumption that science makes is that effects in the natural world all have natural causes, rather than supernatural ones, a position known as **materialism** (**FIGURE 2.8**). Notice that all the hypotheses Semmelweis developed were based on natural causes. He did not hypothesize that childbed fever was caused by demonic possession or by divine retribution for the mother's sins. Those hypotheses may have seemed reasonable to people two or three

materialism :: the idea that effects in the natural world all have natural causes, rather than supernatural ones

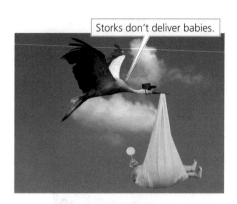

FIGURE 2.8 According to the concept of materialism, all effects in the natural world have natural causes, rather than supernatural ones.

centuries ago, but the rise of science in Western culture over the past centuries often has resulted in replacing supernatural causes with natural ones.

The assumption of materialism is *required* for science to work. As mentioned, science studies cause-and-effect relationships that are repeatable, consistent, and conform to general rules of nature. Events with supernatural causes need not have any of these characteristics; they can appear quite unpredictable. Imagine Isaac Newton's difficulty in studying gravity if sometimes the apple fell straight down from the tree, while other times it floated, rose, or shot off horizontally, all seemingly at random. The scientific method would not be useful in such a case.

The scientific method has proven useful. That alone suggests that the assumptions it makes about the nature of the world are valid. Cause and effect do hold, and events have natural causes and are generally repeatable. Many scientists are so convinced by science's success that they adopt a completely materialistic view: *all* events, they believe, have natural causes, and the supernatural does not exist. Here, science can come into conflict with other approaches to knowledge, particularly religion. However, other scientists think instead of the supernatural as a limit to science: it is something that science simply cannot study. Either point of view has more to do with philosophy than science.

2.4 What Are the Principal Features of Science?

As we have seen, the scientific method gives scientists a procedure to follow when they perform their investigations. Science also makes a number of assumptions about how nature works. Many ways of learning use a similar logic, but a few other features combine to make science unique. Science is based on empirical evidence, it can be tested, and it can be applied more generally to other situations. Semmelweis relied on these features when he investigated the cause of childbed fever (**TABLE 2.3**).

Empirical Evidence

empirical evidence :: information that one gets from direct observation, from experience, or from the results of experiments and other tests of hypotheses

Science is based on evidence rather than on opinion or belief. **Empirical evidence** includes information that one gets from direct observation, from experience, or from the results of experiments and other tests of hypotheses. Empirical evidence in the form of background information helps a scientist develop a hypothesis, a possible mechanism to explain a particular fact or

Table 2.3 :: Principal Features of Science, as Illustrated by Semmelweis's Studies

Feature	Semmelweis's Studies
Empirical evidence	Observed that spread of cadaverous particles was associated with childbed fever
Testability	Tabulated death rate before and after the initiation of hand washing
Generality	Determined that findings were relevant to all maternity cases, not just those in the Allgemeines Krankenhaus

observation. The results of a scientific investigation produce more empirical evidence that other scientists can use.

Most important, every scientific hypothesis and theory is evaluated strictly by how well it explains the existing empirical evidence. No matter how popular an idea is, how logical, or how well it conforms to ideology or belief, if it does not fit in with the available evidence, it is wrong. Semmelweis probably expected that childbed fever was caused by miasma, since this was the prevailing view of disease at the time. But the evidence did not support miasma as a cause—the evidence prevailed, and the miasma hypothesis failed.

Testability

A commitment to **testability**—or a procedure for determining the evidence in support of a hypothesis—is the central tenet of science. When scientists hypothesize a cause for a particular fact or observation, they want to know if they are right. In science, that means their hypothesis is supported by evidence and makes a successful prediction. Science is self-correcting because tests like these can be repeated. Controlled experiments are the main form of testing, but in cases where experiments are not feasible, other types of observations can provide a test.

testability :: a procedure for determining the evidence in support of a hypothesis

Regardless, for a hypothesis to be testable, it must be **falsifiable**; it must be able to be proved wrong. As we have seen, a successful test or experiment provides only *support* for a hypothesis, not proof that it is right. But an unsuccessful test, one in which the predicted results do not occur, gives conclusive evidence that the hypothesis is false (provided the experiment was performed properly). False hypotheses eventually are identified and discarded, leaving only those hypotheses that are supported by empirical results. Semmelweis developed several hypotheses about the cause of childbed fever and tested each one. Falsifying his hypotheses—showing the hypotheses were incorrect—though certainly frustrating, helped him rule out many possible causes and eventually led him to find the true cause of the disease.

falsifiable :: able to be proved wrong

Generality

Most scientific investigations are narrow in scope and apply to a particular event, process, or situation. But scientists are interested in the **generality** of their work, how widely it applies to situations other than the specific ones they tested. When scientists present or publish their work, for example, they are careful to point out how their work relates to the work of other scientists. Scientific ideas are tested under a variety of conditions, at different times, and at different places. Over time, scientists develop an understanding of the general principles that underlie these individual results. These general principles are embodied in scientific theories that explain how the world works and the rules by which it operates. Discovering these principles of nature is one of the

generality :: how widely a scientific investigation applies to situations other than the specific ones scientists tested

Technology Connection

Throat Cultures

Although childbed fever is no longer the terrible problem it once was, the bacterium that causes it, *Streptococcus pyogenes*, still infects humans. Strep throat, for example, is due to one type of *S. pyogenes*. This illness is characterized by a very sore throat, red pharynx, and fever. If you have an unusually sore throat—so painful that it can be difficult to swallow—it is important to get tested for strep. While most untreated strep infections eventually clear up, rheumatic fever develops in 3% of the cases. This can lead to heart and kidney problems, as well as arthritis.

How is strep throat detected? First, the healthcare provider will examine the patient's throat for redness, swelling, or patches of pus. He or she will gently touch the patient's throat, particularly the tonsils and pharynx, with a sterile cotton swab. With the rapid strep test, the swab is dipped in a test tube containing chemicals that will extract molecules from the bacteria if it is present. Next, an antibody that binds to molecules specific to *S. pyogenes* is added, along with a third substance that will change color if it detects the binding of antibodies to *S. pyogenes* molecules. The rapid test takes only minutes to get results. It is about 80% sensitive, meaning

In a positive throat culture for strep throat, the presence of hemolysis of blood agar can be detected. *S. pyogenes* causes the destruction of red blood cells in the agar, resulting in a clearing of the red color.

the test will fail to detect 20% of the positive cases. It is 95% accurate, which means it will read positive or negative incorrectly only 5% of the time. If the rapid test gives a positive result, antibiotic treatment begins immediately so that *S. pyogenes* will be eliminated from the patient.

Swab samples are also taken from the patient's throat so that a culture test can be performed. With this technique, the sample from the patient is cultured on a specific type of microbiological medium called blood agar. *S. pyogenes* is *hemolytic*, meaning it will destroy the red blood cells in its vicinity, thus forming a white halo around the bacterial colony. It can take a couple of days to get results from the culture test. Nevertheless, healthcare providers will always run it to make sure they catch the cases of strep throat that were undetected by the rapid test. The culture test, although more time consuming, is highly sensitive (~95%) and accurate (95%). Even though most cases of strep throat do not produce serious health problems, the infection is so contagious that it is essential to detect and treat it, so that people with compromised immune systems or weakened health are not exposed and placed at risk. ::

A   **B**

Strep throat is detected through an exam in which the patient's throat is examined for redness, swelling, or patches of pus (A). Then a cotton swab is used to touch the patient's tonsils and pharynx (B).

major goals of science. The results Semmelweis found could be applied beyond the divisions of the Allgemeines Krankenhaus. Wherever his procedures were applied they saved lives, and today we understand the general truth that microorganisms often cause disease. (See *Technology Connection: Throat Cultures* for a discussion of one method that scientists use to test for infectious microorganisms today.)

In this chapter, we have discussed a number of features of science, but these features are not unique to science. Many other disciplines rely on one or more of them, and you have likely used some of them in your own work. Taken collectively, however, this particular combination of features distinguishes science from other ways of acquiring knowledge and allows it to stand alone as a separate human enterprise.

2.5 # How Does Science Differ from Other Ways of Knowing?

The search for truth or the answers to questions is not unique to science. We humans seek to understand ourselves and the world we inhabit. As you have seen, science explores particular questions about the natural world with a specific set of practices. And science, as a means to learn about the natural world, has been breathtakingly successful.

Does this mean that science is superior to other ways of knowing? Not at all. In fact, there are some questions that science and its tools are not equipped to help us answer. Other areas of knowledge attempt to address other types of questions, and they do so using methods appropriate for their subject area. For example, we can analyze a poem for its aesthetic value, explore what it means to live a good life, or seek to understand God. These are the tasks of literary analysis, philosophy, and theology, respectively.

In some cases, the distinction between science and another field is easy to describe. For example, science and theology could hardly be more different. Scientists never invoke the supernatural or refer to miracles to explain how the natural world works. Divine revelation is never a source of scientific understanding regarding the natural world. In contrast, many theologians do believe in a spiritual, nonmaterial world and in revealed knowledge. Some interpret this distinction as a conflict between science and religion. It does not need to be. Science and religion usually seek answers to completely different questions. Scientists are not going to answer questions about metaphysical or spiritual domains. Neither is religious scripture the place to go to learn about the material world (**FIGURE 2.9**).

Science cannot address the question of whether there is…

A divine being

an afterlife

or reincarnation

Religion cannot explain why…

A tumor forms

ice caps melt

male pattern baldness exists

FIGURE 2.9 Scientists are not going to answer questions about metaphysical or spiritual domains. Neither is religious scripture the place to go to learn about the material world.

It can be difficult to draw a clear line between science and some fields. For instance, cognitive psychologists may have a great deal in common with biologists who study the features of the brain. Similarly, historians who try to piece together long-ago events share this approach with evolutionary biologists who research fossils. This rich cross-fertilization among disciplines should come as no surprise. In fact, collaboration among different specialists is a key for tackling big problems like the health of our planet, human well-being, and our collective futures.

2.6 How Does Science Differ from Pseudoscience and Quackery?

pseudoscience :: fake science

quackery :: promoting the use and/or purchase of remedies even when there is no scientific evidence or plausible rationale for their effectiveness

Being able to recognize the distinctions that separate science from other areas of knowledge is interesting and valuable. However, knowing the difference between science and **pseudoscience** (fake science) or **quackery**—promoting the use and/or purchase of remedies even when there is no scientific evidence or plausible rationale for their effectiveness—is essential for making informed decisions about many aspects of life. Indeed, lives can be negatively affected when people choose to follow the "latest" health remedy advertised on television or when they organize their lives according to the predictions of astrological charts or tarot readings.

Pseudoscience attempts to look like actual science so that its assertions might appear valid. However, unlike science, pseudoscience begins with a claim and looks only for things that support it. Controlled experiments are never done. In fact, direct tests of any kind, even if possible, are generally avoided. Pseudoscience is indifferent to facts and tries to persuade with

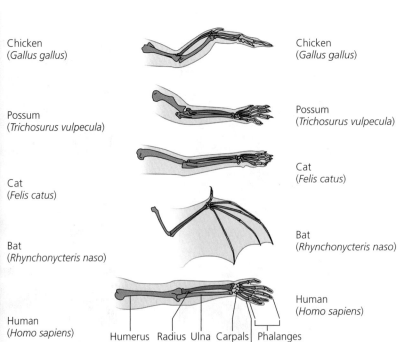

FIGURE 2.10 Evolution predicts that species change and that new species have formed during billions of years of life on Earth.
The geological record, the fossil record, evidence from embryos, and molecular studies all support this theory.

FIGURE 2.11 Creationists claim that biblical scripture supports the notion that dinosaurs and humans inhabited the Earth at the same time, despite evidence to the contrary. In the landmark federal trial *Kitzmiller v. Dover* (2005), the federal district court of Harrisburg, Pennsylvania, found that intelligent design was a form of creationism (religious theory) and therefore could not be taught in American public schools.

appeals to emotion, sentiment, or distrust of established knowledge. Unlike real science, pseudoscience does not progress; nothing is revised or learned.

Let's consider creationism as an example of pseudoscience. Creationism is the idea that a divine being created all species in a single event. That may be a valid theological notion, but it is not a scientific hypothesis. Creationism does not allow for controlled studies to test it, and predictions from it have never been observed. All too often, evidence that refutes creationism as a scientific event is ignored. For example, evolution predicts that species change and that new species have formed during billions of years of life on Earth. The geological record, the fossil record, evidence from embryos, and molecular studies all confirm just that (**FIGURE 2.10**). Moreover, controlled studies in evolution involving selection and selective breeding over many years have produced stickleback fish that can withstand colder temperatures, tame foxes, and mice with increased activity levels.

Creationists disregard this mountain of information. Sometimes, they make up "facts." In contrast to claims made by some creationists, evidence from the fossil record clearly shows that dinosaurs and humans did *not* inhabit the Earth together at the same time (**FIGURE 2.11**). Finally, creationism is a static idea, not an inquiry into nature and a way of learning more. There is no research agenda, no revision of ideas, and no progress. Nothing new is learned.

As bad as pseudoscience is, quackery is even worse. Quackery takes advantage of people when they are sick and vulnerable. The rising cost of health care and the inability of many to get access to medical help have made this a golden age for quacks. We see the claims in the media daily: a vitamin or nutritional supplement supposedly will cure fatigue; a botanical will cure depression; copper bracelets will reduce arthritis pain (**FIGURE 2.12**); a hormone

FIGURE 2.12 Quackery takes advantage of people when they are sick and vulnerable. Are copper bracelets really the best "remedy" for arthritis compared with anti-inflammatory medications?

pill promotes weight loss; a shark cartilage extract will fight cancer . . . the list goes on and on.

How can we avoid being tricked? Your best protection is to know more about biology and the nature of science. By the end of this book, you will have plenty of practice. You should be skeptical about claims that appear too good to be true. You should also trust your health to knowledgeable caregivers who rely on tested and verified treatments.

 # Biology in Perspective

There are a number of ways to learn about how the world works. However, the best way we have discovered so far to learn about the world of nature is through science. Put simply, science works. The scientific method has been used successfully to find causes, treatments, and cures of diseases, and to find ways to improve our health and increase our longevity. Throughout this text, you will encounter many other examples of scientific success.

Science assumes that the events in the world occur through a chain of cause and effect, that the causes are natural rather than supernatural, and that as a result the world operates in a consistent and repeatable fashion. The validity of scientific ideas depends on the evidence that supports them and on how well they explain events in the natural world. Ultimately scientists search for theories that integrate broad areas of science and articulate fundamental natural laws. While science is not the only way that humans gain knowledge, it is the most successful approach for understanding the natural world.

Taken by itself, the scientific method is simply a procedure for learning about nature. The procedure is neither good nor bad, ethical or unethical. However, science is practiced by human beings, and decisions about how (or whether) to use the knowledge gained from scientific investigations are made by humans. This is why it is important for citizens to understand how science works and use their knowledge to ensure that science is used in ways that enrich our lives. It is also essential to use this knowledge to recognize and reject pseudoscientific claims and quack remedies.

Chapter Summary

2.1 How Would a Scientist Investigate Childbed Fever?

Outline the steps that Ignaz Semmelweis took to find the cause of childbed fever.

- Semmelweis identified several potential causes for the deaths of women following childbirth, including miasma, birthing position, rough handling by students, and "cadaverous particles" unintentionally carried by doctors from autopsies to the maternity ward.

- Semmelweis tested each potential cause by changing conditions one at a time; the only change that had any effect on deaths was having doctors wash hands and instruments in diluted bleach prior to entering the doctors' wing.
- Procedures instituted by Semmelweis reduced women's death rate in the doctors' wing to the level of the midwives' wing.

2.2 How Does Science Work?

Determine whether a given statement is an observation, hypothesis, or theory, and use the steps of the scientific method to design and evaluate a simple experiment.

- The scientific method involves making observations and compiling facts about the natural world, constructing hypotheses, testing the predictions made by hypotheses, and evaluating evidence to determine whether the hypotheses are supported.
- One method of testing hypotheses is through controlled experiments.
- Scientific hypotheses and theories are both explanations of the natural world that make testable predictions. Both hypotheses and theories are only supported by evidence, never proven.

2.3 What Assumptions Does Science Make About Nature?

Use the assumptions of science to determine whether a statement is scientific.

- Science makes some important assumptions about the nature of the world.
  - o There is a link between cause and effect.
  - o Events are consistent and repeatable.
  - o Effects in the world have natural causes rather than supernatural ones.
- While some scientists believe the assumptions of science reflect reality, other scientists see the assumptions as limiting the types of questions that can be answered through science.

2.4 What Are the Principal Features of Science?

Explain the principal features of science.

- Scientific hypotheses and theories are evaluated strictly by how well they explain existing empirical evidence.

- Testability is a central tenet of science; in order for scientific hypotheses and theories to be testable, they must be falsifiable.
- Although most scientific investigations are narrow in scope, they gain value when results can be applied beyond the particular experiment in which they were conducted.

2.5 How Does Science Differ from Other Ways of Knowing?

Distinguish science from other ways of knowing.

- Science uses a particular set of methods, assumptions, and features to answer questions. Other disciplines use different sets of practices.
- Science can be strongly contrasted with theology. Acceptable methods specific to theology include invocation of the supernatural, divine inspiration, and reliance on scriptural evidence.
- Many fields, such as history and psychology, can incorporate scientific methods into their discipline.

2.6 How Does Science Differ from Pseudoscience and Quackery?

Determine whether a claim is likely science, pseudoscience, or quackery.

- Pseudoscience looks like science, but it begins with a claim and then selects only evidence that supports the claim.
- Creationism is an example of pseudoscience because it disregards evidence contrary to its predictions, it makes up "facts," and it refuses to change with new evidence.
- Quackery uses pseudoscience to support unfounded claims of medical cures and treatments.

Key Terms

cause and effect 46
controlled experiment 43
empirical evidence 48
fact 40
falsifiable 49
generality 49

hypothesis 42
materialism 47
miasma 36
microorganism 37
observation 40
prediction 42

pseudoscience 52
quackery 52
scientific method 40
test 42
testability 49
theory 45

Review Questions

1. What causes did Semmelweis think could be responsible for the spread of childbed fever when he first began to pursue possibilities?

   a. Miasma, sunlight, delivery method

   b. Bacteria, sunlight, delivery method

   c. Miasma, delivery method, rough treatment of patients

   d. Bacteria, delivery method, rough treatment of patients

   e. Miasma, bacteria, sunlight

2. What was Semmelweis's evidence for his cadaverous particle hypothesis?

3. What are the major steps in the scientific method?

   a. Observations and facts, hypotheses and predictions, hypothesis testing, evaluation and interpretation of results

   b. Observations and facts, hypothesis testing, evaluation and interpretation of results, repeatability

   c. Hypotheses and predictions, hypothesis testing, evaluation and interpretation of results, consistency

   d. Observations and facts, hypothesis testing, cause and effect, evaluation and interpretation of results

   e. Observations and facts, hypothesis and predictions, materialism, hypothesis testing

4. What is the relationship (or difference) between a hypothesis and a prediction?

5. What is a controlled experiment?

   a. An experiment in which a scientist tests all possible factors at the same time

   b. An experiment in which four possible factors are under investigation

   c. An experiment conducted in a temperature-controlled environment

   d. An experiment in which a scientist keeps all possible factors that could affect the outcome of the test the same except for the one under investigation

   e. An experiment that is consistent and repeatable

6. List two reasons why a scientist might repeat an experiment.

7. What are the differences between a hypothesis and a theory?

8. What is materialism?

9. What are three main features of science that help to distinguish science from other forms of knowledge?

10. What does it mean for a hypothesis to be falsifiable?

    a. The hypothesis is material.

    b. The hypothesis is false.

    c. The hypothesis does not match the prediction.

    d. The hypothesis is right.

    e. The hypothesis must be able to be proved wrong.

11. What is inductive reasoning?

12. Which of the following is NOT true of pseudoscience?

    a. It is fake science.

    b. It relies on controlled experiments.

    c. It begins with a claim and looks only for things that will support it.

    d. Controlled experiments are never done, and direct tests of any kind are generally avoided.

    e. It tries to persuade with appeals to emotion, sentiment, or distrust of established knowledge.

13. What is quackery? Describe an example of quackery that you have seen advertised in the media.

The Thinking Citizen

1. Why do scientists reject supernatural explanations as a way to explain natural phenomena?

2. Suppose scientists developed a new drug that they claimed cured the common cold. What tests and experiments would need to be done to support or refute their claim?

3. Two scientists are planning to study prey capture by wolves in Yellowstone National Park. The first scientist develops this hypothesis: the smartest wolves will capture the most prey. The second scientist's hypothesis is this: the fastest wolves will capture the most prey. Which, if either, of these is a valid scientific hypothesis? One of the four "features of science" discussed in the text will help you evaluate these hypotheses for their scientific validity.

4. How would you describe the relationship shown in **FIGURE 2.13** between global average temperature and number of pirates in the world (e.g., "As the number of pirates goes down, _____")? Is this a cause-and-effect relationship? If so, explain the cause and the effect it produces. If not, how do you explain the fact that these two variables appear to be correlated?

FIGURE 2.13 This figure depicts the global average temperature versus the number of pirates. These data should not be taken too seriously!

The Informed Citizen

1. Science is expensive. Many (though not all) scientific investigations done in the United States are funded by the government—that is, through taxpayers' dollars. Scientists apply to government organizations like the National Institutes of Health (NIH) and the National Science Foundation (NSF) for money to perform their research. Given that taxpayers pay for this research, is it reasonable for these organizations to prioritize funding, allocating most for research projects that will be of direct, immediate benefit to people (sometimes called *applied research*)? Are there ways to justify funding research designed primarily to simply discover more new things, such as the behavior of a marine organism or how a plant attracts its pollinators (sometimes referred to as *basic research*)?

2. Describe an example (real or hypothetical) of scientific research that you think is unethical. Is there a distinction between *unethical research* and *unethical uses* to which the research is applied? What sorts of guidelines would you suggest to distinguish ethical research from unethical? Who do you think is in the best position to develop such guidelines, and what are your suggestions for how such guidelines could be enforced?

CHAPTER LEARNING OBJECTIVES

After reading this chapter, you should be able to answer the following questions:

3.1 What Are the Units of Life?
Explain the consequences of cell theory for embryogenesis.

3.2 What Cell Structures Play a Role in Embryo Development?
Describe the cell structures that play a role in embryo development.

3.3 How Do Eggs and Sperm Form?
Show the possible chromosomes in gametes produced through meiosis in a given cell.

3.4 What Happens in Fertilization?
Describe the process of fertilization.

3.5 How Does an Embryo Form and Ultimately Become a Fetus?
Outline the stages of mitosis and the ways that cells become specialized.

3.6 What Are the Key Events of Pregnancy?
Differentiate fetal from embryonic development, and identify key points in development during pregnancy.

3.7 What Happens in Labor and Delivery?
Summarize the stages of labor and important events that are likely to occur in each stage.

3.8 How Do Twins Form?
Distinguish the development of identical from fraternal twins.

3.9 What Can Conjoined Twins Tell Us About Biology and Ourselves?
Assess the value of conjoined twin research and what conjoined twins can tell us about ourselves.

How do we come to be living, breathing, thinking, and feeling individuals with distinct limbs, organs, and the character that makes us who we are? That is the extraordinary riddle of human development.

CHAPTER 3

Human Development

How Do Cells Make a Person?

Becoming an Informed Citizen . . .

Examining human development reminds us of the diversity of life, including human life. Although we understand a great deal about what makes a body, we know much less about what makes a person. And while biology certainly plays a part, there is so much more.

Ow do we come to be living, breathing, thinking, and feeling individuals? How does an embryo become us—with distinct arms, legs, organs, and the individual character that makes us who we are? That is the extraordinary riddle of **human development**—*the process by which an individual grows and matures from a single cell embryo inside a mother's womb to a baby that can survive on its own.*

human development :: the process by which an individual grows and matures from a single cell embryo inside a mother's womb to a baby that can survive on its own

The answers to this puzzle begin even before the embryo forms, with the most basic unit of life, the cell. One biologist, Albert Claude, calls the cell "the mansion of our birth," not just our first home, but all of "our acquired wealth." As we see in this chapter, a single cell already includes the genetic heritage that we bring into the world from our parents. But to develop into humans, cells must communicate with one another and reproduce. We shall see how that small population of cells forms first an embryo, then the more mature stage called a fetus, and, ultimately, a baby. Each part in the journey relies on cell communication, cell movement, and the ability to regulate how genetic information is used.

No two people develop in exactly the same ways—not even identical twins. How can development have such a tremendous impact on a person's life? Consider some *unexpected* development in the story of the Hensel twins.

case study

Unusually Close Sisters

Abigail and Brittany Hensel entered the world on March 7, 1990. The words out of the obstetrician's mouth said it all: "They've got one body and two heads." Still under sedation, the mother, Patty, heard the word "Siamese" and asked, "I had cats?" But no, she had had what people often called Siamese twins, two babies with their bodies connected even after birth.

conjoined twin :: identical twins whose bodies are physically attached at some location

Abigail and Brittany are **conjoined twins**—their bodies are physically attached at some location (**FIGURE 3.1**). Each has her own head, neck, nervous system, heart, esophagus, and gall bladder. Yet they are *parapagus*, or joined at the side, and share many organs. Together they have three kidneys, four lungs (two of which are joined), one small intestine, one large intestine, one pelvis, one bladder, one set of reproductive organs, one liver, and one ribcage. Their spines are joined together at the pelvis, and they have just two legs and two arms. (A third undeveloped arm was removed from their chest when they were infants.)

Even more astonishing, Abigail and Brittany can live much of a normal life. The sisters learned to walk by the age of 15 months, a remarkable achievement given that each girl controls just one side of her body. To do something as ordinary as learning to walk, Abigail and Brittany had to agree on which direction to go; if not, they went in circles. In fact, they are a model of cooperation: they play piano (Abigail the right hand and Brittany the left). They swim, ride horses, play basketball, type, and respond to e-mails—all without having to speak. Each girl passed her driving test and is licensed to drive.

FIGURE 3.1 Abigail and Brittany Hensel are conjoined twins. Each has her own head, neck, nervous system, heart, esophagus, and gall bladder, but they are physically connected at the side and share many organs.

Abigail controls the pedals, radio, and defogger, while Brittany is in charge of the lights and turn signals. They coordinate steering.

Although they share more than other twins do, Abigail and Brittany are quite different in personality, likes and dislikes, and ambitions. Their tastes in food and clothes are different, they style their hair differently, and they experience separately the need to sleep. In school, Abigail and Brittany were good at different subjects. They buy tickets for two seats at the movies and celebrate their birthdays on separate days.

What makes Abigail and Brittany so different? How do humans develop, and what does that variation tell us about the essence of being an individual? At what point during human development does someone become a person? As we will see in this chapter, conjoined twins are only one of the many potential variations in human development.

FIGURE 3.2 Before the invention of microscopes, the humoral theory held that a healthy body required four humors: blood, phlegm, yellow bile, and black bile. When these fluids were not in balance, illness resulted.

cell :: the most basic unit of life

embryogenesis :: The development of cells into embryos and bodies that are able to perform the tasks necessary for life

cell theory :: the realization that all living organisms are made of cells; it was first proposed in 1839 by Matthias Schleiden and Theodor Schwann

3.1 What Are the Units of Life?

We start with the most basic unit of life, the **cell**. From cell structure and function, we can then understand how embryos develop to produce a body, like yours or that of the Hensel sisters, able to perform the tasks necessary for life. That process of development is called **embryogenesis**. Scientists who study it are embryologists.

Before the invention of microscopes in the 17th century, scientists had limited understanding of the substance of life. According to one early idea, a healthy body required four humors: blood, phlegm, yellow bile, and black bile (**FIGURE 3.2**). When these fluids were not in balance, illness resulted. Blood-letting, or the removal of blood, was used well into the 19th century to treat fever and other ailments. Not only did this fail to work, it was potentially harmful. Repeated bleedings of George Washington in response to a headache probably killed him.

Surgical procedures and autopsies revealed the existence of organs (such as the heart) and organ systems (such as the one for blood circulation), and what scientists and physicians described as "tissues and fibers." Later it became apparent that human tissues and fibers are composed of an even smaller unit, the cell.

Of course, that discovery relied on an important tool of science, the microscope. Robert Hooke, in 1663, first observed well-defined chambers in slices of cork (**FIGURE 3.3**). Anton van Leeuwenhoek enthusiastically explored these cells through the early 1700s (see *Scientist Spotlight: Anton van Leeuwenhoek*). As microscopes improved dramatically in the 19th century, it turned out that *all* organisms, plant and animal alike, are composed of cells. There are even single-celled organisms, such as bacteria.

The realization that all living organisms are made of cells is called the **cell theory**. The theory, first proposed in 1839 by Matthias Schleiden and Theodor Schwann, reinforces the common ancestry of life. Knowledge about cells also opens the door to understanding how our bodies function in health and disease.

FIGURE 3.3 In 1663, Robert Hooke first observed well-defined chambers (cells) in slices of cork wood. This is a color-enhanced version of the illustration Hooke provided in his book *Micrographia*, published in 1667.

Scientist Spotlight

Anton van Leeuwenhoek (1632–1723)

Anton van Leeuwenhoek started and finished his life as a tradesman. He was born in Holland in 1632, was apprenticed to a linen drapers shop in 1648, and went into business for himself as a fabric merchant in 1654. In 1668, however, he learned how to grind lenses and make simple microscopes, and he began to look at *everything*. Some historians speculate that he was inspired by Robert Hooke's popular book *Micrographia*, which had illustrations of magnified objects.

In all, van Leeuwenhoek made over 500 microscopes. They opened up the previously unknown world of the very small. What did he examine and observe? To name a few examples, he scraped teeth, collected saliva, and observed the bacteria present. He examined pond water and saw tiny invertebrates, protozoa, and single-celled green algae. After cutting himself shaving, he examined the blood on his razor and saw red blood cells. Van Leeuwenhoek wrote to the Royal Society of London, describing his discoveries. Despite his long, rambling letters and lack of formal education, the society recognized the merit of his work. He was elected as a full member in 1680, and he continued sending descriptions of his observations until the last days of his life.

Van Leeuwenhoek risked social censure with perhaps his most important discovery—sperm cells in animals. At first, he did not want to study the question of what is in human semen; he was too concerned about its propriety. Ultimately, he observed his own semen—"acquired not by sinfully defiling himself," he testified, but "by hastening from the marriage bed to the microscope." Later he showed that sperm are present in the semen of other animals. He also observed large numbers of sperm in the uterus and fallopian tubes of recently mated dogs and rabbits. These observations were essential to our understanding of the cellular basis of reproduction.

Anton van Leeuwenhoek discovered sperm cells in animals using microscopes of his own construction.

This Dutch biologist realized how unusual his drive and curiosity were:

> Some go to make money out of science, or to get a reputation in the learned world. But in lens-grinding and discovering things hidden from our sight, these count for naught. And I am satisfied too that not one man in a thousand is capable of such study, because it needs much time . . . and you must always keep thinking about these things if you are to get any results. And over and above all, most men are not curious to know: nay, some even make no bones about saying, what does it matter whether we know this or not? (translation from "Leeuwenhoek, Anton van [1632–1723]," in *Encyclopedia of World Biography* [Gale, 1998])

Anton van Leeuwenhoek exemplifies the intense desire to observe, learn, and understand that characterizes all good scientists. ::

3.2 What Cell Structures Play a Role in Embryo Development?

Each of us developed from a single cell (**FIGURE 3.4**), as did twins like Abigail and Brittany. As we saw in Chapter 1, our bodies are composed of *eukaryotic* cells, with an enclosed nucleus and specialized structures called **organelles**. (Prokaryotic cells, such as bacteria, do not possess organelles, and we consider them in greater detail in Chapter 10.) Organelles play a role in the structure, function, and development of all multicellular organisms, including humans. The common features of all eukaryotic cells are striking evidence that all species share a common ancestor. All eukaryotic cells include a cell membrane, nucleus, mitochondria, endomembrane system, and cytoskeleton.

organelle :: specialized structure in eukaryotic cells

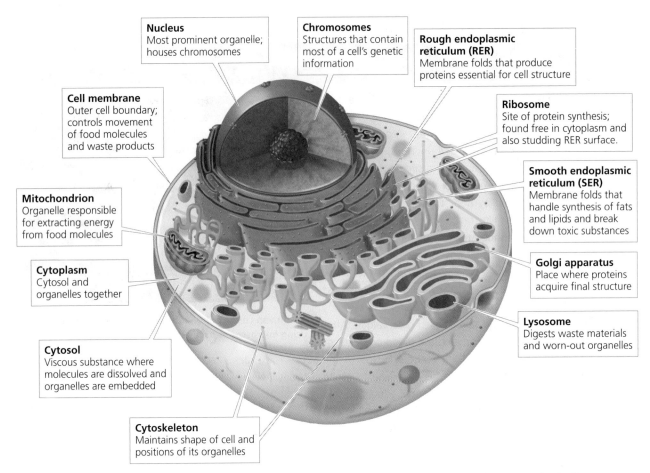

Nucleus
Most prominent organelle; houses chromosomes

Chromosomes
Structures that contain most of a cell's genetic information

Rough endoplasmic reticulum (RER)
Membrane folds that produce proteins essential for cell structure

Cell membrane
Outer cell boundary; controls movement of food molecules and waste products

Ribosome
Site of protein synthesis; found free in cytoplasm and also studding RER surface.

Smooth endoplasmic reticulum (SER)
Membrane folds that handle synthesis of fats and lipids and break down toxic substances

Mitochondrion
Organelle responsible for extracting energy from food molecules

Golgi apparatus
Place where proteins acquire final structure

Cytoplasm
Cytosol and organelles together

Lysosome
Digests waste materials and worn-out organelles

Cytosol
Viscous substance where molecules are dissolved and organelles are embedded

Cytoskeleton
Maintains shape of cell and positions of its organelles

FIGURE 3.4 The most basic unit of life is the cell. Shown here is an animal cell.

Cell Membrane

cell membrane :: outer boundary of the cell body, where the cell encounters its environment

cytosol :: a viscous substance in a cell where small molecules are dissolved and organelles are embedded

cytoplasm :: in a cell it is the cytosol and organelles together

The **cell membrane**, or outer boundary of the cell body, is where the cell encounters its environment. It regulates what enters and leaves the cell. For example, the membrane controls the movement of food molecules into the cell and the removal of waste products. The cell membrane also maintains the environment *inside* the cell—essential for the many thousands of chemical reactions that occur in cells. The membrane also encloses a viscous substance called the **cytosol**. Here, small molecules are dissolved and organelles are embedded. The cytosol and organelles together are called the **cytoplasm**.

The cell membrane plays many roles in human development: it can detect chemical signals from other cells, starting with communication between a human egg and sperm cell. It can also stick to other cells to make multi-cellular structures like tissues and organs.

Nucleus

nucleus :: the most prominent organelle inside the cell; it houses the chromosomes that contain the genetic information

chromosome :: structure that contains most of the genetic information of the cell; housed in the nucleus

The **nucleus** is the most prominent organelle inside the cell. The nucleus houses the **chromosomes**, the structures that contain most of the genetic information of the cell. These genetic instructions are written in the language of the molecule DNA. The instructions are organized into genes. The DNA of genes directs the production of molecules that are important for a cell's structure, function, and even identity. During embryogenesis, gene expression—which genes are turned off or on—is carefully regulated so that the correct types of cells form where they should.

Mitochondria

Mitochondria are the organelles responsible for extracting energy from food molecules. Muscle cells have many mitochondria, which is no surprise given the energy demands on muscles. Mitochondria are also abundant in the cells of developing embryos, because biological development takes energy.

mitochondrion :: organelle responsible for extracting energy from food molecules

Endomembrane System

Like a factory, the **endomembrane system** of the cell produces important molecules and delivers them to key locations within and outside the cell. In human development, it provides the molecules essential to such organs as brain, lung, liver, or skin. It also collects, packages, and removes waste materials. Otherwise, those materials would lead to improper brain development and other defects. The endomembrane system consists of three structures—the endoplasmic reticulum, the Golgi apparatus, and lysosomes.

endomembrane system :: produces important molecules and delivers them to key locations within and outside the cell; consists of the endoplasmic reticulum, the Golgi apparatus, and lysosomes

Located near the nucleus, the **endoplasmic reticulum (ER)** is a series of membrane folds. The *smooth endoplasmic reticulum* (SER) handles the synthesis of fats and lipids including those that compose the cell membrane. The SER also breaks down and eliminates some toxic substances. Liver cells are especially rich in SER, since this organ detoxifies the blood.

endoplasmic reticulum (ER) :: a series of membrane folds located near the nucleus; smooth endoplasmic reticulum handles the synthesis of fats and lipids, and rough endoplasmic reticulum produces proteins

The *rough endoplasmic reticulum* (RER) produces proteins, molecules essential for cell structure and regulation. The RER gets its name from the **ribosomes** that stud its surface. The ribosomes thread the protein into the interior of the RER as they are making it. Once inside, some proteins are packaged in bubble-like vesicles, which then travel to other destinations in the cell, delivering their protein cargo.

ribosome :: studs the surface of the rough endoplasmic reticulum (RER) and threads the protein into the interior of the RER as it is being made

In the **Golgi apparatus**, proteins acquire their final structure. Some proteins, for example, become digestive enzymes, molecules capable of breaking down other molecules. Vesicles also leave the Golgi apparatus and travel to the cell membrane. There, they release their contents to the space outside the cell. Cells in the respiratory system, for example, secrete mucus in this way.

Golgi apparatus :: the place where many proteins acquire their final structure

Lysosomes digest waste materials and worn-out organelles. For example, when mitochondria age and no longer function, they are marked for destruction and delivered to lysosomes. The lysosomes also recycle the molecules so that they can be reused. Similarly, red blood cells, which live for no more than three months, are destroyed by certain cells of the immune system. These immune cells engulf the old red blood cells and digest them in lysosomes present within these immune cells.

lysosome :: digests waste materials and worn-out organelles and recycles the molecules so they can be reused

Cytoskeleton

The **cytoskeleton (FIGURE 3.5)** maintains the shape of the cell and the positions of its organelles. Its fine fibers also allow for cell motion. The cytoskeleton has three major components: *microtubules, microfilaments,* and *intermediate filaments.* Microtubules are hollow rods that determine cell shape and help with cell movement. Flagella, the motile tails on swimming sperm cells, are made largely of microtubules. Microfilaments are thin, strong, and versatile and also aid in cell movement. A ring of microfilaments, along with the help of some other proteins, pinches the cell membrane so that cells may divide in two. Cell division happens a great deal in human development. Intermediate filaments strengthen cells and areas of the embryo, such as the covering of the cell nucleus.

cytoskeleton :: maintains the shape of the cell and the positions of the organelles

Microfilaments
Versatile, thin, and strong; help to pinch the cell membrane so that cells may divide in two

Microtubules
Hollow rods that determine cell shape and help with cell movement

Intermediate filaments
Strong fibers that help strengthen cells and reinforce the physical connections of cells to each other

The cytoskeleton aids in the movement of the amoeba (left) and sperm (right).

FIGURE 3.5 The cytoskeleton maintains the shape of the cell and also aids in cell movement. It has three major components: microtubules, microfilaments, and intermediate filaments.

gamete :: sex cell

meiosis :: a special type of cell division that produces eggs or sperm

homologue :: one in a pair of chromosomes

testis :: male sex organ where sperm are produced

ovary :: female sex organ where eggs are produced

diploid cell :: cell with two complete sets of chromosomes

haploid cell :: a cell with just one set of chromosomes

interphase :: the stage in cell division when cells prepare materials that will be necessary for meiosis or mitosis to occur

3.3 How Do Eggs and Sperm Form?

All organisms that undergo sexual reproduction make **gametes**, or sex cells. In many species, including humans, males make sperm and females make eggs (**FIGURE 3.6**).

A special type of cell division, called **meiosis**, produces eggs or sperm. In humans, body cells contain 46 chromosomes organized into two sets of 23 (**FIGURE 3.7**). Pairs of chromosomes are called **homologues**. Meiosis halves the number of chromosomes. At the same time, each gamete receives a complete set of genetic instructions to pass on to the next generation.

Sperm cells form in male sex organs, the **testes**, egg cells in female sex organs, the **ovaries**. The process involves the repackaging of chromosomes so that DNA can move without being physically damaged. Critical events along the way make it appear as if meiosis is a series of discrete steps, or *phases*. In fact, phases progress smoothly from one to the next.

Meiosis entails two rounds of cell division (**FIGURE 3.8**). It starts with a single **diploid cell**, a cell with two complete sets of chromosomes. It ends with four **haploid cells**, each with just one set of chromosomes.

Meiosis I: The First Round of Cell Division (**FIGURE 3.9**)

- *Interphase:* Cells in **interphase** are hard at work preparing materials that will be necessary for meiosis to occur. Although sometimes called a "resting stage," it is a very busy time. The chromosomes replicate, all organelles continue to function, and the cell increases in size. At this stage, the chromosomes are loosely packaged and

FIGURE 3.6 All organisms that undergo sexual reproduction make gametes, or sex cells. In many species, including humans, males make sperm and females make eggs.

FIGURE 3.7 In humans, body cells contain 46 chromosomes, which are organized into two sets of 23. Pairs of chromosomes are called homologues.

individual chromosomes are hard to see even with a powerful light microscope.

- *Prophase I:* During **prophase I**, the chromosomes become shorter and thicker, and they line up with their homologues. The membrane surrounding the nucleus comes apart, leaving the chromosomes in the cytoplasm.

Meanwhile, the rest of the cell is busy. The microtubules of the cytoskeleton disassemble and form the **spindle**, a structure that will later help to distribute the chromosomes into new cells. In its completed form, it looks like a starburst of fibers, emanating from the two ends or poles of the cell. These fibers extend into the center of the cell, where they can "catch" a chromosome. This continues until they have snagged each of the paired chromosomes.

- *Metaphase I:* In **metaphase I**, specific proteins associated with the fibers of the mitotic spindle pull the chromosomes toward both poles.

prophase I :: in meiosis, the stage in the first round of cell division when the chromosomes become shorter and thicker and line up with their homologues

spindle :: formed from the microtubules of the cytoskeleton, this is a structure that helps to distribute chromosomes to new cells

metaphase I :: in meiosis, the stage in the first round of cell division when paired chromosomes line up in single file down the middle of the cell

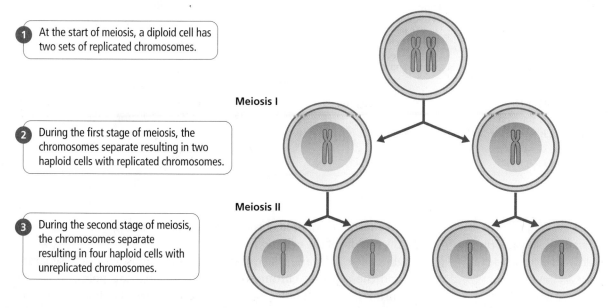

① At the start of meiosis, a diploid cell has two sets of replicated chromosomes.

Meiosis I

② During the first stage of meiosis, the chromosomes separate resulting in two haploid cells with replicated chromosomes.

Meiosis II

③ During the second stage of meiosis, the chromosomes separate resulting in four haploid cells with unreplicated chromosomes.

FIGURE 3.8 Meiosis, a special type of cell division, produces eggs or sperm. It starts with a single diploid cell, a cell with two complete sets of chromosomes, and ends with four haploid cells, each with just one set of chromosomes.

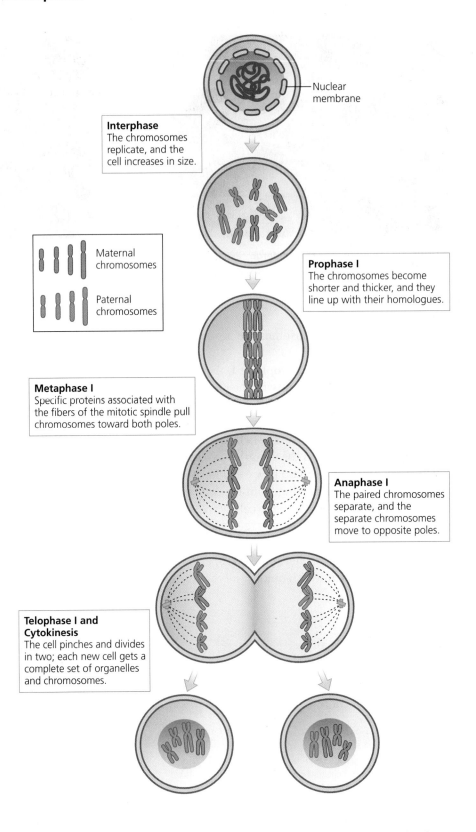

Interphase
The chromosomes replicate, and the cell increases in size.

Nuclear membrane

Maternal chromosomes

Paternal chromosomes

Prophase I
The chromosomes become shorter and thicker, and they line up with their homologues.

Metaphase I
Specific proteins associated with the fibers of the mitotic spindle pull chromosomes toward both poles.

Anaphase I
The paired chromosomes separate, and the separate chromosomes move to opposite poles.

Telophase I and Cytokinesis
The cell pinches and divides in two; each new cell gets a complete set of organelles and chromosomes.

FIGURE 3.9 Meiosis I is the first round of cell division; it ends with two haploid cells. The chromosomes are still in their replicated form, with two chromatids each.

anaphase I :: in meiosis, the stage in the first round of cell division when paired chromosomes separate and move to opposite poles, leaving a haploid set of chromosomes on each side of the cell

As a result, the paired chromosomes line up in single file, right down the middle of the cell.

- *Anaphase I:* In **anaphase I**, the paired chromosomes now separate, and the separate chromosomes move to opposite poles, using the shrinking microtubules as tracks. There is now a haploid set of chromosomes on each side of the cell.

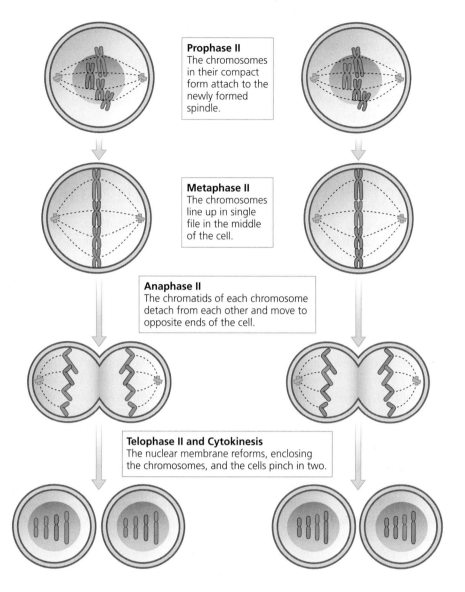

Prophase II
The chromosomes in their compact form attach to the newly formed spindle.

Metaphase II
The chromosomes line up in single file in the middle of the cell.

Anaphase II
The chromatids of each chromosome detach from each other and move to opposite ends of the cell.

Telophase II and Cytokinesis
The nuclear membrane reforms, enclosing the chromosomes, and the cells pinch in two.

FIGURE 3.10 Meiosis II is the second round of cell division; the two attached sister chromatids separate. Meiosis II ends with four haploid cells, and each chromosome has one chromatid.

- *Telophase I and cytokinesis:* In **telophase I and cytokinesis,** the cell at last pinches and divides in two. Each new cell gets a complete set of organelles and chromosomes.

Meiosis I ends with two haploid cells. Meiosis is not over yet, though, because each of these chromosomes has *two* copies of its genetic information called **chromatids.** At the end of meiosis I, each chromosome is composed of two identical chromatids attached to each other. The task of meiosis II is to separate them.

Meiosis II: The Second Round of Cell Division (FIGURE 3.10)

- *Prophase II:* Each stage of meiosis II occurs in both cells produced in meiosis I. At the start of **prophase II,** the chromosomes, still in their compact form, attach to the newly formed spindle.
- *Metaphase II:* In **metaphase II,** chromosomes are captured by one or more microtubules from each side. They line up in single file in the middle of the cell.
- *Anaphase II:* In **anaphase II,** the two chromatids of each chromosome detach from each other and move to opposite ends of the cell.

telophase I and cytokinesis :: in meiosis, the stage in the first round of cell division when the cell pinches and divides in two; each new cell has a complete set of organelles and chromosomes

chromatid :: one of two copies of the DNA and proteins that make up a replicated chromosome; replicated chromosomes are composed of two chromatids

prophase II :: in meiosis, the stage in the second round of cell division when chromosomes, still in their compact form, attach to the newly formed spindle

metaphase II :: in meiosis, the stage in the second round of cell division when chromosomes line up in single file in the middle of the cell

anaphase II :: in meiosis, the stage in the second round of cell division when the two chromatids of each chromosome detach from each other and move to opposite ends of the cell; each half of the cell has a complete haploid set of chromosomes

telophase II :: in meiosis, the stage in the second round of cell division when the nuclear membrane reforms, enclosing the chromosomes

cytokinesis :: the stage in cell division when the cells are pinched in two

Each half of the cell has a complete haploid set of chromosomes. Each chromosome is composed of a single chromatid.

- *Telophase II and cytokinesis*: In **telophase II**, the nuclear membrane reforms, enclosing the chromosomes. The chromosomes return at last to their relaxed, decondensed form. **Cytokinesis** now pinches the cells in two.

Since meiosis II began with two cells, the final products are four haploid cells.

What Happens in Fertilization?

fertilization :: the process of a sperm cell from the father being fused with an egg cell from the mother

zygote :: a one-celled embryo produced as a result of fertilization

After meiosis, we now have the egg and sperm. In **fertilization**, a sperm cell from the father fuses with an egg cell from the mother. The brand-new one-celled embryo is the **zygote**. For fertilization to occur, though, eggs and sperm must find and recognize each other. They must then fuse to form a new diploid cell, with two full sets of chromosomes (see *How Do We Know? Eggs and Sperm Are Both Needed for Fertilization*).

In sea urchins, fertilization occurs outside the body: males and females release gametes into the ocean. In humans, fertilization occurs within the woman's body. Nevertheless, sperm still face an arduous journey to the fallopian tubes where they will meet the egg (**FIGURE 3.11**). Fortunately, sperm are excellent swimmers. The female helps transport sperm too. Contractions of muscles in the uterus make it possible for sperm to travel from the vagina to the fallopian tubes within 30 minutes, much faster than they could get there just by swimming. Once in the fallopian tubes, sperm are directed by chemical signals secreted by the egg. When the sperm find themselves near the egg, they swim even more vigorously.

In addition to its cell membrane, the egg has another coating, the *zona pellucida* (**FIGURE 3.12**). This protective covering shields the egg cell from damage and also ensures that it will not be fertilized by sperm from a different species. The sperm cell membrane and *zona pellucida* have proteins that bind

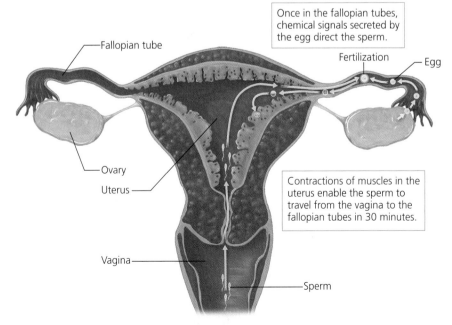

FIGURE 3.11 In humans, fertilization occurs within the woman's body. Sperm face an arduous journey to the fallopian tubes, where the egg may be fertilized.

FIGURE 3.12 The *zona pellucida* is a protective covering that shields the egg cell from damage. Shown here is a scanning electron micrograph (SEM) of the surface of the *zona pellucida*.

Eggs and Sperm Are Both Needed for Fertilization

It takes a male and a female for human or animal reproduction, and both contribute to the development of the newborn. But exactly what do they contribute? For centuries, no one knew. Ideas ranged from physical material to energy—or more mysterious forces.

In 1667, Anton van Leeuwenhoek discovered sperm cells in the semen of humans and other animals. In fertilization, he proposed, sperm enter the egg. Unfortunately, he could not provide evidence for his hypothesis. Mammalian eggs were not even discovered until 1797.

Already, however, scientists were out to test van Leeuwenhoek's hypothesis. In 1784, the Italian scientist Lazzaro Spallanzani placed little trousers on male frogs and let them try mating with female frogs. Despite their best efforts, no offspring developed. Spallanzani also noticed drops of semen on the frog pants. When he combined these drops with frog eggs, fertilization took place.

In 1824, the French scientists J. L. Prevost and J. B. Dumas demonstrated that sperm cells, not just semen, were necessary to trigger development in amphibian eggs. Peering into a microscope in 1854, the British scientist George Newport finally watched frog sperm enter frog eggs.

In 1876 the German scientist Oskar Hertwig looked within these cells. He demonstrated that as the egg and sperm fuse in sea urchin, their nuclei fuse as well. The new nucleus indeed has genetic information from each parent. ::

1. Spallanzani placed trousers on male frogs.

2. Spallanzani let the male frogs try mating with female frogs.

3. No offspring developed.

4. Spallanzani noticed drops of semen on the frog pants.

5. Spallanzani combined the drops of semen with frog eggs.

Fertilization took place.

In 1784, the Italian scientist Lazzaro Spallanzani placed little trousers on male frogs and let them try mating with female frogs. Despite their best efforts, no offspring developed.

together and attach the two cells to each other. Once bound, the sperm releases enzymes that digest a hole in the *zona pellucida*, and another set of proteins triggers cell fusion. The egg cell now springs into action releasing enzymes to remove all sperm-binding proteins from the *zona pellucida*. This ensures that only one sperm fuses with an egg. The zygote starts development with exactly one set of chromosomes from each parent—none missing and none extra. If fertilization is not successful, the egg is absorbed by the body.

3.5

How Does an Embryo Form and Ultimately Become a Fetus?

After fertilization, embryogenesis begins. It starts with **cleavage**, the division into new cells that will form the earliest stage of development, or the **embryo**. Cleavage involves **mitosis**, a second type of cell division, which is somewhat

cleavage :: the start of embryogenesis when the cells divide into new cells

embryo :: the earliest stage of development

mitosis :: a type of cell division during which all of the chromosomes in a cell are copied or replicated, and each new cell receives genetic information identical to the parent cell different from meiosis. During mitosis (**FIGURE 3.13**), all of the chromosomes in a cell are copied or replicated, and each new cell receives genetic information identical to the parent cell. In humans, cells have 46 chromosomes, meaning two sets of 23. After mitosis, each cell has 46 chromosomes, just as the parent cell did. Mitosis is therefore less complex than meiosis: it involves only one round of cell division.

Interphase
The chromosomes replicate, and the cell increases in size.

Prophase
The chromosomes become shorter and wider, and the mitotic spindle forms.

Late prophase
The spindle fibers attach to the chromosomes.

Metaphase
The chromosomes line up in single file.

Anaphase
The chromatids of each chromosome separate; one full set, composed of single chromatids, moves to one pole, and another to the opposite pole.

Telophase and Cytokinesis
A nucleus reforms in each half of the cell, enclosing the chromosomes; the cell pinches and divides in two.

FIGURE 3.13 During mitosis, all of the chromosomes in a cell are copied or replicated, and each new cell receives genetic information identical to the parent cell.

Mitosis

Mitosis governs all cell division except for making an egg or sperm. You'll recognize many of the steps from similar stages in meiosis:

- Mitosis begins with *interphase*. Here the chromosomes replicate, and the cell makes other materials necessary for cell division. The cell increases in size.
- In *prophase*, the chromosomes become shorter, wider, and easier to see with a microscope. The nuclear membrane disappears, and the mitotic spindle forms. As in meiosis, microtubules extend from opposite poles of the cell and bind to chromosomes. Each chromosome is composed of two chromatids.
- In *metaphase*, the chromosomes line up in single file, as specific proteins associated with the microtubules pull on them from each pole.
- In *anaphase*, the chromatids of each chromosome separate. One full set of chromosomes, composed of single chromatids, moves to one pole, and one set moves to the opposite pole. As a consequence, each half cell has a full complement of genetic instructions.
- During *telophase and cytokinesis*, a nucleus reforms in each half of the cell, enclosing the chromosomes. The chromosomes resume their relaxed form, and the cell pinches and divides in two. In humans, the newly formed cells again possess two sets of 23 chromosomes.

FIGURE 3.14 In humans and other mammals, the first couple of cleavages create a cluster of cells, none of which has specialized.

In humans and other mammals, the first couple of cleavages create a cluster of cells, none of which has specialized (**FIGURE 3.14**). By the time the embryo reaches the eight-cell stage, the cells compact together into a tight group. At the 16-cell stage, there are two populations of cells present: the outer ones will help to make the **placenta**, the organ that connects the developing fetus to the mother's uterine wall, and other structures that are needed to support the development of the embryo and fetus. The inner cell population will eventually form all of the cells, tissues, and organs of the embryo and fetus. At this point in development, the inner cells are **embryonic stem cells**: they will produce all of the types of cells in the body.

placenta :: an organ that connects a developing fetus to the mother's uterine wall and provides for an exchange of nutrients and waste elimination

embryonic stem cell :: cell that can produce any type of cell in the body

Gastrulation and Organ Formation

During **gastrulation** (**FIGURE 3.15**), cells and tissues move to new locations, where they will grow into organs. For example, the cells that give rise to the skin remain on the exterior surface of the embryo, while the cells that will ultimately form the heart move into the interior.

Organ formation involves many of the same types of cell and tissue rearrangement seen in gastrulation. For example, the brain and spinal cord of the central nervous system start their development as a flat sheet of cells. This tissue rolls up to form a **neural tube**. The anterior, or front part, of the tube expands, ultimately making the brain. The posterior portion, or rear part, becomes the spinal cord (**FIGURE 3.16**).

As with all aspects of development, mistakes can have serious consequences. In the case of the central nervous system, failure to close the anterior region of the neural tube will likely result in a fetus that has no brain. Survival is not possible. If the posterior region of the neural tube does not close, spina bifida occurs (**FIGURE 3.17**). The health of an individual with spina bifida depends on the location and size of the opening.

gastrulation :: a stage in embryonic development in which cells and tissues move to new locations, where they will grow into organs

neural tube :: the tissue that starts from a flat sheet of cells and rolls up, ultimately becoming the brain and spinal cord

FIGURE 3.15 During gastrulation, cells and tissues move to new locations, where they will grow into organs.

As for Abigail and Brittany, or other conjoined twins, it is not entirely clear how errors in setting up the organization of their bodies led to the particular patterns of organ formation observed in them. Ideas about how conjoining occurs are discussed later.

Differentiation

For an embryo to become a fetus, its cells must take on new roles on the way to becoming human organs like the heart and lungs. Hans Spemann, an embryologist, captured the wonder of development this way: "We are standing and walking with parts of our body which could have been used for thinking had they developed in another part of the embryo."

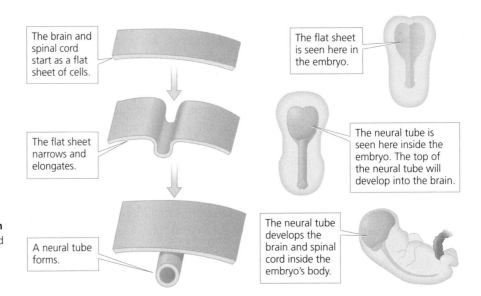

FIGURE 3.16 Organ formation involves many of the same types of cell and tissue rearrangement seen in gastrulation. For example, the brain and spinal cord of the central nervous system start their development as a flat sheet of cells.

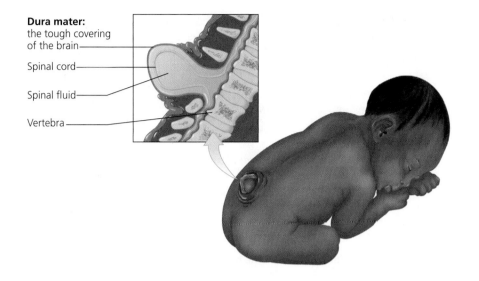

Dura mater:
the tough covering
of the brain

Spinal cord

Spinal fluid

Vertebra

FIGURE 3.17 Mistakes in the development of the central nervous system can have serious consequences. If the posterior region of the neural tube does not close, *spina bifida* occurs; in serious cases, part of the spinal cord may protrude through the bones.

The zygote, the very first embryonic cell, can produce any type of cell in the body. Cells in an adult body no longer have that unlimited potential. A human adult, in fact, has more than 200 types of specialized cells. The change from cells that have unlimited potential to cells that are specialized is called cell **differentiation**. In the process, cells take on a specific structure, set of behaviors, and function.

Cell differentiation (**FIGURE 3.18**) takes place through a series of commitments. Think of a marriage. Marriage does not generally happen moments after a couple meets. Instead, it comes at the end of a series of choices and an increasing level of commitment. When two individuals meet, there needs to be some "chemistry" between them. They might go out a few times to see whether the relationship has a future. If all goes well, they agree to see one another exclusively. Eventually they may decide to become engaged. Finally, they make the big commitment, marriage. They are fully "differentiated."

Returning to cells, let's consider the differentiation of the muscle cells in the heart (**FIGURE 3.19**). Again, we start with the zygote, which has unlimited potential: it has made no commitment to become any one type of cell. During cleavage, inner and outer cell populations arise, and muscle cells originate from the inner cells. During gastrulation, three major types of embryonic cells emerge: *ectoderm, mesoderm,* and *endoderm.* These three types are not yet fully specialized, but each will give rise only to specific cells. Muscle cells derive from the mesoderm alone. The mesoderm produces several different

differentiation :: the change from cells that have unlimited potential to cells that are specialized

FIGURE 3.18 Cell differentiation takes place through a series of commitments that is in some ways analogous to courtship and marriage.

Meeting	First date	Additional dates	Dating exclusively	Engagement	Marriage
Cells in the early embryo have undergone a small number of divisions, but no decisions have been made about anything.	Cells can be found in one of two embryonic cell populations: outer or inner.	Cells narrow their future developmental choices.	Cells can give rise to a limited set of cell types.	Cells can form only a particular cell type, but they have not done so yet.	Cells complete differentiation and commit to being a particular type of cell.

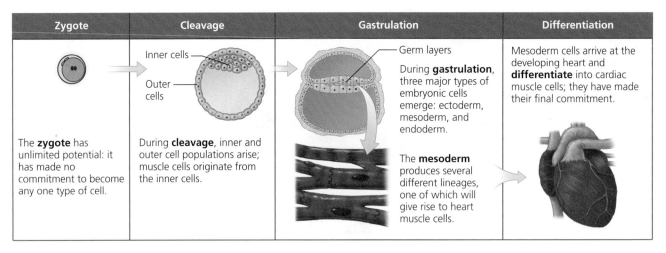

FIGURE 3.19 This figure illustrates how cells differentiate into cardiac muscles of the heart.

mesoderm lineages, one of which will give rise to heart, or cardiac, muscle cells. Finally, these mesoderm cells arrive at their final destination (the developing heart) and differentiate into cardiac muscle cells. They have made their final commitment.

Gene Expression

What guides the process? What ensures that a healthy individual gets the right complement of healthy organs? First, in complex organisms like humans, cell differentiation depends on cell communication. Chemical signals between neighboring cells tell them when and where to differentiate. This "conversation" enables the construction of elaborate, precisely functioning organs like the eye.

Second, cell differentiation depends on using just the right genetic information, a process called **gene expression** (**FIGURE 3.20**). All cells contain the

gene expression :: the process during which the information encoded in genes is used

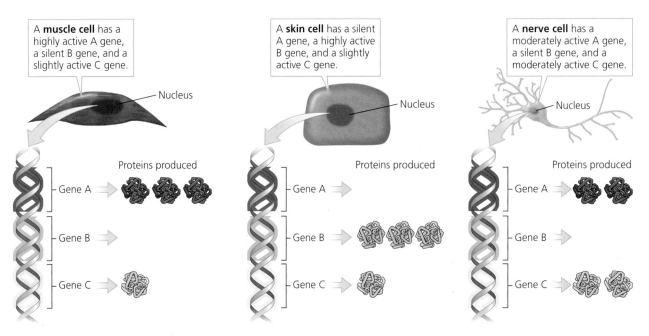

FIGURE 3.20 Cell differentiation depends on cell communication and gene expression. Chemical signals between neighboring cells tell them when and where to differentiate, and different cells use different combinations of the same genes to specialize or differentiate.

same genes, but different genes are used in different cells. Some genes are uniquely expressed in brain cells and a different set in liver cells. If you think of all your genes as a fully loaded MP3 player, then all of your cells possess all of the songs. Gene expression is like listening to just the songs you need.

Cell differentiation results in highly specialized cells—with their own shapes, internal structure, and behavior. For example, the white blood cells of the immune system are amorphous blobs; these cells crawl around the body, ever vigilant for infection. In contrast, muscle cells are long and thin, connected to other muscle cells in parallel. They can thus coordinate their individual contractions so that the entire muscle contracts.

3.6 What Are the Key Events of Pregnancy?

Now we know how we get from an egg and sperm to an embryo and the early steps of making organs. Let's take a closer look at pregnancy, starting with conception and ending with labor and delivery. All this has to happen to produce a healthy newborn.

Embryonic Development: Conception to Eight Weeks (FIGURE 3.21)

It might seem that becoming pregnant is the easiest thing in the world. After all, a great deal of effort goes into avoiding unwanted pregnancy. In fact, the odds of conceiving during a given cycle of fertility are only 15–25%, depending on age, health, and other factors. The ability to conceive decreases starting at age 30. A couple who is trying to conceive but has not been successful will not be considered infertile until at least close to a year of effort.

During the first week after conception, cleavage produces the cells that will form the embryo and chorion, a component of the placenta. In week 2, the embryo, still only the size of the period at the end of this sentence, implants in the wall of the **uterus**—also known as the womb. Gastrulation occurs, and cell specialization is underway: ectoderm, mesoderm, and endoderm are formed.

uterus :: the womb of a mother

The third week includes **neurulation**, or formation of the earliest stages of the central nervous system, and the beginning of heart development. In week 4, heart development continues and limb buds, or swellings that will become limbs, can be seen. The embryo is barely larger than a letter on this page. In week 5, a discernible nose, eyes, and ears appear. By weeks 6 and 7, fingers and toes are evident, and the embryo has a skeleton made of cartilage. By the end of week 8, the embryo is approximately 1.5 inches long, bone is beginning to replace cartilage, and all organ systems are developing. At this point, embryonic development is finished and fetal development begins.

neurulation :: formation of the earliest stages of the central nervous system

Fetal Development: Three Months to Nine Months (FIGURE 3.22)

During the rest of development, organ systems are refined, and the **fetus**—the stage of development that follows the embryo—grows dramatically. During the third month, the gender of the fetus can be discerned and fingernails form. Also, the placenta is fully developed by the 10th week. In the fourth month, the fetus is approximately 6 inches long and weighs 6 ounces. The skeleton is visible using **ultrasound** imaging (see *Technology Connection: Ultrasound*, p. 79). Eyelashes, eyebrows, and hair on the head are all present. If the mother has had a previous pregnancy, she can feel and identify fetal movement.

fetus :: the stage of development that follows the embryo

ultrasound :: a device that relies on high-frequency sound waves to send back images of internal organs

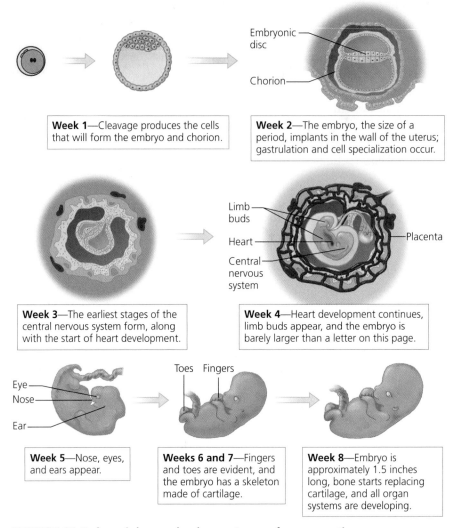

Week 1—Cleavage produces the cells that will form the embryo and chorion.

Week 2—The embryo, the size of a period, implants in the wall of the uterus; gastrulation and cell specialization occur.

Embryonic disc

Chorion

Limb buds

Heart

Central nervous system

Placenta

Week 3—The earliest stages of the central nervous system form, along with the start of heart development.

Week 4—Heart development continues, limb buds appear, and the embryo is barely larger than a letter on this page.

Eye
Nose
Ear

Toes   Fingers

Week 5—Nose, eyes, and ears appear.

Weeks 6 and 7—Fingers and toes are evident, and the embryo has a skeleton made of cartilage.

Week 8—Embryo is approximately 1.5 inches long, bone starts replacing cartilage, and all organ systems are developing.

FIGURE 3.21 Embryonic human development occurs from conception to eight weeks.

Month 3 (12–15 weeks)	Month 4 (16–19 weeks)	Month 5 (20–23 weeks)	Months 6 and 7 (24–31 weeks)	Months 8 and 9 (32–39 weeks)
The gender of the fetus can be discerned, and fingernails form.	The fetus is approximately 6 inches long and weighs 6 ounces; the skeleton is visible using ultrasound imaging, and eyelashes, eyebrows, and hair are present.	A stethoscope placed on the mother's abdomen can detect a fetal heartbeat, and a mother can feel fetal movement.	Organ systems continue to mature, and the eyes open.	The fetus increases in size, lungs mature, and body hair disappears; a typical newborn weighs around 7 pounds and is approximately 20 inches in length.

FIGURE 3.22 Human fetal development occurs from three months to nine months.

Technology Connection

Ultrasound

Not long ago, a "baby book" opened with a picture of the newborn infant. In many families, that place of honor is now occupied by a black-and-white, slightly fuzzy image of a tiny fetus. Ultrasound imaging has become a routine part of pregnancy care in developed countries.

Just as bats use sound to "visualize" their world and fishermen use fish finders, an ultrasound examination relies on high-frequency sound waves. As the waves echo back from an object, they are sent through a recording device called a transducer and transformed into an image on a computer. By measuring the waves, we can determine the shape of an object, its distance, its size, and whether it is liquid, solid, or both. Ultrasound imaging operates in real time. You see a live image.

During pregnancy, ultrasound is used to monitor fetal growth, development, and the health of the placenta. It can also identify many developmental problems 20 weeks into pregnancy—or even sooner. It can detect anencephaly (the lack of a brain), dwarfism, spina bifida (failure to close the neural tube), cleft palate, and heart abnormalities, for example.

Ultrasound also has other medical uses. It can image such organs as the heart and its major blood vessels, liver, breasts, spleen, gallbladder, pancreas, kidneys, bladder, uterus, ovaries, and scrotum. Ultrasound may even guide biopsies, or tissue samples drawn by a needle. It can assess damage after a heart

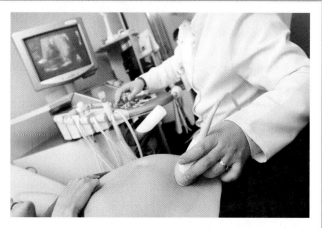

Ultrasound imaging has become a routine part of pregnancy care in developed countries.

attack, verify the presence of blood clots, measure the size of blood vessels to see whether there is a buildup of fatty deposits, and visualize tumors. A modification, called Doppler ultrasound, can measure blood flowing through the heart.

Because it is versatile, noninvasive, and safe, the use of ultrasound is growing. New breakthroughs are leading to the next generation of ultrasound scanners. These include 3D ultrasound, in which a three-dimensional image is generated from successive two-dimensional images, and 4D ultrasound—or 3D images in real time. ::

In the fifth month, a stethoscope placed on the mother's abdomen can detect a fetal heartbeat. Even a first-time mother can feel fetal movement. The sixth and seventh months are marked by a continual maturing of organ systems. The body of the fetus is covered by a layer of fine hair called lanugo. By the seventh month, the fetus is 12 inches long and weighs around 3 pounds. The eyes open, and if the fetus is male, the testes will have descended into the scrotum.

By this point, the fetus is positioned head down (**FIGURE 3.23**), the orientation appropriate for delivery. A fetus born after 7 months has roughly a 90% chance to survive. The eighth and ninth months are characterized by increases in size, maturation of the lungs, disappearance of body hair, and deposits of fat for insulation. A typical newborn weighs around 7 pounds or so and is approximately 20 inches in length.

Because human embryos and fetuses develop inside their mothers, they are protected from such things as heat, cold, rain, and wind. Yet that location makes them vulnerable to other assaults. Whatever the mother puts in her body also winds up in her offspring. If she drinks alcohol during pregnancy, the consequences for her baby can be grave and include the

Uterus

Cervix

Vagina

FIGURE 3.23 By the end of seven months, the fetus is positioned head down, the orientation appropriate for delivery. A fetus born after 7 months has roughly a 90% chance of survival.

Fetal Alcohol Syndrome

Fetal alcohol spectrum disorders (FASDs) encompass a range of health problems experienced by children who were exposed to alcohol prenatally. The most severe form, fetal alcohol syndrome (FAS), is a lifelong condition.

Children with FAS may suffer from a variety of symptoms—including poor growth, abnormal facial features, hyperactivity, poor reasoning skills, vision problems, hearing problems, and mental disabilities. FAS occurs in from 0.2 to 1.5 out of every 1000 births in the United States, while the less severe form is present in up to 6 out of every 1000. These children are at a higher risk for psychiatric problems and criminal behavior. Although there is no treatment, a supportive home and education can achieve a more positive outcome.

FASD is completely preventable: women should not drink alcohol during pregnancy. There is no known safe amount. Even occasional drinking can lead to FASD.

While prevention is quite simple in principle, it can be challenging in practice. Many women have no idea that they are pregnant soon after conception. If a woman is sexually active and drinks alcohol, she should use an effective contraceptive to prevent pregnancy. If a woman is trying to start a family, she should not drink. If she learns that she is pregnant and drinks, she should stop—the sooner, the better.

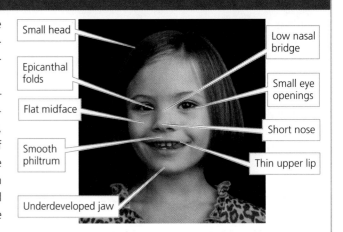

Small head · Epicanthal folds · Flat midface · Smooth philtrum · Underdeveloped jaw · Low nasal bridge · Small eye openings · Short nose · Thin upper lip

Children with fetal alcohol syndrome (FAS) may suffer from a variety of symptoms—including poor growth, abnormal facial features, hyperactivity, poor reasoning skills, vision problems, hearing problems, and mental disabilities.

Prenatal exposure to alcohol is not the only problem. Exposure to other drugs—including cocaine (crack or powder), heroin or other opiates, barbiturates, and PCP—can have terrible consequences. ::

---

fetal alcohol spectrum disorder (FASD) :: encompasses a range of health problems experienced by children who were exposed to alcohol prenatally

risk of **fetal alcohol spectrum disorders** (**FASDs**), the most severe of which is **fetal alcohol syndrome** (see *Life Application: Fetal Alcohol Syndrome*).

 # What Happens in Labor and Delivery?

fetal alcohol syndrome (FAS) :: the most severe FASD; symptoms may include poor growth, abnormal facial features, hyperactivity, poor reasoning skills, vision problems, hearing problems, and intellectual disabilities

Normally the fetus triggers labor once the lungs are fully developed. A surfactant is a substance that coats the tissues lining the inside of the lungs so that they don't stick together. Without adequate surfactant, the sticky tissues make breathing very difficult. In fact, immature lungs are one of the most serious health challenges facing infants born prematurely. When a fetus is ready to be born, it secretes surfactant, which causes an immune response in certain fetal cells. These cells migrate to the uterus, where they secrete a chemical signal that causes the uterus to contract.

Light uterine contractions occur throughout pregnancy. The contractions that mark labor are different. At the onset of labor, contractions occur every 15–20 minutes, each lasting around a minute. As delivery nears, contractions occur every 1–2 minutes, each lasting a minute.

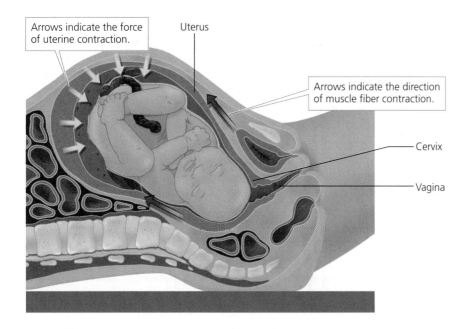

Arrows indicate the force of uterine contraction.

Uterus

Arrows indicate the direction of muscle fiber contraction.

Cervix

Vagina

FIGURE 3.24 During labor, the cervix stretches and causes a uterine contraction, which stimulates release of the hormone oxytocin, which in turn stimulates more uterine contraction and pushes the fetus downward.

Although the timing of events varies, most labors involve the same scenario: the lower end of the uterus, or **cervix**, stretches and causes a uterine contraction (**FIGURE 3.24**). This contraction stimulates release of the hormone oxytocin, which in turn stimulates more uterine contraction. As these contractions push the fetus downward, pressure from the fetus makes the cervix stretch more. This cycle repeats until eventually the cervix dilates (or expands) to give an opening 4 inches (10 centimeters) in diameter.

Labor and delivery are divided into three stages (**FIGURE 3.25**). In stage 1, uterine contractions pull the lower part of the uterus up toward the fetus's

cervix :: the lower end of the uterus

Stage 1
Uterine contractions pull the lower part of the uterus up toward the fetus's head and push the fetus down.

Umbilical cord

Fetus

Uterus

Vagina

Dilated cervix

Stage 2
Strong contractions occur every minute; the mother experiences an urge to push, and the baby emerges.

Crowning of the head

Stage 3
The placenta is delivered.

Placenta

Umbilical cord

Umbilical clamp

FIGURE 3.25 Labor and delivery are divided into three stages.

head. As described earlier, the contractions also push the fetus down, thus helping cervical dilation. The **amnion**, or sac in which the embryo is suspended, breaks, releasing amniotic fluid. (This event is often referred to as breaking water.) Stage 1 ends when the cervix is fully dilated. Stage 2 is sometimes called "hard labor." Contractions are very strong, occurring every minute or so. The mother experiences an almost irresistible urge to push. The baby emerges. Finally, in stage 3 the placenta is delivered.

Unlike some animals, human newborns are essentially helpless and remain so for quite some time. Scientists hypothesize that humans are born less fully developed than many other mammals because of the size of a human head. If the fetus grew much further, its head would be too big to deliver without surgery. Even now, some women have babies too big to deliver vaginally and have to resort to a Caesarian, or surgical, delivery.

3.8 How Do Twins Form?

Abigail and Brittany are twins, although of a very unusual kind. Even so, most pregnancies produce only one baby. Twins are born in a little more than 1% of pregnancies, unless the woman has been treated with hormones for infertility; the odds of twins then increase to 3%.

Identical twins come from a single fertilized egg that splits into two embryos sometime during very early development (**FIGURE 3.26**). About a third of twins are identical. The types of identical twins can be divided into three categories.

- In 33% of identical twins, the embryos have completely separate chorions (a component that forms in the placenta). Because the tissue that develops this support structure for the embryo itself forms at five days post-conception, it is clear that twinning occurs prior to five days after conception.
- In 66% of identical twins, the embryos share a chorion but have separate amnions, the fluid-filled sac that surrounds and cushions the embryo. Twinning in these cases occurs after five days—the embryonic tissue that will form the chorion is present—but before nine days post-conception, the time when the amnion forming tissue would have developed.

1 chorion   2 amnions   2 chorions   2 amnions   1 chorion   1 amnion
1 placenta   Fused placentas   1 placenta

In 66% of identical twins, the embryos share a chorion but have separate amnions.

In 33% of identical twins, the embryos have completely separate chorions.

In 1% of identical twins, the embryos share both a common chorion and amnion.

FIGURE 3.26 Identical twins come from a single fertilized egg that splits into two embryos sometime during very early development. About a third of twins are identical.

- In 1% of identical twins, the embryos share *both* a common chorion and a common amnion. Separation of the twins must happen later than nine days after conception. When twins separate this late during embryogenesis, they are at risk for being conjoined.

Fraternal twins, in contrast, occur when the mother releases two eggs, and each is fertilized by a separate sperm. Although they develop in the same uterus, fraternal twins are no more closely related genetically than siblings born to the same parents. A set of fraternal twins can consist of two boys, two girls, or one boy and one girl.

3.9 What Can Conjoined Twins Tell Us About Biology and Ourselves?

Amazingly, conjoined twins form in 1 out of every 50,000 pregnancies—and their births are energetically reported. Given that the world population is more than 6 billion, we would expect there to be at least 60,000 sets of conjoined twins alive on Earth today. Why does conjoining seem so rare?

Not all pregnancies are successful. Especially when the embryo or fetus has serious genetic or physical defects, the pregnancy may abort or end spontaneously. This event is called a miscarriage. Approximately 40–60% of all conceptions result in miscarriages. Although 1 out of every 50,000 pregnancies involves conjoined twins, only 1 out of every 200,000 births produce conjoined individuals. In a majority of these cases, the fetus dies, thus resulting in a stillbirth. Even when conjoined twins are delivered alive, their chances of survival are lower than 25%. For some types of conjoining, survival is less than 5%.

Abigail and Brittany are especially rare. They are joined side by side, sharing many organs and structures. About three-fourths of conjoined twins are joined at the chest, abdomen, or both. Joining occurs at the buttocks in 18% of twins and at the ischium (a part of the hip) in 6%. Finally, head-to-head joining happens in 2% of conjoined twins. The angle of joinings may also vary, as may the sizes of each twin—anything from slightly different sizes to a situation in which one of the twins is "parasitic," or not fully formed and often not completely conscious.

Explanations for Conjoining

Conjoining has had many explanations, few taken seriously today. A 16th-century French physician, Ambroise Paré, offered several: God was showing his wrath or his power; the pregnant woman looked at a terrible sight; her uterus was too tight; she sat the wrong way during pregnancy. Explanations like this do not count as scientific hypotheses. Why? It is difficult to think of ways to test them.

The true causes of conjoining are unknown, and the few clues so far don't lead to a clear avenue of inquiry. Conjoining is much more common in females: 75% of conjoined twins are sisters. It also occurs more often in India and Africa, less often in China and the United States. Some scientists argue that the early embryos in conjoined twins fail to separate properly. Yet an increasing number of scientists think that this "improper fission" cannot be the whole story. How can it explain which parts are attached and at what angle? One possibility is the fusion of embryos. Twin embryos develop in very close proximity, within a single amnion. Perhaps they bump into each other and stick together.

Whatever their cause, conjoined twins show how tightly the body regulates a growing embryo. Conjoined twins are the result of the best effort to follow a normal developmental program. Although their body plan sure looks different from one we are used to seeing, conjoined twins are beautiful in their own ways.

What Conjoining May Tell Us About Biology

Scientists often learn more from studying rare events than they do from everyday occurrences. For example, research on cancer cells tells us an enormous amount about the structure and function of normal cells. Similarly, conjoined twins can help us understand the mechanisms by which embryos form a body, because development has followed an unexpected path.

Conjoining is not unique to humans; it can occur in other organisms, too. It can also be produced experimentally in many organisms, so that we can study it more closely. Hans Spemann's lab first produced conjoined embryos in the 1920s in amphibians—creatures that can live both on land and in water, like frogs and toads (**FIGURE 3.27**). Spemann, a pioneer in experimental embryology, had assigned a research problem to a graduate student, Hilde Mangold: how do tissues interact and communicate to direct embryogenesis? Mangold removed small pieces of tissue from one set of embryos and transplanted them to another set of embryos. To Spemann's amazement, when one piece of tissue was transferred, it actually directed the cells of the other embryo to form a second body axis. The result was tadpoles, or newborn frogs, joined at the abdomen.

This basic experiment has been repeated thousands of times in many different organisms. Scientists have learned about the molecular signals, cell interactions, and changes in gene expression that accompany both conjoining and normal development. Spemann received the Nobel Prize in Physiology or Medicine in 1935 for his contributions to experimental embryology. Sadly, Mangold did not share the award, because she had died in a tragic accident, and Nobel Prizes are never awarded after death.

What Conjoining May Tell Us About Ourselves

As much as conjoining reveals about biology, it tells us even more about ourselves as people. Conjoining, like other physical differences, forces us to confront how we define "normal." It also forces us to ask how we treat people who are different from ourselves. Throughout history, the answer has been influenced by culture, religious views, and individual experience.

① Spemann's graduate student Hilde Mangold removed small pieces of tissue from one set of embryos and transplanted them to another set of embryos.

② Once the piece of tissue was transferred, it actually directed the cells of the other embryo to form a second body axis.

The result was tadpoles, or newborn frogs, conjoined at the abdomen.

Piece from embryo 1

Developing embryo 1   Developing embryo 2   Developing embryo 2   Resulting embryo

FIGURE 3.27 Conjoining is not unique to humans; it can also be produced experimentally in many organisms. Hans Spemann's lab first produced conjoined embryos in the 1920s in amphibians.

In ancient times, conjoined twins might be worshipped as gods, banished, or killed. In one of the earliest reported cases, Eliza and Mary Chulkhurst were born in England in 1100, joined at the hip. When one of the sisters died, the other refused to attempt separation. She died shortly thereafter. Upon their deaths, the Chulkhurst sisters, also known as the Biddenden maids, left 20 acres of land to the poor. Ever since, cakes imprinted with their images are baked and distributed to the poor at Easter.

The Biddenden maids have been held in high regard for almost 900 years. Much more often, conjoined twins have been treated as freaks and displayed in sideshows or circuses. In some cases, the experience was profoundly humiliating, and in others, the twins managed to eventually take charge of their situation and have successful careers. There are probably few if any examples from the historical record in which the twins did not suffer exploitation for at least part of their lives. The seemingly inevitable display of conjoined twins for some manner of entertainment was prevalent from at least the 15th century into the 20th century.

Perhaps the most famous conjoined twins in history, Chang and Eng Bunker (**FIGURE 3.28**), were born in Siam (now Thailand) in 1811. Billed as the "Siamese Twins," the Bunkers enjoyed a lucrative career in show business, touring around the world with an act that included acrobatics and other physical feats. They eventually settled in North Carolina, owned and ran a store, and married two sisters. Amazingly, they fathered 21 children between the two of them. (None were twins.) Many of their numerous descendants continue to live in the area of North Carolina where the brothers made their home.

Violet and Daisy Hilton, born in 1908 in England, were joined at the buttocks. The twins were sold to their single mother's boss, who saw the girls as a way to make money. The sisters were forced into show business and held against their will for almost 20 years. They finally escaped and sued for their independence. The Hilton sisters were then able to chart their own professional course and performed: singing, dancing, and playing saxophone and violin. They also appeared in some motion pictures, including *Freaks* and *Chained for Life*. They did well for quite a while, but after 20 or so years of touring and performing, interest in them decreased. After their agent abandoned them in Charlotte, North Carolina, the sisters got a job in a grocery store, made themselves a home in Charlotte, and generally fit into the community.

Finally, Clara and Altagracia Rodriguez were born in 1973, joined at the abdomen. By then most people's notions of conjoined twins had shifted from oddities to individuals. Medical science, too, had developed, so that surgical separation was now feasible in many cases. The Rodriguez sisters were separated successfully in 1974 by a team of 23 doctors headed by Dr. C. Everett Koop (**FIGURE 3.29**).

When conjoined twins are born now, efforts are generally undertaken to separate the individuals. However, some conjoined twins share a critical organ, such as the heart, and separation would kill one twin or leave them both profoundly disabled. Such would be the case for the Hensel twins, for example.

Conjoining and other types of physical difference challenge the concept of normal. Abigail and Brittany Hensel are living full lives with interests, ambitions, friends, and loved ones. Chang and Eng Bunker probably experienced fulfilling lives as well. In contrast, the Hilton sisters had difficult journeys. Some of the greatest challenges faced by those with a "disability" are the ways that others treat them.

Conjoining brings into sharp relief questions about what it means to be a person. What makes *anyone* an individual? Although identical twins,

FIGURE 3.28 Perhaps the most famous conjoined twins in history, Chang and Eng Bunker, were born in 1811. Billed as the "Siamese Twins," the Bunkers enjoyed a lucrative career in show business before settling in North Carolina, marrying two sisters, and fathering 21 children between the two of them.

FIGURE 3.29 Clara and Altagracia Rodriguez were born in 1973, joined at the abdomen. The sisters were separated successfully in 1974 by a team of 23 doctors headed by Dr. C. Everett Koop.

conjoined or not, possess identical genetic material, each is clearly an individual. As we saw at the start of this chapter, Abigail and Brittany are quite different from each other in many ways. Without a doubt, there is more to our development as individuals than our genes alone.

With respect to the question of when personhood is established, no one knows for certain. Since gastrulation does not occur until 14 days, and twinning can take place at least up until then, it is hard to argue that we are defined as individuals before this time. Another way to consider this question is to ask: when does human life begin? This question is surprisingly difficult to answer in a biological sense or even in a religious manner, since different faith traditions have different views. We will return to a full consideration of this issue in Chapter 8.

Biology in Perspective

An understanding of human embryogenesis and development opens a door to the informed and intelligent consideration of many of the challenges and choices we face as a society with regard to reproduction. Learning about human development also reveals the fascinating biological story of how each of us came to be.

Examining human development reminds us of the diversity of life, and it challenges us to confront our views about what is normal. Even though we are all the same species, the bodies we have are so very variable. A look around any room filled with people will demonstrate how different we are from one another. Conjoined twins are just one dramatic example. They force us to ask how we engage with others who are different from us. The answer says a lot about each of us as individuals and all of us as a society. Although we understand a great deal about what makes a body, we know much less about what makes a person. Biology certainly plays a part, but there is so much more.

Chapter Summary

3.1 What Are the Units of Life?
Explain the consequences of cell theory for embryogenesis.
- Cell theory states that cells are the most basic unit of life, all living things are composed of cells, and all living things come from existing cells.
- Embryogenesis is the process of development from single cells to whole organisms.

3.2 What Cell Structures Play a Role in Embryo Development?
Describe the cell structures that play a role in embryo development.
- Humans are composed of eukaryotic cells with membrane-enclosed nuclei and organelles.

- Cell membranes are the outer boundary of cells that regulate what enters and leaves cells and detect chemical signals.
- The nucleus houses the chromosomes and most of the genetic information for the cell as DNA.
- Mitochondria extract energy from food molecules.
- The endomembrane system consists of three membrane-bound structures: the endoplasmic reticulum (ER), including the smooth endoplasmic reticulum (SER) and rough endoplasmic reticulum (RER); the Golgi apparatus; and lysosomes.
- The cytoskeleton maintains the shape of the cell and is responsible for cell movement and organelle placement.

3.3 How Do Eggs and Sperm Form?

Show the possible chromosomes in gametes produced through meiosis in a given cell.

- Almost all human cells are diploid with 23 chromosome pairs (46 chromosomes total).
- Human eggs and sperm (gametes) are haploid; they have only one set of 23 chromosomes and are formed through meiosis.
- Meiosis entails two rounds of cell division and results in four haploid cells.
- Meiosis I includes interphase, prophase I, metaphase I, anaphase I, and telophase I and cytokinesis; it ends with two haploid cells. The chromosomes are each composed of two chromatids.
- Meiosis II includes prophase II, metaphase II, anaphase II, and telophase II and cytokinesis; it results in four haploid cells. The chromosomes are each composed of one chromatid.

3.4 What Happens in Fertilization?

Describe the process of fertilization.

- Fertilization occurs when a sperm fuses with an egg; the one-celled embryo is called a zygote.
- In humans, sperm must swim to the fallopian tubes to reach a fertile egg; they are aided by contractions of the uterus and chemical signals secreted by the egg.
- A coating outside the cell membrane of the egg (*zona pellucida*) contains proteins that protect the egg; sperm with the correct protein bind to the egg cell and release enzymes that digest a hole to the egg membrane; the egg removes the sperm-binding proteins from the *zona pellucida* so that only one sperm will fuse with the egg.

3.5 How Does an Embryo Form and Ultimately Become a Fetus?

Outline the stages of mitosis and the ways that cells become specialized.

- Embryogenesis begins with cleavage.
- Mitosis involves a single division resulting in two cells with exactly the same chromosomes as the mother cell; the phases are similar to meiosis: interphase, prophase, metaphase, anaphase, and telophase and cytokinesis.
- When the zygote divides into the 16-cell stage, the cells separate into two groups: the outer cells make the chorion, and the inner cells become the embryo.
- During gastrulation, cells in the embryo move into three layers, which will give rise to specific cell types.
- The neural tube forms following gastrulation; the anterior part becomes the brain and the posterior becomes the spinal cord.
- Cell differentiation is controlled by chemical signals between cells.

3.6 What Are the Key Events of Pregnancy?

Differentiate fetal from embryonic development, and identify key points in development during pregnancy.

- During the first week of conception, cleavage produces cells that will form the embryo and the chorion.
- In week 2, the embryo implants in the wall of the uterus and gastrulation occurs.
- In week 3, the earliest stages of the nervous system and heart begin.
- Weeks 4–7 involve further differentiation and development of some external features such as ears, nose, fingers, and toes.
- At week 8, fetal development begins: organ systems are refined and the fetus grows dramatically.
- In the third month, sex can be determined.
- In the fifth month, the heartbeat can be heard with a stethoscope.
- By the seventh month, the fetus is positioned head down, appropriate for delivery.
- In the eighth and ninth months, size increases and the lungs mature.

3.7 What Happens in Labor and Delivery?

Summarize the stages of labor and important events that are likely to occur in each stage.

- When a fetus's lungs are fully developed, they secrete surfactant, which also causes the uterus to contract.
- At onset of labor, the contractions occur every 15–20 minutes; near delivery, contractions occur every 1–2 minutes.
- Stretching of the cervix releases oxytocin, which causes contractions that push the fetus downward, further stretching the cervix; this repeats until the cervix dilates to a 4-inch diameter.
- Labor and delivery consist of three phases: in stage 1, the fetus is pushed down and the amnion breaks, releasing amniotic fluid; during stage 2, strong contractions occur and the mother experiences an almost irresistible urge to push; in stage 3, the placenta is delivered.
- If there are complications caught early in a labor, the doctor may opt for surgical delivery.

3.8 How Do Twins Form?

Distinguish the development of identical from fraternal twins.

- Identical twins come from a single fertilized egg that splits into two embryos early in development.
  - 33% of identical twins have separate chorions.
  - 66% of identical twins share a chorion but have separate amnions.
  - 1% of identical twins share both a chorion and an amnion; this category of twins is at risk of becoming conjoined.
- Fraternal twins occur when two separate eggs are fertilized by separate sperm. Fraternal twins are no more closely related than regular siblings.

3.9 **What Can Conjoined Twins Tell Us About Biology and Ourselves?**

Assess the value of conjoined twin research and what conjoined twins can tell us about ourselves.

- The survival rate of conjoined twins is very low.
- Conjoined twins can be produced experimentally in many organisms; these experiments help us understand the process of development.

- Conjoined twins were previously often exploited as oddities, but now conjoined twins are more likely to be treated as individuals with individual personalities.
- Although it is now possible to surgically separate many conjoined twins, some conjoined twins cannot be separated without causing significant harm to one or both individuals.

Key Terms

amnion 82
anaphase I 68
anaphase II 69
cell 62
cell membrane 64
cell theory 62
cervix 81
chromatid 69
chromosome 64
cleavage 71
conjoined twin 60
cytokinesis 70
cytoplasm 64
cytoskeleton 65
cytosol 64
differentiation 75
diploid cell 66
embryo 71
embryogenesis 62

embryonic stem cell 73
endomembrane system 65
endoplasmic reticulum (ER) 65
fertilization 70
fetal alcohol syndrome (FAS) 80
fetal alcohol spectrum disorder (FASD) 80
fetus 77
gamete 66
gastrulation 73
gene expression 76
Golgi apparatus 65
haploid cell 66
homologue 66
human development 60
interphase 66
lysosome 65
meiosis 66
metaphase I 67
metaphase II 69

mitochondrion 65
mitosis 71
neural tube 73
neurulation 77
nucleus 64
organelle 63
ovary 66
placenta 73
prophase I 67
prophase II 69
ribosome 65
spindle 67
telophase I and cytokinesis 69
telophase II 70
testis 66
ultrasound 77
uterus 77
zygote 70

Review Questions

1. What is the cell theory?
   a. The idea that bloodletting can be used to maintain balance within cells
   b. The realization that all living organisms are made up of cells
   c. The idea that cells have specialized structures called organelles
   d. The realization that all cells have a nucleus
   e. The idea that all cells must divide through the process of mitosis

2. Identify the principal function of each organelle or cell structure: (a) nucleus; (b) mitochondria; (c) rough endoplasmic reticulum; (d) smooth endoplasmic reticulum; (e) Golgi apparatus; (f) lysosome; (g) cell membrane.

3. Give an example of the function of each component of the cytoskeleton: (a) microfilaments; (b) microtubules; (c) intermediate filaments.

4. Starting with a diploid cell that has six chromosomes (two chromatids per chromosome), diagram meiosis I and meiosis II.

5. Starting with a diploid cell that has six replicated chromosomes (meaning two chromatids per chromosome), diagram mitosis.

6. What protects the egg from being fertilized by sperm of another species?
   a. Fallopian tubes
   b. Meiosis
   c. Gametes
   d. Homologues
   e. *Zona pellucida*

7. What are the main events of the first two months of human development?

8. What are the main events of fetal development, and when does each occur?

9. How is labor triggered by the fetus?

   a. When a fetus is ready to be born, the cervix will shrink for a time, causing a contraction that stimulates the release of the hormone oxytocin.

   b. When a fetus is ready to be born, it will secrete phlegm, which will block the lungs and send a signal to the uterus to contract.

   c. When a fetus is ready to be born, it will kick the wall of the uterus, causing a contraction.

   d. When a fetus is ready to be born, it will secrete surfactant, which causes certain fetal cells to migrate to the uterus, where they secrete a chemical signal that causes the uterus to contract.

   e. When a fetus is ready to be born, neurulation will occur, causing contractions.

10. Describe the three stages of labor and delivery.

11. Describe the difference between the formation of fraternal and identical twins.

   a. Identical twins arise from a single fertilized egg, which splits into two embryos sometime during very early development; fraternal twins arise when the mother secretes two eggs, and each egg is fertilized by a separate sperm.

   b. Identical twins arise when the mother secretes two eggs, and each egg is fertilized by a separate sperm; fraternal twins arise from a single fertilized egg, which splits into two embryos sometime during very early development.

   c. Identical and fraternal twins both arise from a single fertilized egg, but in the case of fraternal twins, the egg is fertilized by separate sperm.

   d. Identical and fraternal twins both arise when the mother secretes two eggs, but in the case of fraternal twins, the eggs are fertilized by the same sperm.

   e. There is no significant difference between the development of identical and fraternal twins.

12. Why is there a greater risk of conjoined twins forming if twins separate later than nine days after conception?

13. What are two ideas regarding how conjoined twins form?

The Thinking Citizen

1. Why is it necessary to produce gametes by meiosis rather than mitosis? What would the consequence be if gametes were instead formed by mitosis?

2. Given the high rate of spontaneous miscarriage (40–60% of all conceptions), why do you think we have not declared this situation a public health crisis?

3. Human and other mammalian embryos are attached to the mother's circulatory system and obtain nutrients and oxygen delivered from the placenta. Birds do not develop in their mothers' bodies. Where do bird embryos get their nutrients and oxygen?

4. Even though identical twins have identical genes, they are not the same person. Why is this so? What is responsible for their individualities?

The Informed Citizen

1. Many thoughtful people disagree about when human life begins. What information would you need in order to figure out when an embryo or fetus becomes a person?

2. People do not generally have funeral ceremonies to mourn miscarried embryos or fetuses. Should we? Why or why not?

3. Many infants born dramatically premature have serious health problems for the rest of their lives. Do you think that heroic medical efforts should always be made to save these infants, no matter the cost? Why or why not?

4. If a baby is born with a physical difference, it is generally the parents who decide whether to allow surgery. Is it always appropriate for parents to make this choice? Why or why not? Can you think of any examples in which no decision should be made about surgery until the child is old enough to make his or her own choice?

5. What are your thoughts about reality television programs, movies, or live shows that are focused on people with physical differences?

At its most basic, a gene is the instruction to make a specific protein. The notion that there is a gene "for" eye color misses the mark; rather, we should think of genes as *used for* or *used in* eye development.

CHAPTER 4

Inheritance, Genes, and Physical Characteristics

Does Disease Have a Genetic Basis?

Becoming an Informed Citizen …

We all want to know where we came from and why we look and act the way we do. Science is making great strides in understanding genes and even manipulating what they can do, but those advances also have real consequences for individuals and society.

We all want to know where we came from. Why do we look and act the way we do? How much of who we are can be traced to our genes?

Science is making tremendous advances in understanding **genes**—and even manipulating what they do. In this chapter, we explore how inheritance works. We see how the genetic information passed down from your parents plays a role in your unique development. To do so, we must learn the basic rules of inheritance. We need to know how the molecules DNA, RNA, and protein produce what is coded in our genetic instructions.

gene :: hereditary unit consisting of DNA

The function of a gene is widely misunderstood. At its most basic, a gene is simply the instructions to make a specific protein. The notion that there is a gene "for" eye color, or height, or intelligence, or aggression, or sexual orientation, or any other characteristic misses the mark entirely. Instead of a gene *for*, we should think of genes as *used for* or *used in*: for example, we can speak of the gene used in eye development. And many factors influence that development—including the environment. Genes act within a community of other genes as well, not as solitary actors.

In light of all this complexity, a single gene can have far-ranging consequences. Take the case of two devastating diseases—sickle cell disease and malaria.

case study

Sickle Cell Disease, Malaria, and Human Evolution

FIGURE 4.1 Malaria is caused by the parasite *Plasmodium falciparum*, which is carried by the *Anopheles* mosquito. When one of these mosquitoes bites an individual, the parasitic cells take up residence in the blood, destroying red blood cells, the cells that carry oxygen through the body.

malaria :: a disease caused by a parasite carried by the *Anopheles* mosquito; it is often fatal

protozoan :: a simple, single-celled organism

red blood cell :: a cell that carries oxygen through the blood

Tony Allison grew up in Kenya. His father, a farmer, had relocated the family from England in 1919. As a boy, Tony went on long excursions with professional naturalists to observe and help collect birds for the Natural History Museum in London. He also visited the archeological excavation site of Louis Leakey, the preeminent anthropologist of his time, and became intrigued with human evolution and the relationships among the various tribes he saw in Kenya.

During one of the Allisons' holidays on the beaches of Malindi in Kenya, Tony contracted malaria; he was only 10 years old. **Malaria** is a terrible disease and often fatal. It is caused by a **protozoan**, a simple single-celled organism, and carried by a specific type of mosquito. When one of these mosquitoes bites a person, it injects this protozoan, *P. falciparum*, into the individual (**FIGURE 4.1**). These parasitic cells take up residence in the blood and destroy **red blood cells**, the cells that carry oxygen through the blood. There is currently no vaccine against malaria. Tony's experience with malaria led him to switch gears: rather than becoming a naturalist, he decided to become a physician.

After earning his undergraduate degree in South Africa, Tony moved to England to finish his medical training at Oxford University. Although he enjoyed medical school, he had not lost his keen interest in human evolution. Tony was convinced that there had to be a way to measure human evolutionary relationships more precisely than anthropologists did by looking only at bones.

In 1949, when Oxford University sent a group of scientists to Kenya to survey and study plants and animals all over the country, Tony jumped at the opportunity to revisit the land of his childhood and to study human populations. He accompanied the expedition team on their travels throughout

Kenya, and at each location he collected blood samples from individuals of each of the various tribes, including the Kikuyu, Luo, and Masai tribes. He tested the samples for red blood cell membrane proteins to see whether there might be a pattern that would reveal genetic relationships. Tony also tested for the presence of **sickle cells** (**FIGURE 4.2**), a type of abnormal red blood cell. Prior to embarking on the trip, Tony had learned from other scientists that sickle cells, common in some areas of Africa and inherited, produced **sickle cell disease**—a painful and, in the early twentieth century, generally fatal disorder.

To Tony's disappointment, the analysis of red blood cell membrane proteins failed to reveal a pattern. In contrast, the sickle cell results presented an unexpected surprise. Tony observed the sickle cell trait in more than 20% of the individuals in tribes living along the coast or near Lake Victoria. Yet fewer than 1% of people in tribes living in dry land areas or at high altitudes exhibited the trait. Tony wondered why there was such a big difference in its prevalence and why it seemed to depend on where a person lived.

Sometimes a good idea comes like a bolt out of the blue. Tony knew that the density of mosquito populations varied in different locales, as did the incidence of malaria. In lowlands and wet areas, the mosquitoes were abundant, malaria was prevalent, and the incidence of the sickle cell trait was high. In the dry highlands, mosquitoes were few and far between, malaria was not a problem, and the sickle cell trait was rare. Was there a relationship? Tony hypothesized that possession of the sickle cell trait provided some resistance to infection with *P. falciparum*, the malaria parasite, but he could not test the idea because he had to return to England to finish medical school.

FIGURE 4.2 Normal red blood cells appear disc shaped, while sickle cells appear in the shape of a sickle blade.

sickle cell ∷ a type of abnormal red blood cell that is shaped like a sickle

sickle cell disease ∷ an inherited genetic disorder in which red blood cells sickle easily, leading to potentially serious physical consequences

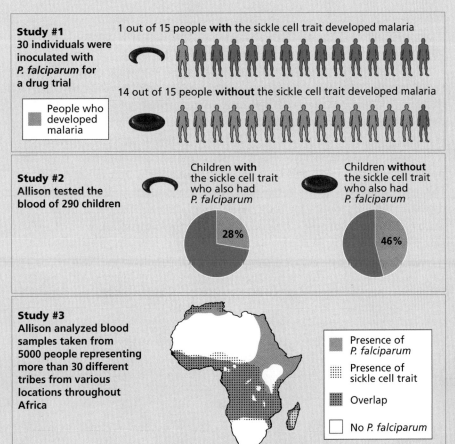

Study #1
30 individuals were inoculated with *P. falciparum* for a drug trial

1 out of 15 people **with** the sickle cell trait developed malaria

14 out of 15 people **without** the sickle cell trait developed malaria

☐ People who developed malaria

Study #2
Allison tested the blood of 290 children

Children **with** the sickle cell trait who also had *P. falciparum*

Children **without** the sickle cell trait who also had *P. falciparum*

28%

46%

Study #3
Allison analyzed blood samples taken from 5000 people representing more than 30 different tribes from various locations throughout Africa

■ Presence of *P. falciparum*
▦ Presence of sickle cell trait
▨ Overlap
☐ No *P. falciparum*

FIGURE 4.3 In 1953, Tony Allison designed three studies to see whether the inheritance of the gene associated with sickle cell disease would confer resistance against malaria.

Tony completed medical school in 1952 and received the financial and technical support he needed to return to Africa in 1953. As Dr. Allison, his goals were straightforward: he wanted to see whether the inheritance of the gene that is associated with sickle cell disease would indeed confer resistance against malaria.

Dr. Allison designed three studies to address his research question (**FIGURE 4.3**). In the first study, he piggybacked onto a drug trial being conducted by a pharmaceutical company. Scientists injected *P. falciparum* into healthy volunteers. When subjects developed malaria, the scientists administered their test drug to see whether it was effective. Allison had access to all 30 of the individuals in the study. He took blood samples from all and looked for a relationship between those possessing the sickle cell trait and those who developed malaria. The results were dramatic. Fourteen out of 15 infected subjects who did not have the sickle cell trait developed malaria. Only 1 out of 15 subjects who had the sickle cell trait did so.

Allison was excited about these results, but he knew that adults living in mosquito-infested areas probably had been exposed to *P. falciparum* earlier in their lives. The results may have been due to immune responses remaining from previous infections. To address this possibility, Allison undertook a second study. He offered free medical checkups to children at the local farmers' markets. He administered physical exams and collected a small amount of blood from each child. He examined the blood samples for two things: the presence or absence of the sickle cell trait and whether *P. falciparum* was in the blood.

Eventually Allison collected data for 290 children. Once again, the results were dramatic. *P. falciparum* was present in only 28% of children who had the sickle cell trait. In contrast, 46% of children who did not have the trait were infected and the number of parasites present was greater than that observed in the children with the sickle cell trait.

In Allison's final study, he traveled throughout Africa collecting blood samples from 5000 individuals representing more than 30 tribes. He observed that the frequency of the sickle cell trait ranged from 0% to 40%. In all cases, the sickle cell trait was rare or absent in locations where malaria was absent and common in locations where malaria was a serious problem (**FIGURE 4.4**).

Through his three studies, Allison demonstrated unequivocally that possession of the sickle cell trait afforded some protection against malaria. However, many questions remained. How does a person develop sickle cell disease? What accounts for the relationship between sickle cell disease and malaria? For our purposes, let's start with a simpler question: what exactly is sickle cell disease?

Distribution of malaria

Distribution of sickle cell trait

■	> 20%
■	15–20%
■	1–10%
■	< 1%

Incidence of
sickle cell trait

FIGURE 4.4 In Tony Allison's final study, he discovered that the sickle cell trait was rare or absent in locations where malaria was absent and common in locations where malaria was a serious problem.

What Is Sickle Cell Disease?

In 1910, Dr. James Herrick of Chicago took a drop of blood from a dental student and examined it with a microscope. The student suffered from **anemia** and was deficient in **hemoglobin**, which is the protein that carries oxygen in red blood cells. Symptoms of anemia may include yellowing of eyes, fatigue, dizziness, shortness of breath, and chest pain, to name just a few. Herrick saw something quite unusual: instead of being disc shaped, the red blood cells were shaped like commas or sickles (curved blades generally used for harvesting crops). The cause of the sickling was not known. In 1917, Dr. Victor Emmel placed blood from an individual with sickle cell disease onto a slide and sealed it from the air. He observed that the red blood cells sickled as the amount of oxygen available to them decreased.

By 1923, it was clear to scientists that sickle cell disease was genetic. A person could inherit the disorder from parents. And in 1949, scientists realized that there were two forms of the disease. In the severe type, red blood cells sickle easily; this is sickle cell disease (or *sickle cell anemia*). In the less severe version, red blood cells do not sickle unless they are completely without oxygen. Someone with this less severe form is referred to as "being a carrier" or "possessing the sickle cell trait."

anemia :: a disease in which a person is deficient in hemoglobin

hemoglobin :: the protein that carries oxygen in red blood cells

How Sickling Happens

Sickle cell disease is the most commonly inherited blood disorder in the United States. Approximately 72,000 Americans have this disease. One out of every 500 African Americans is affected.

While normal red blood cells are flexible, sickle cells are rigid and can get stuck in narrow blood vessels, thus forming blood clots (**FIGURE 4.5**). Cells, tissues, and organs "downstream" from a blood clot will be deprived of oxygen. If the lack of oxygen persists, tissue and organ damage are certain. The range of physical consequences is quite broad and potentially serious: muscle cramps, intense pain, anemia, swollen joints, kidney failure, and heart failure are some examples (**FIGURE 4.6**). A sickling incident can be triggered by many events such as physical exertion or other health issues such as infections, heart disease, or respiratory ailments.

FIGURE 4.5 Normal blood cells are flexible, but sickle cells are rigid and can get stuck in narrow blood vessels, forming blood clots. The range of physical consequences is quite broad and potentially serious.

Normal red blood cells

Normal hemoglobin

Cross section of RBC

Normal red blood cell

Disc-shaped, flexible red blood cells move easily through the blood vessel.

Sickled (abnormal) red blood cells

Abnormal hemoglobin forms fibers that result in the sickled shape of the red blood cell.

Cross section of sickled cell

Rigid and pointy sickle cell

Sickle cells get stuck in narrow blood vessels.

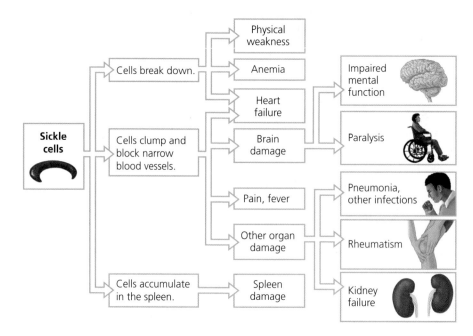

FIGURE 4.6 Sickled cells can lead to a range of serious physical consequences caused by lack of oxygen, including physical weakness, intense pain, anemia, heart failure, kidney failure, impaired mental function, and paralysis.

HbA :: normal hemoglobin that carries oxygen molecules throughout the body

HbS :: abnormal hemoglobin; although it can carry oxygen, it forms long fibers once it has given up its oxygen, causing red blood cells to sickle

The principal biochemical defect of sickle cell disease is in the molecule hemoglobin. Located inside red blood cells, normal hemoglobin (**HbA**) carries oxygen molecules throughout the circulatory system, delivering them to cells and tissues within the body. Sickle cell hemoglobin (**HbS**) is abnormal. Although it can carry oxygen, unlike HbA, HbS forms long fibers once it has given up its oxygen, causing the red blood cells to become sickle shaped; deoxygenated HbA does not form fibers (**FIGURE 4.7**).

Sickle Cell Disease and Inheritance

The inheritance pattern of sickle cell anemia is well understood (**FIGURE 4.8**). If a person inherits genes that encode for HbA protein, he or she will not have sickle cell disease. Inheritance of one normal gene and an altered one that directs the synthesis or production of HbS protein means that the person will be a carrier of the disorder. Such a person will have *both* HbA and HbS proteins in his or her red blood cells, but he or she will not develop full-blown sickle cell disease. Finally, individuals who inherit two altered genes will have only HbS protein in their red blood cells and will definitely suffer from sickle cell disease.

FIGURE 4.7 Both normal hemoglobin (HbA) (A) and sickle cell hemoglobin (HbS) (B) carry oxygen molecules throughout the circulatory system, delivering them to cells and tissues within the body. However, when oxygen is low, HbS forms long fibers because of the attraction of the HbS molecules to one another, causing cells to become sickle shaped.

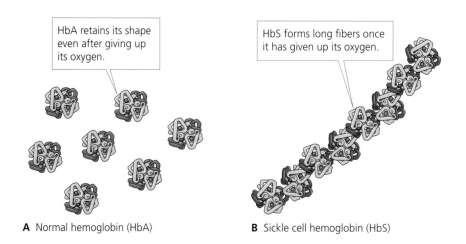

HbA retains its shape even after giving up its oxygen.

HbS forms long fibers once it has given up its oxygen.

A Normal hemoglobin (HbA)

B Sickle cell hemoglobin (HbS)

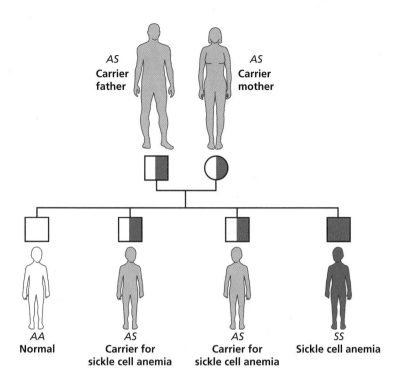

FIGURE 4.8 Only individuals who inherit genes that encode for HbS protein from both a mother and a father will suffer from sickle cell disease. In the figure, *A* stands for the HbA allele, and *S* stands for the HbS allele.

Individuals will also differ sharply in their susceptibility to malaria. People who have only HbA protein are vulnerable to infection, while those with both HbA and HbS proteins are resistant. Having only HbS provides malaria protection but guarantees sickle cell disease. Evidently, individuals who have the gene that encodes HbA protein *and* the gene that directs the synthesis of HbS protein have the best-case scenario possible in a malarial area: they have resistance to a deadly disease, malaria, but don't develop a different potentially fatal problem, sickle cell disease.

How does the sickle cell trait confer protection against malaria? Scientists have shown that sickled cells are more difficult for *P. falciparum* to enter. Even parasites that manage to get into the red blood cells while they are still shaped normally will be killed by a sickling event.

Even though the inheritance, biochemistry, and molecular biology of sickle cell disease are well understood, we have no effective treatments. Physicians can only prevent symptoms from developing and relieve them when they occur. Public health efforts to screen for carriers and children with sickle cell disease have been only partially successful. The problem remains serious: at least 8% of all African Americans are carriers of the sickle cell trait.

Could Molecular Medicine Prevent Sickle Cell Disease?

Research to understand the biochemical basis of sickle cell disease began almost accidentally. In 1945, Linus Pauling (**FIGURE 4.9**), one of the most important chemists of the 20th century, was visiting Harvard University to deliver a lecture. While there, he met with other scientists to share research findings and questions. In his meeting with professor William Castle, Pauling learned about sickle cell disease and how red blood cells sickle in the absence of oxygen.

FIGURE 4.9 Linus Pauling revealed that sickle cell disease was due to a mutated form of hemoglobin.

Technology Connection

Electrophoresis

"Who done it?" On television and in the movies, DNA tests often hold the answer. You know the scenario. DNA is collected from the saliva on a cigarette butt carelessly left behind at the crime scene. If the police have a suspect and there is enough evidence, a court order can require the accused to provide a DNA sample. But how do we know whether the samples match?

One of the first steps is to chop up the DNA into manageable fragments. Chemicals called enzymes can cut a DNA molecule at specific nucleotide sequences. For example, the enzyme EcoR1 cuts the molecule between A and G everywhere the sequence GATTC appears. Because each of us has unique DNA sequences, the enzymes will produce DNA sequences of different lengths. By comparing the DNA fragments from different samples, scientists can determine whether the DNA came from the same person.

Electrophoresis supplies the comparison. First developed in the 1930s by Arne Tiselius, this technique is used widely for forensic investigation. It also identifies genes for paternity determination or disease screening and diagnosis.

Electrophoresis works much like the chromatography you might have done in elementary school. Back then, you might have taken a piece of paper towel or filter paper and dotted it with food coloring. When you dip the edge of the paper in water, it absorbs fluid, which travels up the paper past the colored dots, and the colors separate. The green dot, for example, reveals that it is actually a mixture of yellow and blue. Like chromatography, electrophoresis separates mixtures of substances by the movement of molecules. Unlike chromatography, it relies on their movement through an electrical field.

With electrophoresis, a mixture of DNA fragments is placed in a well cut into the gel, a solid matrix similar in consistency to gelatin. The gel, submerged in a conductive liquid, is subjected to an electric field. Because DNA is negatively charged when in solution, it will migrate toward the positively charged end of the field. Small fragments migrate the fastest, large ones the slowest. Eventually, the mixture of DNA pieces separate into a ladder of fragments that can be dyed and thus visualized as stripes called bands. By comparing the bands produced by different DNA mixtures, the forensic scientist can tell whether the DNA on the cigarette did indeed come from the suspect. ::

① Restriction enzymes cut DNA into segments of different sizes.

② DNA segments are placed into wells or slots in a porous gel.

③ When an electric current passes through the chamber, DNA fragments move toward the positively charged end.

④ Smaller DNA segments migrate the furthest and the fastest; large ones migrate the least far and the slowest.

Gel electrophoresis is used to separate mixtures of molecules such as proteins, RNA, and DNA. It is one of the most important tools in biomedical, biotechnological, conservation biological, and forensic science laboratories.

Pauling seized on this information and set out to study the HbA and HbS proteins. What exactly was different between the two molecules? He wanted to show that disease could be studied—and even prevented or cured—at the biochemical level. Pauling's big idea was that molecular medicine was the path to the future.

Pauling and his coworkers examined the HbA and HbS proteins using a technique known as **electrophoresis** (see *Technology Connection: Electrophoresis*). They found that the two types of hemoglobin were different chemically: HbA and

electrophoresis :: a widely used method for separating mixtures of molecules, including proteins, DNA, and RNA

4   5   6   7

Normal amino acid sequence with
glutamic acid in position 6

4   5   6   7

Valine replaces glutamic acid in position 6
of amino acid sequence

FIGURE 4.10 In HbS, the amino acid
that is in position 6 on the chain is
valine instead of glutamic acid (as it
is in HbA). A change in just one amino
acid produces an abnormal protein that
causes the disease.

HbS had a different distribution and localization of electrical charges on their
surfaces. Also, they demonstrated that the red blood cells of individuals who
were carriers contained an approximately equal mix of HbA and HbS.
Finally, Pauling and his colleagues revealed that the sickling that occurs in
low oxygen is because HbS forms long filaments under those conditions, thus
deforming the cells.

Although biochemical information was important, Pauling's insights
about the results were game changing. First, he realized that the type of
hemoglobin present in a person was related to the genetics of the disease. A
person with two normal hemoglobin genes showed no evidence of the disease
at all. Inheritance of one normal gene and one sickle cell gene produces a
carrier. An individual who suffers from sickle cell disease has inherited two
sickle cell genes, one from each parent.

Second, he demonstrated that one could trace the cause of a disease to an
alteration in the molecular structure of a protein. Finally, Pauling argued that
since sickle cell disease is inherited, genes must determine the structure of
proteins. This was years before anyone knew what genes did or what they
were made of. Pauling ushered in the age of **molecular medicine**, although it
took decades for other scientists to catch up to this idea that understanding
the way genes, proteins, and other molecules work could be the key to diagnosing and treating disease.

What precisely was different between the HbA and HbS molecules? In
1957, Vernon Ingram tackled this problem using a more sensitive technique
he had developed. He examined HbA and HbS structures and showed that
they differed in only one tiny place.

Proteins are composed of subunits called **amino acids**. The most basic
level of protein structure is simply the types and order of amino acids hooked
together like snap beads. In HbS, the amino acid that is in position 6 on the
chain is the wrong one. Instead of glutamic acid, as is present in HbA, HbS
has valine in that spot (**FIGURE 4.10**). Evidently, a change in a gene results in
the alteration in amino acid sequence, thus producing an abnormal protein
that causes a disease. And as if these connections are not sufficiently amazing,
here is one more of a different type: the source of the HbS protein for Ingram's
study was none other than Dr. Tony Allison!

molecular medicine :: a field of
medicine that aims to understand the
relationship between health and genes,
proteins, and other molecules, in order
to diagnose and treat disease

amino acid :: a subunit of protein

Where Is Our Genetic
Information Stored?

Newspapers, magazines, television, and the Internet bombard us with stories about inheritance. We hear that genes are related to everything from the
risk of disease to intelligence, musical talent, athleticism, and even sexual

orientation. How can we assess whether these claims are accurate? Is how we look, act, and live primarily determined by the genes we inherit? For an answer, let's first consider how genetic information is passed from one generation to the next.

Humans have been interested in how heredity works since at least the dawn of agriculture. Plant breeding for crop improvement and animal breeding to increase milk yield, strength, or meat production were practiced long before anyone had a clue about the mechanisms responsible for these successes. People developed significant practical knowledge regarding breeding, and for most, this was all a person needed to know. However, as is true in all time periods and places, there were some individuals who were scized by the need to know *how* a thing works. They wanted to understand the exact mechanisms of inheritance.

One early idea that persisted for more than 2000 years was **pangenesis**. According to this notion, each part of the body produced a characteristic

pangenesis :: an idea that persisted for more than 2000 years and that held that each part of the body produced a characteristic seed that traveled to reproductive organs; this idea was debunked by August Weismann in the 1880s

❶ Weismann measured the tails of a population of mice and then cut off their tails.

❷ Next, he had these mice mate, and when the offspring were born, grown, and ready to mate, he measured their tails and then had these mice breed with each other.

❸ When these grandchildren mice were born, he again let them grow, measured their tails, cut off the tails, and mated the mice among their generation…

❹ Weismann did this for 23 generations, and the tails never got shorter, even though it was tailless mice that were breeding generation after generation.

If pangenesis had been accurate, the tails of the mice should have been shorter as the "tail pangenes" were lost every generation!

FIGURE 4.11 In the 1880s, August Weismann performed an experimental test of pangenesis using mice. Weismann concluded that the information to construct the tails of the mice did not reside in the tail itself.

"seed" that traveled to reproductive organs. When mating occurred, the "seeds" were released by males and females and thus combined. As a consequence, the seeds from the parents formed the respective parts of the offspring. For example, the nose seeds, arm seeds, and leg seeds of the parents mixed, and the result was a blending of characteristics. It is easy to understand why pangenesis seemed like a realistic mechanism. If you look at offspring, human or other, they often resemble a blend of their parents' characteristics. Also, there was not a better explanation available yet.

Finally in the 1880s, August Weismann performed an experimental test of pangenesis (**FIGURE 4.11**). He took a population of mice and removed their tails (with their hypothetical seeds). He then bred offspring from these mice and measured whether the tails were shorter.

When Weismann measured the tails, he found they were just as long as the tails of the parents. Unfortunately for the mice, Weismann was a patient and methodical man. He repeated this experiment for seven years. Mice of this 23rd generation had normal-sized tails. Weismann concluded that the information to construct the tail does not reside in the tail itself. So where is it? The answer is the nucleus of each cell, in a molecule called DNA. The rest of this chapter tells the story.

How Did Mendel Discover the Rules of Inheritance?

The path to answering the questions of where genetic information resides and how inheritance works is a long one. In fact, it began 20 years before Weismann's experiment. Gregor Mendel (**FIGURE 4.12**), an Augustinian monk from the Brunn Monastery (in what is now the Czech Republic) undertook a series of carefully planned experiments. In doing so, he revealed the basic rules of inheritance.

Thanks to one of those accidents of circumstance, Mendel happened to be assigned to a monastery with a long tradition of scientific pursuits. It was equally fortunate that he was well trained in science and mathematics. The religious order's original plan was for Mendel to be a science teacher. Luckily for science, Mendel turned his formidable mind to the problem of inheritance.

Mendel's Experiments

Mendel made some important decisions as he designed his experiments (**FIGURE 4.13**):

1. He chose garden peas as the organism to study. Peas are simple to cultivate, and a person can easily control the matings by manually applying pollen to the female parts of the flowers. Mendel's initial genetic experiments were done with mice, but the church hierarchy pulled the plug on this work. Apparently all of that mouse sex was deemed inappropriate. It appears that the fathers did not understand that plants mate too.

2. Mendel focused on one **trait**, or characteristic, at a time, rather than just recording the overall appearance of the offspring of a **cross** (mating). For example, he looked at traits that were discrete and easy

FIGURE 4.12 Gregor Mendel was an Augustinian monk who conducted a series of carefully planned experiments that revealed the basic rules of inheritance.

trait :: characteristic

cross :: mating

① **Choose experimental organism carefully.** Mendel chose garden peas, which are simple to cultivate and whose matings may be controlled by manually applying pollen to the female parts of another flower using a paintbrush.

② **Focus on one trait at a time.** Rather than just recording the overall appearance of the offspring of a cross, Mendel focused on one trait (such as flower color, pod color, or stem length) at a time.

Flower color

Pod color

Stem length

③ **Follow the cross for more than one generation.** Mendel followed the crosses beyond the first set of offspring.

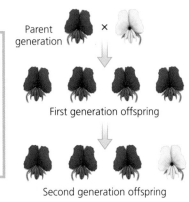

Parent generation

First generation offspring

Second generation offspring

④ **Collect quantitative data.** Mendel counted the number of progeny possessing a trait and kept detailed records.

FIGURE 4.13 Mendel followed four steps that led to his success.

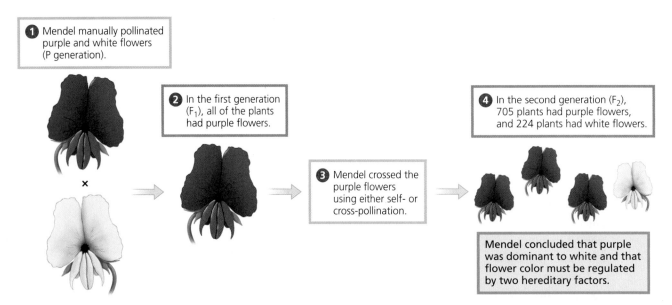

① Mendel manually pollinated purple and white flowers (P generation).

② In the first generation (F₁), all of the plants had purple flowers.

③ Mendel crossed the purple flowers using either self- or cross-pollination.

④ In the second generation (F₂), 705 plants had purple flowers, and 224 plants had white flowers.

Mendel concluded that purple was dominant to white and that flower color must be regulated by two hereditary factors.

FIGURE 4.14 Mendel crossed pea plants that had purple flowers with pea plants that had white flowers. In the first generation of offspring, all of the flowers were purple, but in the second generation, Mendel noted a mix of purple flowers and white flowers.

to identify, such as yellow peas versus green peas or white flowers versus purple flowers. In humans, examples of discrete traits include attached or unattached earlobes, left- or right-handedness, and red blood cells with HbA or HbS protein.

3. Mendel followed the crosses for more than one generation. This made it possible to see whether a characteristic that disappeared in the children of a cross would reappear in the grandchildren.

4. Finally, Mendel collected numerical data: he counted the number of progeny possessing a particular trait.

Mendel began his experimental crosses with pure-breeding parental strains that always produced offspring just like themselves; self-pollinating plants with purple flowers had offspring that had purple flowers and plants with white flowers had offspring with white flowers. In one of Mendel's experimental crosses, he mated pure-breeding plants with purple flowers with other pure-breeding plants that had white flowers (**FIGURE 4.14**). After these flowers formed seeds (peas), the seeds were planted. The plants that grew produced flowers. In this first cross, all of the flowers were purple. Next, Mendel crossed these purple flowers among themselves, planted the resulting peas, and observed the flowers produced by these "grandchildren" plants. He reported that 705 plants produced purple flowers and 224 had white.

A lot of information is packed into this experiment. First, Mendel observed that the "white" trait disappeared from view in the progeny of the first cross but reappeared in the next cross. This result showed that purple was dominant to white. However, the white reappeared. This demonstrated that whatever was responsible for the production of this color still existed.

Second, Mendel figured out that flower color must be regulated by two hereditary factors. He reasoned that one factor is received from each parent, because each must contribute hereditary instructions. He knew that the offspring in the first cross had a purple factor; after all, the flowers were purple. Because white showed up in the grandchildren, Mendel argued that a white factor must exist in the children. People just could not see the expression of its information. Therefore, there must be *two* factors in the children: one white and one purple. When the children mated, the purple and white factors from each made very specific combinations in their offspring.

Gametes and Monohybrids

We now know that these factors sort into different gametes, or sex cells (see Chapter 3). In the cross of the children, the gametes from the mother had *either* a purple *or* a white factor. Since the grandchildren are each produced by the union of one male gamete and one female gamete, only the following combinations of factors are possible: purple, purple; purple, white; and white, white (**FIGURE 4.15**).

Mendel knew nothing about cells, cell division, chromosomes, or meiosis, but he did understand statistics. He realized that if you had two independent factors sort into pairs, the following ratios would occur: 1 purple, purple; 2 purple, white; and 1 white, white. Mendel observed that plants with a purple factor and a white factor produced flowers that looked identical to those produced by plants with two purple factors. If we consider only how the flowers look, Mendel's hypothesis predicted that the ratio of purple to white flowers in the grandchildren should be 3:1. Mendel's results were 705 purple to 224 white flowers, or a ratio of 3.15:1, amazingly close.

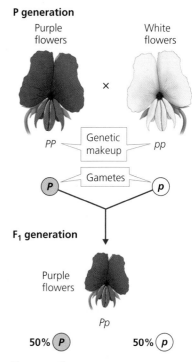

P generation

Purple flowers × White flowers

PP — Genetic makeup — *pp*

Gametes — *P* *p*

F₁ generation

Purple flowers

Pp

50% *P*   50% *p*

F₂ generation

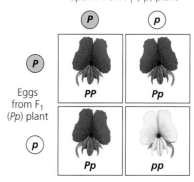

Sperm from F₁ (*Pp*) plant

	P	*p*
Eggs from F₁ (*Pp*) plant *P*	*PP*	*Pp*
p	*Pp*	*pp*

3 purple flowers : 1 white flower

FIGURE 4.15 This figure depicts Mendel's genetic cross examining flower color, with symbols representing the inherited factors *P* (purple) and *p* (white).

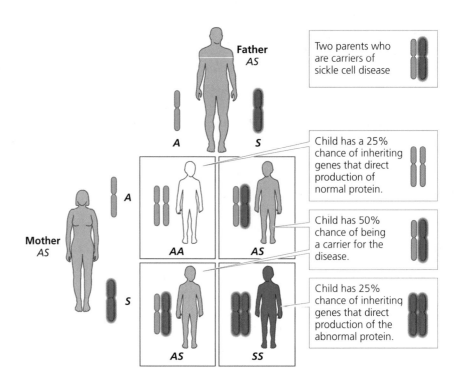

FIGURE 4.16 Mendel's rules apply to all organisms that mate and with respect to all monohybrid crosses, including sickle cell disease.

monohybrid :: a type of cross in which scientists pay attention to one trait at a time

factor :: a gene that possesses the information for the production of a specific protein

The type of cross in which we pay attention to one trait at a time is called **monohybrid**. Mendel repeated his analysis of monohybrid crosses in peas for six more traits: seed shape, seed color, pod shape, pod color, stem height, and flower position. He established that each trait is encoded by two factors, one from each parent. While Mendel did not understand the mechanism, other scientists have shown that the **factors** are genes that possess the information for the production of a specific protein. Meiosis sorts these genes or factors into gametes. When gametes fuse upon fertilization, genes from the mother and father are brought together again.

Although we have considered the details of the monohybrid cross and inheritance with respect to peas, the rules apply to all organisms that mate. For example, sickle cell disease follows the same pattern (although we can't undertake a controlled mating experiment with people, even volunteers) (**FIGURE 4.16**). A person who has sickle cell disease has two genes, each of which has the information for the synthesis of HbS protein. Someone with only normal hemoglobin has two genes that encode for HbA protein. And if an individual has one gene that encodes for HbA and another for HbS, he or she is a carrier of the disease.

Two parents who produce normal hemoglobin will have children that do so too. Two parents with sickle cell disease will have children who suffer from this disorder. The children of parents who are both carriers have a 25% chance of making HbA protein only, a 25% chance of making only HbS and having sickle cell disease, and a 50% chance of making HbA and HbS and being carriers. Mendel's rules predict just that.

Mendel's Rules

Mendel's research on monohybrid crosses revealed much of the basic information about inheritance:

allele :: a different form of a gene

1. Genes can come in more than one form, or **allele**. For example, the pea flower color gene can be either a purple or white allele. Similarly, the gene that encodes for hemoglobin can be the normal or sickle cell allele, which encodes for HbA or HbS protein, respectively.

2. The alleles of a particular gene sort individually into gametes during meiosis. In other words, a pea plant with a purple allele and a white allele would produce gametes each possessing one of these alleles. A person who is a carrier of sickle cell disease will make gametes, 50% of which have the normal hemoglobin allele and 50% the sickle cell allele.

3. Finally, pea flower color is a trait that exhibits a dominant/recessive relationship between the purple and white alleles. Purple is the **dominant** trait, or the one seen when both alleles appear together. When present together, the two different alleles produce a flower just as purple as one developed by a plant with two purple alleles. White, which is then not seen, is the **recessive** trait.

dominant :: a trait that is seen even when one allele is present

recessive :: a trait that is not seen when only one allele is present

4.5 How Much Do Mendel's Rules Explain?

Do Mendel's rules explain inheritance completely? Not exactly. Suppose that we compare the physical outcomes for individuals with two normal alleles, two sickle cell alleles, or one normal and one sickle cell allele. We find that there is a lot more to the story than Mendel might have predicted. First, alleles can interact, and so can genes. Second, genes do not always affect just one characteristic. Finally, gene expression depends on the environment.

Alleles Can Interact, and So Can Genes

Sickle cell disease inheritance is more complicated than a simple dominant/recessive relationship between alleles. One copy of the sickle cell allele cannot totally dominate the effect of the normal HbA allele. Instead, the alleles interact (**FIGURE 4.17**). That is why a carrier has a mild form of the disease. Carriers do not have completely normal red blood cells, but they are not afflicted with sickle cell disease. As you saw earlier, carriers make both HbA and HbS proteins, and each red blood cell contains both types of hemoglobin. Red blood cells of carriers don't sickle unless the oxygen levels are very low. Interactions like these are actually much more common than dominant or recessive alleles.

Genes can interact with each other in other ways, too. For example, several different alleles influence whether you are blond, brunette, black haired, or red

Red blood cell	Hemoglobin	Sensitivity to oxygen concentration	Resistance to malaria
Normal	HbA	Not sensitive	Not resistant
Carrier	HbA and HbS	Intermediate sensitivity	Resistant
Sickled	HbS	Very sensitive	Resistant

FIGURE 4.17 There is a predictable relationship between the type of hemoglobin present in a person's red blood cells and whether he or she will experience sickling or resistance to malaria.

FIGURE 4.18 Genes can interact with one another in different ways. If an individual has the genes to make a certain set of skin pigments but also one gene that says "make no pigment at all," the skin color directions will be irrelevant; the individual will end up with albinism.

cystic fibrosis :: a disease in which a thick, abnormal mucus accumulates in lungs and the digestive system

haired. To make things more complicated, another gene is associated with baldness. Suppose someone has two "no hair" alleles. He might also have two blond alleles, but he would still have a bald head. In another example, if a person has a "no skin pigmentation" allele, it will not matter if she also has alleles for light or dark skin. She will end up with albinism, a condition in which individuals have no pigmentation in their skin, hair, or eyes (**FIGURE 4.18**). Sickle cell disease depends on other genes as well. Genes can influence the degree to which cells are sticky and form blood clots, even during a sickling event.

Genes May Affect More Than One Characteristic

Most genes also affect more than one characteristic. The sickle cell gene illustrates this point well. It also makes red blood cells less susceptible to infection by *P. falciparum*.

A similar situation is seen in **cystic fibrosis** (CF), a disease in which a thick, abnormal mucus accumulates in lungs and the digestive system. The CF allele encodes for an abnormal chloride channel in cell membranes. Interestingly, carriers of the CF allele are resistant to cholera and other serious infectious intestinal diseases.

Because a single gene can affect multiple traits, it could be wrong to remove genes even to cure disease. You could solve one problem but cause an even worse one. The consequences are unpredictable and potentially grave.

Gene Expression Depends on the Environment

Finally, how genes contribute to physical characteristics can depend on the environment. Flowers of *Hydrangea* vary in color depending on the acidity of the soil in which they grow (**FIGURE 4.19**). Some animal embryos develop differently depending on whether they are exposed to predators during their development. For example, in the absence of predators, carp are a slender fish. In contrast, exposure to predators during development results in a fatter, hunchbacked shape—no doubt more difficult to eat. Our development, too, is influenced by exposure to environmental factors, especially *in utero* (in the womb).

Carp are generally slender fish.

The more acidic the soil, the deeper and more vivid the blue.

Carp exposed to predators during development will be larger.

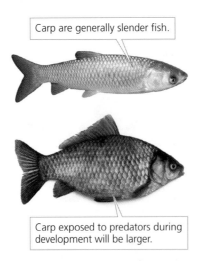

Chemicals
Alcohol
Cortisone
Diethylstilbestrol (DES)
Lead
Mercury
Retinoic acid (Accutane)
Streptomycin
Tetracycline

Chemicals and microorganisms such as these can be harmful to human embryos.

Microorganisms
Herpes simplex
Parvovirus
Rubella
Toxoplasma gondii (toxoplasmosis)
Treponema pallidum (syphilis)

A   **B**   **C**

FIGURE 4.19 Gene expression can depend on the environment. Hydrangea vary in color due to the acidity of the soil (A), carp vary in size depending on whether or not they are exposed to predators during their development (B), and humans can be affected by exposure to pathogens and chemicals *in utero* (C).

What Are Genes Made Of?

Suppose you went to a genetic counselor to find out whether you are at risk for passing on a genetic disorder to your children. The counselor needs some basic information:

- What genes are on your chromosomes, and how are the alleles for a particular gene sorted into gametes during meiosis?
- What gametes could each parent contribute during fertilization?

The resulting probability of having children with specific alleles can then be calculated in many cases using a technique known as **pedigree analysis**. (See *How Do We Know? Pedigree Analysis*.)

pedigree analysis :: a family history study; it shows a pattern of inheritance

How Do We Know?

Pedigree Analysis

In 1866 the physician Paul Broca wondered whether the breast cancer from which his wife suffered was hereditary. He observed a cluster of cases in his wife's own family: 10 out of 24 women had breast cancer. Broca mapped a family history, or pedigree, to see if he could figure out the pattern of inheritance. The idea that some traits "run in the family" is a very old one. Broca saw that they also run in patterns.

Today, pedigree analysis helps genetic counselors determine the risk of developing or passing on traits, such as disease. First, a counselor interviews the individual about the occurrence of the disease in other family members. The counselor then constructs a chart and applies the rules of inheritance.

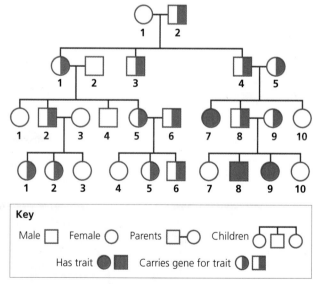

The pattern of inheritance of sickle cell disease in a family can be seen in a pedigree analysis.

For a disease like sickle cell disease or cystic fibrosis, in which a single gene is responsible, successful interpretation is likely. In other cases, such as ovarian cancer, disease depends on more than one gene as well as the environment. Still, a pedigree will point to the risks of developing the disease relative to someone with a different genetic background. It can also reveal unexpected patterns of inheritance across generations. ::

For instance, if a man who is a carrier of the sickle cell allele and a woman with normal hemoglobin genes have children, there is a 50% chance that these children will be carriers of the sickle cell trait, making both HbA and HbS proteins, and a 50% likelihood that they will make only HbA protein.

Sometimes, however, a person wants to be tested to see whether he or she carries a specific allele of a gene—one that does not show an obvious pattern of inheritance. In that case, these methods won't suffice. The techniques to identify specific genes for medical or forensic reasons all rely on knowledge about the molecular basis of genes. Thanks to our understanding of DNA, genes can be identified, manipulated, and even changed. To identify genes, unravel the causes and mechanisms of many diseases, and design drugs to treat these ailments, we need to know the basics of DNA structure. We need to know what genes are made of.

DNA (deoxyribonucleic acid :: a type of nucleic acid located in the cell nucleus; it is the physical material of which genes are made

nucleotide :: a subunit of a DNA molecule

DNA (deoxyribonucleic acid) is a type of nucleic acid located in the cell nucleus. DNA is the physical material of which genes are made. Chromosomes are composed principally of DNA (**FIGURE 4.20**). DNA holds our genetic information.

DNA molecules are composed of subunits called **nucleotides**. There are four different nucleotides in DNA: adenine (A), thymine (T), guanine (G), and cytosine (C). Each has similar components: a sugar (deoxyribose), a phosphate group, and a nitrogen-rich ring structure called a *base*. To form the DNA molecule, the nucleotides hook together in a specific manner: the sugar of one nucleotide connects to the phosphate of another nucleotide. In turn, its sugar group attaches to a phosphate of another nucleotide and so on until

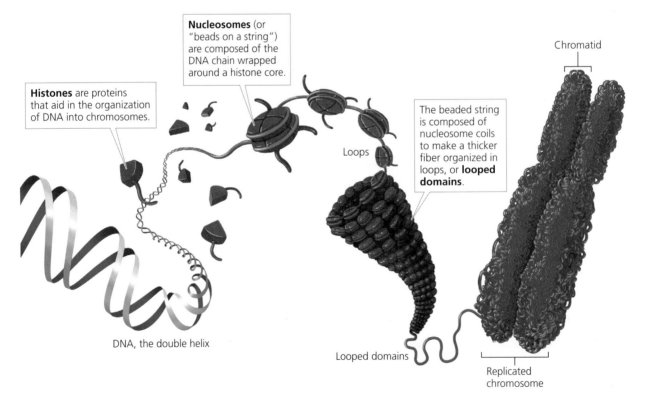

Histones are proteins that aid in the organization of DNA into chromosomes.

Nucleosomes (or "beads on a string") are composed of the DNA chain wrapped around a histone core.

Loops

The beaded string is composed of nucleosome coils to make a thicker fiber organized in loops, or **looped domains**.

Chromatid

DNA, the double helix

Looped domains

Replicated chromosome

FIGURE 4.20 DNA holds our genetic information. Chromosomes are composed of DNA.

a long strand of DNA is made. As a consequence, DNA has a sugar-phosphate backbone with bases sticking out from one side (**FIGURE 4.21**).

This is only the first aspect of DNA structure. The final form is a double helix: two single strands of DNA bound to each other in a shape like a twisting staircase. The attachment of the strands occurs thanks to the complementary base-pairing interactions between pairs of nucleotides. The base pair combinations are specific; A and T bind to each other and G and C to each other. No other combination works.

Let's put all of this information together for a look at DNA structure as a whole. The nucleotides of a particular DNA sequence bind to each other, forming a sugar-phosphate backbone. The bases stick out from this linear structure—in, for example, the sequence AATTGG. The DNA strand that will bind to this sequence is 100% predictable: TTAACC. Once scientists worked out these basic rules, they tackled the question of DNA function. Eventually, this fundamental understanding of DNA structure and complementary base pairing led to the development of the molecular techniques that make possible genetic testing, gene manipulations, biotechnology, drug development, and forensic science. (For a discussion of the discovery of DNA's double-helix structure, see *Scientist Spotlight: Rosalind Franklin.*)

Base pairs

Sugar-phosphate backbone

Adenine · Thymine

Cytosine · Guanine

FIGURE 4.21 DNA has a sugar-phosphate backbone with bases sticking out from one side. The final form is a double helix with two single strands of DNA bound to each other in a shape like a twisting staircase.

 4.7 # How Does DNA Function?

By 1928, scientists had demonstrated the inheritance of genes but did not know much at all about the physical basis of genes. It took scientists two steps to figure it out. First, Frederick Griffith found that a substance in bacteria could transform how other bacteria function to cause disease. Second, Oswald Avery, Colin MacLeod, and Maclyn McCarty showed that the mysterious substance was in fact DNA.

Transformation

Griffith found a clue to DNA's function while studying a species of bacteria that caused pneumonia. These bacteria came in two forms. One, referred to as smooth, grew on petri dishes as glistening, moist colonies. The other, rough, produced colonies that were drier looking and not smooth at all. Griffith knew that the smooth bacteria caused pneumonia, whereas the rough type did not. Griffith was interested in what was different between the two types of bacteria so that one caused a disease and the other was harmless.

Griffith tried a lot of manipulations of the two types of bacteria (**FIGURE 4.22**, p. 111). He would then infect mice to see whether they would develop pneumonia from the experimentally altered bacteria. The results from one of the tests were truly amazing. Griffith killed a population of smooth bacteria by heating them. He combined the dead smooth bacteria with living rough bacteria. Next, he infected mice with rough bacteria alone, dead smooth bacteria alone, or the mixture of dead smooth and live rough bacteria. Griffith monitored the health of the mice carefully. Only one group got sick—the mice infected with the mixture of dead smooth and live rough bacteria. The other two test populations remained completely healthy.

Griffith isolated bacteria from the sick mice and observed that they were all smooth. Somehow the physical characteristics of the formerly rough bacteria had been altered by the dead smooth bacteria. Griffith referred to this process as **transformation** because the bacteria had changed their

transformation :: a term that Frederick Griffith used to describe the process by which dead smooth bacteria altered rough bacteria in experiments; the transforming substance was later found to be DNA

Scientist Spotlight

Rosalind Franklin (1920–1958)

James Watson and Francis Crick are most closely associated with discovering the structure of DNA. One strand of the double helix is even called Watson, and the other is Crick. However, the work of other scientists was essential to their discovery. Maurice Wilkins shared the 1962 Nobel Prize in Physiology or Medicine with Watson and Crick. Another scientist, Rosalind Franklin, received little recognition.

Born in London to a wealthy family, Rosalind Franklin was interested in getting answers to questions, even as a child. She was fortunate to attend one of the few schools in London that taught chemistry and physics to girls, and she decided at age 15 to become a scientist. Franklin passed the entrance exam for Cambridge University in 1938, but her father refused to pay. He was against *any* woman attending university. Fortunately, Franklin's aunt stepped in and offered to fund her niece's education. Franklin's mother also supported her daughter's ambition. Finally, her father relented.

A brilliant student, Franklin completed her degree in 1941 and earned a Ph.D. in 1945. Her scientific talent was evident from the start. She published several research papers before she was 26 years old. Franklin spent 1947–1950 working in a research laboratory in Paris.

While in Paris, Franklin became expert in a technique called X-ray crystallography. In 1951, she was offered a job as a research associate in the laboratory of John Randall at King's College, London. She was assigned to research the structure of DNA.

With X-ray crystallography, a molecule is bombarded with X-rays. The pattern produced, called a diffraction pattern, gives clues to the positions of the atoms. Imagine shining a flashlight on a kitchen chair and figuring out its structure from its shadow.

Maurice Wilkins was another research associate in Randall's lab. Even though Wilkins and Franklin were peers, he initially misunderstood their relationship and thought she was his assistant. To make things worse, women in England were not permitted in campus dining halls, senior common rooms, or many pubs. Feeling isolated, Franklin threw herself into her work.

A

B

An expert in X-ray crystallography, Rosalind Franklin (A) produced a beautiful image of crystallized DNA (B) that was essential for figuring out the structure of this molecule.

She soon discovered that crystals of DNA take two forms. When these forms mix together, it is impossible to get a clear diffraction image. It is like shining the flashlight on a chair that has a table in front of it. However, she then figured out how to separate the forms, and she achieved excellent pictures of the A form. The scientist J. D. Bernal described them as "the most beautiful photographs of any substance ever taken." Finally, her analysis demonstrated that DNA was double stranded. The molecule had a sugar-phosphate backbone on the outside and bases paired on the inside. But how were they paired?

The answer came from Watson and Crick. Unbeknown to Franklin, Wilkins showed them her photograph of DNA. They took what they had learned from their own research, combined it with Franklin's pivotal information, and quickly described DNA's structure.

Franklin never knew exactly what had happened. Her accompanying research paper did not make the huge splash enjoyed by Watson and Crick. Sadly, she did not live to enjoy the accolades that her work deserved. She died of ovarian cancer in 1958 at the age of 37. Franklin was, however, recognized in her day as a first-class scientist with wide-ranging insights into chemical structures. She kept working in the lab until weeks before her death. ::

characteristics dramatically. Moreover, this change was heritable. Once transformed, these "new" smooth bacteria had offspring that were also smooth. Griffith, however, had no idea what this "transforming substance" was, just that it survived heating. Nevertheless, he realized that he had discovered a physical material that could change bacteria in a way that could be inherited.

FIGURE 4.22 In 1928, Frederick Griffith found a clue to DNA's function, which he called transformation, while studying a species of bacteria that caused pneumonia.

The "Transforming Substance"

The next step was to figure out the identity of this transforming material. Sadly, Griffith was unable to complete this work because he was killed, while working in his lab in London, in a bombing raid during World War II. In fact, the next piece of the puzzle was not filled in until the mid-1940s. Like so many basic research questions, this one was put aside because of World War II, a time when scientists focused their attentions on military service or projects related to the war.

Finally in 1944, Avery, MacLeod, and McCarty demonstrated that DNA was responsible for the transformation. These scientists performed a series of experiments modeled after Griffith's (**FIGURE 4.23**). They killed smooth bacteria. They then used enzymes to eliminate carbohydrate, protein, fat, RNA, or DNA from the virulent filtrate (the liquid that is left after you filter something). Next, they combined each filtrate from which they had removed *one* type of molecule with the rough bacteria. Only the mixture containing DNA combined with rough bacteria resulted in the appearance of smooth bacteria. DNA was the transforming substance.

4.8 What Processes Must DNA Accomplish?

Long before the work of Griffith or Avery, MacLeod, and McCarty, it was clear to scientists that the molecule of inheritance has to accomplish three functions:

1. It must replicate itself.
2. It must be able to change, or *mutate*.
3. It must produce the proteins responsible for physical changes.

The rest of this chapter tackles the details of these three processes.

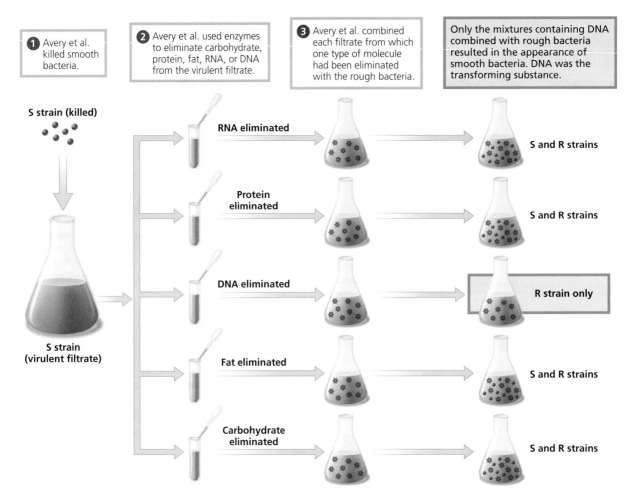

① Avery et al. killed smooth bacteria.

② Avery et al. used enzymes to eliminate carbohydrate, protein, fat, RNA, or DNA from the virulent filtrate.

③ Avery et al. combined each filtrate from which one type of molecule had been eliminated with the rough bacteria.

Only the mixtures containing DNA combined with rough bacteria resulted in the appearance of smooth bacteria. DNA was the transforming substance.

S strain (killed)

S strain (virulent filtrate)

RNA eliminated — S and R strains

Protein eliminated — S and R strains

DNA eliminated — R strain only

Fat eliminated — S and R strains

Carbohydrate eliminated — S and R strains

FIGURE 4.23 In 1944, Oswald Avery, Colin MacLeod, and Maclyn McCarty demonstrated that DNA was the substance responsible for the transformation discovered by Griffith.

Replication

replication :: the process by which the double helix of DNA unzips and new strands form

The first characteristic of DNA is that it can replicate. **Replication** is the process by which the double helix of DNA unzips and new strands form. During mitosis, identical cells form, chromosomes replicate, and a complete set sorts into each daughter cell (see Chapter 3). Similarly, during meiosis, chromosomes replicate, and two rounds of cell division parcel out single complete sets of chromosomes to individual gametes. Both mitosis and meiosis must involve DNA replication.

When James Watson and Francis Crick discovered the molecular structure of DNA in 1953, they suggested a hypothesis for how replication might work. Because of complementary base-pairing rules (A binds to T, G to C), the information for accurate replication is built right into the molecule.

To see why, imagine that the double strands of the DNA molecule separate from each other—like a zipper opening. To form two new molecules, each separate strand must now link to other nucleotides to form a *complementary strand*. As they separate, each strand reveals a sequence of nucleotides, and each nucleotide can pair with one and only one other nucleotide. Correct base-pairing thus ensures that the newly made strands will have the correct sequences (**FIGURE 4.24**).

No one in 1953 knew *how* the information in DNA led to the development of physical traits. Even so, scientists reasoned that copying of genetic instructions

Original DNA

DNA unzips

Old

New

New

Old

New strands form

FIGURE 4.24 One important characteristic of DNA is its ability to replicate. During DNA replication, the double helix unzips and new strands form.

had to be accurate for inheritance to work. Over more than a decade, scientists collected evidence that supports Watson and Crick's idea about DNA replication.

Mutation

The second characteristic of DNA is that it can mutate. A **mutation** is a change in DNA. Genes for a given trait come in different forms, as we have seen for the gene that codes for normal and sickle cell hemoglobin. Genes can be altered or mutated in a way that is inherited. DNA must be able to do this, too.

mutation :: a change in DNA; five examples of mutations are point, deletion, duplication, inversion, and translocation

If DNA sequences and genes remained identical one generation after the next, never changing, so would life. There would be none of the great diversity of life we see all around us. Instead, all living things would be genetically identical. But we *do* see diversity—in millions of species. We see variation even within a species and between parents and children. Why?

The answer is that although DNA replication is accurate, it is not perfect. Sometimes an organism's DNA is damaged. As a consequence, there is a small probability that genes can become altered, or *mutate*. In addition to the point mutation present in the sickle cell allele in which a nucleotide in the DNA is replaced by an incorrect nucleotide (a substitution point mutation), there are four more types of mutations: deletion, duplication, inversion, and translocation. A deletion removes a segment from the chromosome, a duplication repeats a segment, an inversion reverses a segment within a chromosome, and a translocation moves a segment from one chromosome to another (**FIGURE 4.25**). As a result, the nucleotide sequences of the chromosome are altered.

Most of the time, no physical effect can be discerned as a result of mutations. At other times, however, the change can be lethal, as with certain rare cancer-causing genes. In still other cases, the change in DNA, or mutation, produces an allele that confers a selective advantage to those who possess it.

For example, a mutation in the gene *CCR5* prevents binding of the human immunodeficiency virus (HIV) to immune cells. Individuals who possess this allele are resistant to acquired immune deficiency syndrome (AIDS). And, as we have discussed, the sickle cell allele provides protection against malaria.

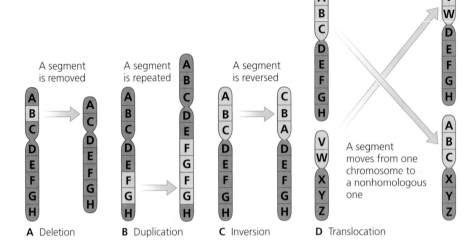

FIGURE 4.25 A second characteristic of DNA is its ability to mutate or change. In addition to the type of mutation seen in sickle cell disease, there are other types of mutations that can involve chromosome changes, including deletion (A), duplication (B), inversion (C), and translocation (D).

A segment is removed

A segment is repeated

A segment is reversed

A segment moves from one chromosome to a nonhomologous one

A Deletion   **B** Duplication   **C** Inversion   **D** Translocation

Organisms have developed ways to protect their DNA or even repair it. However, DNA mutations are sometimes due to faulty DNA repair. Exposure to ultraviolet light from the sun, tanning beds, or radiation, certain industrial chemicals like benzene, and activities like smoking tobacco can all damage DNA. The damage can be relatively minor, such as a change in or loss of one nucleotide. It can also be major, involving the deletion of a large sequence of DNA. In some cases, the cell can recognize relatively small mutations and enzymes can repair the damage by placing the correct nucleotides into position (**FIGURE 4.26**). If the repair is not done accurately, the

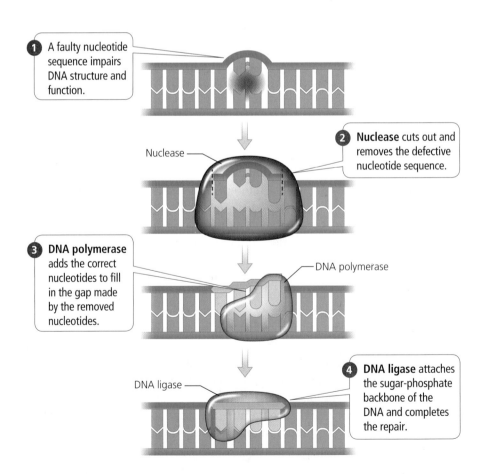

1 A faulty nucleotide sequence impairs DNA structure and function.

2 **Nuclease** cuts out and removes the defective nucleotide sequence.

Nuclease

3 **DNA polymerase** adds the correct nucleotides to fill in the gap made by the removed nucleotides.

DNA polymerase

4 **DNA ligase** attaches the sugar-phosphate backbone of the DNA and completes the repair.

DNA ligase

FIGURE 4.26 In some cases, enzymes can repair the damage of small mutations by placing the correct nucleotides into position.

mutation will persist. In other circumstances, the defect is too large, and repair is impossible.

Sometimes mutations originate during replication. Accurate DNA replication relies on enzymes that "proofread" the newly formed strand to make sure that the sequences are correct. Some mutations are the consequences of errors in this mechanism. If the wrong nucleotide is added during replication and the mistake goes undetected, the incorrect sequence will become a component of the DNA molecule for generations to come.

Protein Production

The last characteristic of DNA is that it can direct the production of specific proteins. Linus Pauling's sickle cell research demonstrated that the normal hemoglobin allele and the sickle cell allele encode for different proteins. As we saw in Section 4.4, individuals with two normal alleles made HbA protein, those with two sickle cell alleles produced HbS protein, and carriers made both types of hemoglobin. The connection between genes and proteins was clear. What remained unknown in 1949 was *how* the information in DNA directs the synthesis of a specific protein.

As was true for deciphering the details of DNA structure, replication, and mutation, it took a legion of scientists from many labs in many countries over a decade to work out the details. They discovered how DNA fulfills the third requirement of genetic material, along with replicating and mutating: how it is involved in the production of proteins and the development of physical characteristics. Encoding for proteins has three aspects—the *genetic code* itself, its *transcription* into RNA, and finally its *translation* into proteins (**FIGURE 4.27**). In this way, the sequence of nucleotides is ultimately translated into a specific sequence of amino acids connected together to make a protein.

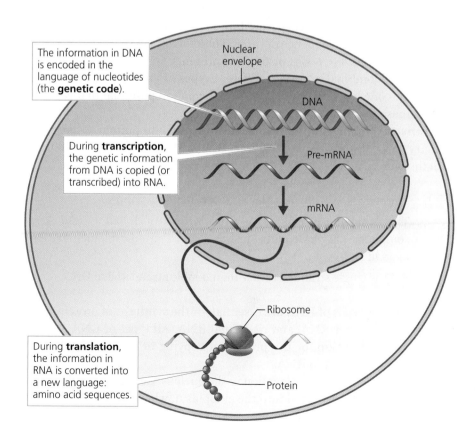

The information in DNA is encoded in the language of nucleotides (the **genetic code**).

Nuclear envelope

DNA

During **transcription**, the genetic information from DNA is copied (or transcribed) into RNA.

Pre-mRNA

mRNA

Ribosome

During **translation**, the information in RNA is converted into a new language: amino acid sequences.

Protein

FIGURE 4.27 A third characteristic of DNA is its ability to direct protein synthesis. Encoding for proteins involves three steps: the reading of the genetic code, the transcription of the genetic code into RNA, and the translation of RNA into proteins.

FIGURE 4.28 The genetic code is the language of nucleotides. The genetic code encodes the information in DNA.

genetic code :: the information encoded in DNA; it is the language of nucleotides

The Genetic Code The information encoded in DNA is in a language of nucleotides called the **genetic code** (**FIGURE 4.28**). The alphabet of the genetic code consists of four letters: A, T, G, and C. The "words" are three nucleotides each. A particular nucleotide triplet corresponds to a specific amino acid that will be incorporated into a protein. Only four nucleotides can be present in any sequence of three. Therefore, the maximum number of triplets possible is $4 \times 4 \times 4$, or 64. The proteins present in organisms are composed of combinations of 20 different types of amino acid. As a result, the genetic code has more than enough information potential to direct protein synthesis.

Transcription DNA is not translated directly into protein. Other intermediary molecules, RNAs, are required. RNA is structurally similar to DNA, with three exceptions:

1. The sugar in the nucleotides is ribose, not deoxyribose.
2. The nucleotide uracil (U) is used instead of T, but U and A are complementary base pairs (just as A and T are complementary base pairs in DNA).
3. RNA is single stranded rather than double stranded like DNA.

Three major types of RNA are essential for the synthesis of protein: messenger RNA, transfer RNA, and ribosomal RNA. All types of RNA are produced by **transcription**, or the process of copying or transcribing genetic information from DNA to RNA.

DNA never leaves its safe location in the nucleus. Consequently, its information has to be transported into the cytoplasm, where proteins are made.

transcription :: the process of copying or transcribing genetic information from DNA to RNA

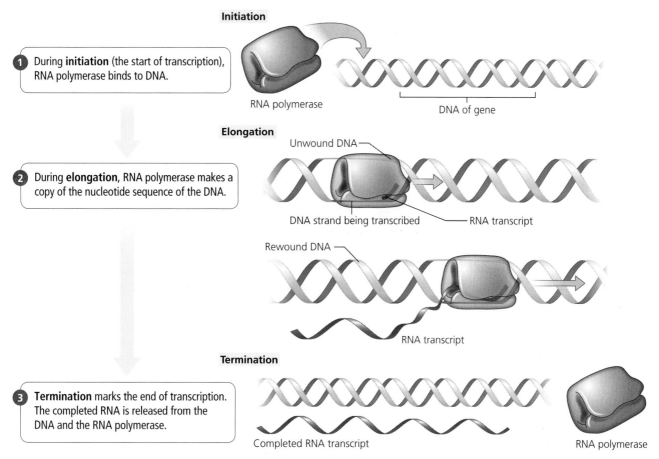

Initiation

① During **initiation** (the start of transcription), RNA polymerase binds to DNA.

RNA polymerase

DNA of gene

Elongation

② During **elongation**, RNA polymerase makes a copy of the nucleotide sequence of the DNA.

Unwound DNA

DNA strand being transcribed — RNA transcript

Rewound DNA

RNA transcript

Termination

③ **Termination** marks the end of transcription. The completed RNA is released from the DNA and the RNA polymerase.

Completed RNA transcript

RNA polymerase

FIGURE 4.29 During the process of transcription, the genetic information in DNA is transcribed (or copied) into RNA.

If you think of the DNA as the blueprints for a building, RNA is the photocopy of those instructions that you can take to the construction site.

Transcription begins in the nucleus when the double-stranded structure of a specific gene opens, so that single-stranded structures are revealed (**FIGURE 4.29**). Much as in DNA replication, the single-stranded regions of DNA bind nucleotides according to complementary base-pairing rules: A and U join together, and G joins to C. Once the entire DNA sequence has been transcribed, the RNA detaches and moves into the cytoplasm. The DNA restores its double-stranded structure.

Translation   The information present in DNA is in "nucleotide language." This information is transcribed into RNA, another molecule that is in nucleotide language. To move the information from the RNA to protein entails **translation** into a new language: amino acid sequences.

Once transcription has been completed, all three types of RNA are located in the cytoplasm, and each plays its role in the synthesis of protein (**FIGURE 4.30**):

- **Messenger RNA (mRNA)** sequences are composed of nucleotide triplets. These **codons** identify precisely which amino acids should be joined together and in what order.
- **Transfer RNA (tRNA)** is a carrier molecule. At one end it has a triplet nucleotide sequence, called the **anticodon**. The other end has a binding

translation :: the process by which the information in RNA is converted to a new language: amino acid sequences

messenger RNA (mRNA) :: codons that identify precisely which amino acids should be joined together and in what order

codon :: a specific mRNA sequence of three nucleotides

transfer RNA (tRNA) :: a carrier molecule; at one end is the anticodon, and at the other is a binding site to which a specific amino acid is attached

anticodon :: a triplet nucleotide sequence at one end of tRNA

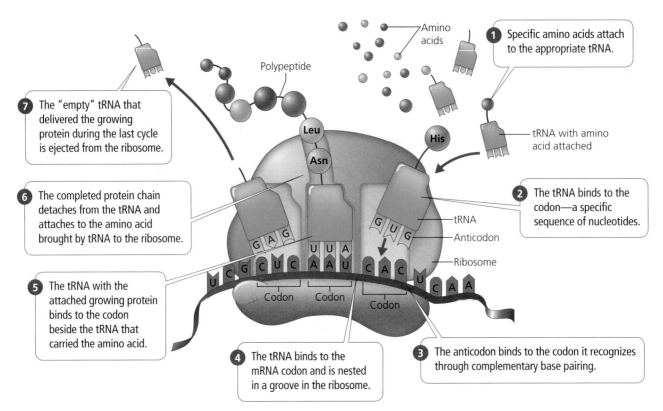

FIGURE 4.30 During translation, the information in RNA is converted into a new language: amino acid sequences.

ribosomal RNA (rRNA) :: forms ribosomes, clamp-like structures that hold mRNA and tRNA in the correct orientation so that the amino acids can be connected together

site to which a specific amino acid is attached. A host of different tRNAs each carry the specific amino acid requested by the mRNA codon.

- Finally, **ribosomal RNAs** (rRNA) form ribosomes, clamp-like structures that hold mRNA and tRNA in the correct orientation so that the amino acids can be connected together.

The overall picture that emerges is incredible. DNA, the repository of genetic information, resides in the nucleus. It possesses information for the production of all proteins in the cell. Transcription is the first step for moving the information present in DNA on its journey to the information built into protein. Finally, translation is efficient and successful because ribosomes, composed of various rRNAs, hold the growing protein in the correct arrangement.

Let's return to hemoglobin to see how this process works. The gene that encodes for the synthesis of hemoglobin opens. The nucleotide sequence that encodes for which amino acids should be hooked together is transcribed into mRNA. The anticodon of tRNAs bear amino acids, and these bind one by one to the codons of mRNA. They thus deliver the appropriate amino acids in the correct order. Because of the work of the ribosomes, the entire mRNA sequence is translated to make HbA protein. In the case of the sickle cell alleles, translation produces HbS protein, not HbA.

As mentioned earlier, these two types of hemoglobin differ only by one amino acid. The alleles themselves differ by only one nucleotide in one triplet of the DNA and hence in just one mRNA codon. Yet the physical consequences are profound, because this tiny difference is responsible for meaningful changes in the structure of the hemoglobin protein. The structural differences between HbA and HbS proteins have huge physical consequences: they may mean the difference between having sickle cell disease or not.

Why Is Protein Structure So Important?

When amino acids join together to produce a protein, the final structure depends on the identities of each amino acid. Because of the differences in their size, shape, and electrical charge, a string of amino acids will fold into the shape that is the most chemically stable. If the wrong amino acid, with different chemical properties, is inserted during protein synthesis, the entire protein may fold incorrectly (**FIGURE 4.31**). As a result, the shape of the protein will be altered, dramatically changing its function. The information present in DNA and RNA is in the language of nucleotides. In contrast, shape is the language of protein structure.

Suppose you ball your hand into a fist. Too belligerent? Suppose instead that you open it fully to reveal your empty palm, close it except for sticking up your thumb, or close it except for pointing one finger. Each shape has a different meaning. The closed fist might mean anger or a threat, the open palm is friendly, the thumbs-up might mean cheerful agreement, and your index finger could be pointing to something of interest. (We won't even ask about your middle finger.) Shape is one way that your hand can convey information.

In the case of hemoglobin, the shape of the molecule allows it to fulfill its primary function: binding oxygen and delivering it to cells and tissues in the

FIGURE 4.31 Shape is the language of protein structure. If the wrong amino acid is inserted during protein synthesis (as when valine is substituted for glutamic acid in sickle cell disease), the protein may fold incorrectly, changing the shape of the protein and, as a result, its function.

Life Application

The Effectiveness of Genetic Screening

Scientists and physicians have the ability to screen for carriers of many genetic diseases. Yet the outcomes of screening programs are quite variable. If we measure success as a reduction in the number of infants born with a particular disorder, screening for Tay Sachs disease has been spectacularly effective, whereas screening for sickle cell disease is disappointing so far. The difference in how well the screening programs have worked shows that it is important to educate patients. It is important, too, for healthcare professionals to be culturally sensitive and aware.

Tay Sachs disease afflicts children who are born with two copies of a mutant allele. The normal allele encodes for a protein called *hexosaminidase A*. This protein breaks down a type of fatty molecule called GM2, which is present in the brain and nervous system. A child with Tay Sachs disease is missing normal hexosaminidase A, and as a consequence GM2 builds up, thus destroying brain and nerve cells. Symptoms begin to appear by the age of 6 months. From then until death, usually before age 5, the nervous system deteriorates.

While not exclusive to a specific ethnic group, the Tay Sachs allele is especially prevalent in people of Ashkenazi ancestry (Eastern European Jews). One out of every 27 people from this community is a carrier of the Tay Sachs trait. Because Jewish teaching explicitly encourages medical intervention to prevent disease and save lives, genetic screening for Tay Sachs is consistent with religious beliefs. Add to this the active engagement of community and religious leaders by physicians and the strenuous efforts to educate members of the community, and the ingredients for success were in place. In fact, the campaign to screen potential parents led to a 90% reduction in the number of infants born with Tay Sachs in North America.

In contrast, the screening program for sickle cell disease reads like a script for how *not* to do something properly. Testing of African Americans began in the early 1970s—but without consent, without community outreach, and without adequate public education. When people learned that they had the sickle cell trait, they were rarely told they were *not* going to be sick. Nor did they learn the chances that their children would develop the disease. As if this were not bad enough, individuals who tested positive were discriminated against and unable to get life or health insurance. Until 1981, the U.S. Air Force Academy refused entrance to any applicant who carried the sickle cell trait. Even after these discriminatory practices were stopped, it was too late to build the trust necessary for this program to succeed. Instead, the program's legacy was an enhanced suspicion that medical genetics has racist intentions. ::

In 2004, the U.S. Postal Service issued a commemorative stamp to raise awareness about sickle cell disease and the importance of testing babies.

body (**FIGURE 4.32**). Each hemoglobin molecule can bind to four oxygen molecules. When the oxygen is released, the hemoglobin, which is in the red blood cell, maintains its structure and picks up a new cargo of oxygen when it returns to the lungs.

In contrast, sickle cell hemoglobin, with its single amino acid difference, behaves differently. Although HbS protein can bind four oxygen molecules, when the oxygen is unloaded the HbS has a slightly different shape from HbA protein. Instead of remaining as individual molecules, HbS molecules bind together to produce fibers. As Tony Allison saw, these fibers can have serious consequences. Because they can distort the red blood cells that carry them, cell sickling results. Depending on where the sickled cells form, blood flow can be blocked anywhere in the body. Tissues downstream from a blood clot will be deprived of oxygen and nutrients. If the clot is not dissolved, tissues will die. When blood flow to vital organs is impaired, the clots can be fatal. And this entire disease scenario was triggered by just one error in protein structure.

The importance of protein structure and shape is not unique to sickle disease and hemoglobin. Most cell components and regulatory molecules are proteins. Whether it is cell communication, cell division, development, cell adhesion, regulation of gene expression, or reception of hormone signals, proteins are the main players. It is not hard to imagine what could result from the presence of abnormal proteins: cancer, birth defects, autoimmune diseases, diabetes—just about anything. (See *Life Application: The Effectiveness of Genetic Screening* for a discussion of the potential cultural complexities of genetic testing.)

Biology in Perspective

Sickle cell disease reveals a great deal about how a genetic mutation can be inherited and result in a disease. It also demonstrates how even a complex syndrome can be caused by the malfunctioning of a single protein.

Sickle cell disease starts with a mutation in DNA. The RNA that is transcribed from this mutated gene is translated into hemoglobin S, a protein that differs from normal hemoglobin A by only one amino acid. Because HbS attains a final structure and set of surface electrical charges different from those of HbA, it forms fibers in red blood cells when oxygen levels are low. The HbS fibers in turn deform red blood cells into a sickle shape. Sickled cells do not behave like normal red blood cells: they get stuck in narrow blood vessels, thus producing clots. Depending on the size and location of the clot, symptoms experienced by an individual with sickle cell disease range from muscle pain to organ damage to death. All of this results from one incorrectly placed amino acid.

It is not correct, however, to think of the sickle cell allele as all bad. Individuals who have one HbS allele and one HbA allele might experience mild sickle cell symptoms under certain conditions. But they are also resistant to developing malaria, another potentially lethal disease.

The sickle cell allele illustrates a general reality about genes and inheritance. Genes do not function in isolation. The path between the information present in DNA and its arrival in protein and the ultimate journey from there to the development of physical characteristics is not so straightforward. Alleles interact with each other, as do genes. A single gene can affect numerous characteristics. The entire genetic background of an individual matters. Finally, the environment plays a role in how genes behave, too.

Red blood cell

Hemoglobin molecules transport oxygen; red blood cells contain millions of them.

Oxygen binds to heme on the hemoglobin molecule.

O₂

O₂

Heme —

Hemoglobin molecule

FIGURE 4.32 The shape of the hemoglobin molecule allows it to fulfill its primary function: binding to oxygen and delivering it to cells and tissues in the body.

Chapter Summary

4.1 What Is Sickle Cell Disease?

Specify the cause and consequences of sickle cell disease and the relationship between sickle cell disease and malaria.

- Sickle cell disease is a severe inherited genetic disorder.
- In sickle cell disease, red blood cells that are normally disc shaped and flexible become rigid and pointy shaped.
- Sickled cells get stuck in blood vessels, which can lead to blood clots and other serious physical consequences, including kidney failure, paralysis, and heart failure.
- The inheritance of the gene associated with sickle cell disease also confers resistance against malaria.

4.2 Could Molecular Medicine Prevent Sickle Cell Disease?

Explain how Linus Pauling's research into sickle cell disease ushered in the age of molecular medicine.

- Linus Pauling and his colleagues examined two types of hemoglobin (HbA and HbS).
- Pauling demonstrated that the cause of sickle cell disease could be traced to an alteration in the molecular structure of a protein.
- Pauling concluded that since sickle cell disease is inherited, genes must determine the structure of proteins.
- It took decades for other scientists to catch up to this idea that understanding the way genes, proteins, and other molecules work could be the key to diagnosing and treating disease.

4.3 Where Is Our Genetic Information Stored?

Describe the study that disproved the concept of pangenesis.

- In the 1880s, August Weismann performed an experimental test of pangenesis.
- Weismann took a population of mice and removed their tails and then bred offspring from these mice and measured whether the tails were shorter.
- The tails of the offspring were just as long as the tails of the parents.
- Weismann repeated this experiment for seven years.
- Weismann concluded that the information to construct the tail does not reside in the tail itself.
- Later scientists discovered that this information resides in the nucleus of each cell in a molecule called DNA.

4.4 How Did Mendel Discover the Rules of Inheritance?

Outline how Mendel conducted his experiments related to inheritance and the resulting set of rules.

- Mendel chose his experimental organism carefully: garden peas, which are simple to cultivate and whose mating can be controlled manually.
- Mendel focused on one trait at a time and followed the cross for more than one generation, collecting quantitative data and keeping detailed records.
- Mendel's research revealed some basic rules of inheritance.
  - Genes can come in more than one form, or allele.
  - Alleles of a particular gene sort individually into gametes during meiosis.
  - Certain traits are dominant, while others are recessive.

4.5 How Much Do Mendel's Rules Explain?

Identify the three reasons that Mendel's rules fail to explain inheritance completely.

- Alleles can interact, and so can genes.
- Most genes affect more than one characteristic.
- How genes contribute to physical characteristics can depend on the environment.

4.6 What Are Genes Made Of?

Detail the structure of DNA.

- DNA is the physical material of which genes are made.
- DNA molecules are composed of nucleotides, and there are four different types: adenine (A), thymine (T), guanine (G), and cytosine (C).
- Each nucleotide has a sugar, a phosphate group, and a nitrogen-rich ring structure called a base.
- To form the DNA molecule, the nucleotides hook together in a specific manner.
- As a consequence, DNA has a sugar-phosphate backbone with bases sticking out from one side.
- The final form is a double helix: two single strands of DNA bound to each other in a shape like a twisting staircase.

4.7 How Does DNA Function?

Describe how scientists uncovered the physical basis of genes.

- Frederick Griffith found a clue to DNA's function while studying a species of bacteria that caused pneumonia.
  - He killed a population of smooth bacteria.
  - He combined the dead smooth bacteria with living rough bacteria and infected mice with either rough bacteria alone, dead smooth bacteria alone, or a combination of the two.
  - Only the mice infected with the combined mixture got sick.
  - After isolating the bacteria from the sick mice, he observed that they were all smooth; he had discovered a physical material that could change bacteria in a way that could be inherited, but he had no idea what the "transforming substance" was.
- Scientists Oswald Avery, Colin MacLeod, and Maclyn McCarty showed that the mysterious substance was in fact DNA.
  - They killed smooth bacteria.
  - They used enzymes to eliminate carbohydrate, protein, fat, RNA, or DNA from the virulent filtrate.
  - They combined each filtrate from which they had removed *one* type of molecule with the rough bacteria.
  - Only the mixture containing DNA combined with rough bacteria resulted in the appearance of smooth bacteria; DNA was the transforming substance.

4.8 What Processes Must DNA Accomplish?

Diagram the processes of DNA replication, mutation, and protein production.

- DNA must accomplish three functions: replication, mutation, and protein production.
- During replication, the double helix of DNA unzips and new strands form.
- During mutation, a change in the DNA occurs; five examples are point mutation, deletion, duplication, inversion, and translocation.
- During transcription, the genetic information is transcribed or copied from DNA to RNA.
- During translation, the information in RNA is converted to amino acid sequences.

4.9 Why Is Protein Structure So Important?

Explain why a protein's structure is so important to its function.

- Protein structure determines a protein's function; a change in protein structure can dramatically alter a protein's function.
- Most cell components and regulatory molecules are proteins.
- Abnormal proteins can result in cancer, birth defects, autoimmune diseases, diabetes—just about anything.

Key Terms

Review Questions

1. What evidence did Dr. Allison use to understand the relationship between malaria and sickle cell disease?

2. What happens to the blood cells in an individual with sickle cell disease? What can trigger a sickling incident?

3. If an individual inherits an HbS protein and an HbA protein, which of the following is likely to be true?

   a. The individual will likely develop full-blown sickle cell disease.

   b. The individual will likely have no resistance to malaria, if exposed.

   c. The individual will likely have some resistance to malaria, if exposed, and will also be unlikely to develop full-blown sickle cell disease.

   d. The individual will likely have no resistance to malaria, if exposed, and will also be unlikely to develop full-blown sickle cell disease.

   e. The individual will likely develop malaria, if exposed, and will also be likely to develop full-blown sickle cell disease.

4. What did Linus Pauling and his colleagues learn about sickle cell disease?

   a. An individual with two normal hemoglobin genes would be unlikely to exhibit sickle cell disease.

   b. An individual with two normal hemoglobin genes would be likely to exhibit sickle cell disease under certain conditions.

   c. An individual who suffers from sickle cell disease has inherited one normal and one sickle cell gene.

   d. Sickle cell disease is not an inherited disease.

   e. There is no such thing as a carrier when it comes to sickle cell disease.

5. What is pangenesis?

   a. The idea that genetic information is stored in DNA.

   b. The idea that certain traits are dominant and others are recessive.

   c. The idea that genes can come in more than one allele.

   d. The idea that each part of the body produces a characteristic seed that is released by males and females during mating, resulting in offspring with blended traits.

   e. The idea that alleles sort individually into gametes during meiosis.

6. What were the principal discoveries that Mendel made from his monohybrid crosses?

   a. Genes can come in more than one form; some traits are dominant and others are recessive; pangenesis is a proven theory.

   b. Genes can come in more than one form; the alleles of a particular gene sort individually into gametes during meiosis; some traits are dominant and others are recessive.

   c. Genes can come in more than one form; pangenesis is a proven theory; the alleles of a particular gene sort individually into gametes during meiosis.

   d. There is no such thing as dominant and recessive traits; genes can come in more than one form; the alleles of a particular gene sort individually into gametes during meiosis.

   e. Pea flower color is a trait that exhibits a dominant/recessive relationship between the purple and white alleles; gene expression depends on the environment; genes always affect more than one trait.

7. What were the reasons that Mendel was successful whereas others who wanted to understand the rules of inheritance were not?

8. What is an allelic interaction?

9. What is a gene interaction?

10. What is an example of the environment influencing the expression of genes?

11. Describe or sketch the general features of the structure of DNA.

12. For the DNA sequence GCATTTGCCAACTGA, please answer the following:

    a. What is the complementary DNA sequence?

    b. What is the mRNA sequence that would be transcribed from it?

    c. What are the specific anticodons that would bind to the codons of the mRNA?

13. Describe the evidence that demonstrates that genes are made of DNA.

14. Describe or diagram the pathway of information flow starting with DNA and ending with a protein.

15. Explain the relationship between protein structure and function.

16. Suppose a man who carries the sickle cell trait marries a woman who also carries the trait. What is the likelihood that their children will have sickle cell disease? What are the chances of the children being carriers?

The Thinking Citizen

1. How could strenuous physical activity and/or high altitude potentially lead to organ failure in a person with sickle cell disease?

2. Predict the outcomes of matings between the following if the theory of pangenesis were correct:

   - A black dog and a white dog
   - A man missing his left ear with a woman with an extra toe on each foot
   - A gerbil with long hair and a gerbil with short hair
   - A pea plant with red flowers and a pea plant with white flowers
   - An Olympic swimmer and a completely out-of-shape couch potato

3. What observations can you make that refute the theory of pangenesis?

4. Why does the incidence of the sickle cell trait in people in a geographic area vary in relationship to the incidence of malaria? What do you predict would occur to the incidence of the sickle cell trait if malaria were to be eliminated entirely? Why?

5. Imagine you checked the news, online or print, and you read that the "gene for alcoholism" had been isolated. Is this claim likely to be correct? Why or why not?

6. Assume that a specific gene is associated with alcoholism. Would a person who has two copies of this gene definitely be an alcoholic? Why or why not?

7. Suppose Mendel wanted to demonstrate that the offspring of a white-flowered and a purple-flowered parent actually has a "hidden" white allele even though it looks purple. What cross could demonstrate that the white allele is indeed present?

8. Even after Avery and his colleagues showed DNA's role in heredity, many scientists were skeptical. DNA is so simple, skeptics said. Its double helix has just four possible base pairs. Surely protein would be a more likely component of our genes. What would you say in reply?

The Informed Citizen

1. Tony Allison supported his theory about the relationship between malaria and the selective advantage of the sickle cell trait by doing three studies that involved human subjects. Is it ethical to do research on humans? If so, what safeguards are necessary? If not, why not? How would you learn about human structure and function?

2. In one of the studies that Allison joined, a pharmaceutical company was testing its drug to see whether it was an effective treatment for malaria. Is it ethical to infect people with a disease to see whether you can cure it? If not, how can you test whether a drug does indeed treat a disease successfully and safely? Is it even ethical to test drugs on people? If so, what safeguards are needed? If not, how should drugs be tested safely in humans?

3. The rights and responsibilities of individuals and the public sometimes seem to come in conflict. Consider genes that are unequivocally associated with sickle cell disease, cystic fibrosis, Huntington's chorea, or muscular dystrophy. Should all individuals be screened? If so, what should be done with this information? Should it be off-limits to insurance companies and employers, or do they have a right to know to minimize their own financial risks? Is it acceptable to refuse screening even though an individual might add to the healthcare burden of society? Is it acceptable to force people to be screened against their will, or would that violate a right to privacy?

4. Suppose it is possible to remove or permanently disable the gene associated with a disease. Should it be done? What are the potential risks and benefits? Does the patient get to decide whether to have the treatment, or is it a legal obligation for the good of society?

Every day approximately 1500 people in the United States are diagnosed with cancer. Some cancers and causes are well understood, while others are less so.

CHAPTER 5

Cancer

How Can It Be Prevented, Diagnosed, and Treated?

Becoming an Informed Citizen ...

Cancer is a disease that is likely to touch each of us in some way at some point in our lives. All cancers are the consequence of alterations or mutations of genes responsible for regulating cell division. But what is cancer exactly? What causes it? Can it be prevented? Are there effective treatments? These are just some of the questions we investigate.

dward R. Murrow was one of journalism's most important figures. He was a pioneer in television journalism. He was well respected for his integrity and commitment to the truth. His face became well known as television developed into an integral source of information for millions.

Murrow's ever-present cigarette was just as familiar. Reminiscing about his 1965 death, the broadcast journalist Alexander Kendrick remarked, "On CBS Radio, the news of Ed Murrow's death, reportedly from lung cancer, was followed by a cigarette commercial."

You might not be surprised about the placement of that commercial. Perhaps you think that knowledge about the dangers of cigarette smoking is relatively recent. Only it isn't. The connection between tobacco and cancer was suspected centuries ago. Moreover, clear evidence of the relationship was published in 1950. Why, then, did a person as intelligent and informed as Edward R. Murrow continue to engage in a behavior that increased his chances of developing cancer, a disease that was often a death sentence in his day?

A disease of uncontrolled cell proliferation, **cancer** impacts the lives of most Americans because they themselves have the disease or a friend or family member does. Every day approximately 1500 people in the United States are diagnosed with cancer. Close to 40% of all Americans will be diagnosed with cancer, and half of these people will die from the disease. Hundreds of millions of dollars are spent on cancer research and treatments each year.

In this chapter, we'll discuss what cancer is and what causes it. We'll also outline the tools of prevention and treatment. Far from being a modern phenomenon, descriptions of cancer date back to the ancient Egyptian, Greek, and Roman empires. Some cancer causes are well understood, while others are less so. And some cancers may be caused or exacerbated by certain behaviors, while others may result from the genes we inherit. Let's take a look at one such genetic disorder that leads to a high incidence of skin cancer.

cancer :: a disease of uncontrolled cell proliferation

case study

Xeroderma pigmentosum

cyst :: a fluid-filled sac

melanoma :: an invasive and deadly skin cancer

***Xeroderma pigmentosum* (XP)** :: a rare disorder in which individuals are extremely sensitive to ultraviolet (UV) light and more easily develop skin cancers if they are exposed to sunlight or other sources of UV light

Thirteen-year-old Jeff Markway finally got around to showing his mother the golf ball–sized lump on his right arm. He demonstrated for her how the lump would move up and down when he wiggled his right thumb. Not surprisingly, Jeff's mother was alarmed by what she saw and rushed her son to the doctor. At first their doctor was not concerned because he thought the lump was simply a **cyst**, a fluid-filled hollow, and not a serious problem at all. When the doctor attempted to drain the cyst, however, instead of finding fluid in the lump, the doctor observed that it was filled with cells. The lump was not a cyst at all; it was cancer. In fact, the tumor formed because Jeff had **melanoma**, a particularly invasive and deadly skin cancer (**FIGURE 5.1**). How could a 13-year-old develop melanoma—or any cancer, for that matter?

Jeff and his family were referred to Dr. Henry Lynch, a physician and scientist whose specialty was the study of genetic factors important in cancer. Jeff was diagnosed with ***Xeroderma pigmentosum* (XP)**, a rare disorder in which people are extremely sensitive to ultraviolet (UV) light and develop skin cancers more easily if they are exposed to sunlight or other sources of UV

Asymmetry   Border irregularity

Color   Diameter

FIGURE 5.1 Melanoma is a particularly invasive and deadly skin cancer. The ABCD characteristics of a mole that should prompt a medical examination include (A) asymmetry, (B) border irregularity, (C) changes in color, and (D) diameter (large size).

light. Lynch examined Jeff's entire family and found that his sisters Patrice and Kathleen and his twin brothers Phil and Gregg all had XP. In fact, of the seven siblings of the Markway family, five suffered from the disease. XP affects only 1 out of every 250,000 people, and yet most of the Markway children had it. Clearly there is a genetic component to this disorder.

Individuals with XP who are under the age of 20 years have a 2000-fold higher incidence of skin cancer than those without XP. Also, the median age of skin cancer for individuals with XP is 8 years, while the median for individuals without XP is 58 years. Evidently, XP increases the sensitivity of skin to sunlight and speeds up the process whereby UV light exposure causes skin cancer.

Dr. James Cleaver learned of the Markway family's extreme sensitivity to UV light from an article he read in the newspaper. Interested in the effects of UV radiation on DNA, Cleaver eagerly contacted Dr. Lynch to see whether it would be possible to study their cells. They willingly agreed.

Cleaver's study unlocked the secret of exactly what went wrong in XP. Normal cells exposed to UV radiation experience DNA damage, but the cells are able to repair it. In XP cells, the UV damage to DNA is permanent (**FIGURE 5.2**). The cells are incapable of doing DNA repair. Consequently, the UV-exposed XP cells accumulate DNA damage, some of which renders the cells more likely to become cancerous. Affected individuals must avoid sunlight and UV light completely. Many go outdoors only at night.

Incoming UV photon

Adjacent thymine (T) subunits of DNA become linked due to UV exposure; DNA cannot function in this form.

FIGURE 5.2 Cells exposed to ultraviolet (UV) light experience DNA damage. Normal cells repair the damage, but in *Xeroderma pigmentosum* (XP) cells, the UV damage is permanent, and the cells are more likely to become cancerous.

5.1 How Does Cancer Make You Sick?

tumor :: a lump that results from the production of extra cells

benign :: harmless

Some growths, such as warts, are benign.

Some growths, such as the bone tumor osteosarcoma, continue to increase in size and may even become malignant, spreading to other places in the body.

FIGURE 5.4 Sometimes the balance between cell birth and death is not properly regulated, which results in the production of extra cells.

In order to function smoothly, the cells of the body live as a carefully regulated society in which each type of cell has its own place and function. As we saw in Chapter 1, cells assemble into tissues such as tendons. Tissues compose organs, such as the heart, and organs with a shared overall objective form organ systems—for example, the digestive system (**FIGURE 5.3**). Cells communicate with one another so that all parts of the body operate in a stable manner. For instance, the amount of cell proliferation is generally carefully balanced with the number of cells that die. In this way, the body can replace many worn-out or damaged cells, thus avoiding problems with tissue, organ, or organ system function.

Sometimes the balance between cell birth and death is not properly regulated. When this results in the production of extra cells, a lump, or **tumor**, might form. In many cases, these growths are **benign**, or harmless, as is true for warts. Benign tumors do not spread from their sites of origin and often do not continue growing. In other situations, the growths are **malignant**, meaning cell division continues in an uncontrolled manner, resulting in cancer (**FIGURE 5.4**).

There are three reasons that the production of extra cells creates such serious health consequences for the body, including death:

1. *Tumors can grow quite large.* In doing so, they impair the function of the organ from which they originate. They may even press down on blood vessels, blocking circulation and reducing or eliminating blood flow to organs. Organs starved of oxygen and nutrients because of impaired circulation can die.

2. *Cancer can become invasive.* It can break out of its place of origin, damaging more than one organ in the process. Individual cancer cells may leave a tumor, enter the bloodstream, and produce secondary tumors at new sites, a process called **metastasis** (**FIGURE 5.5**). This is what people mean when they say that a cancer has spread.

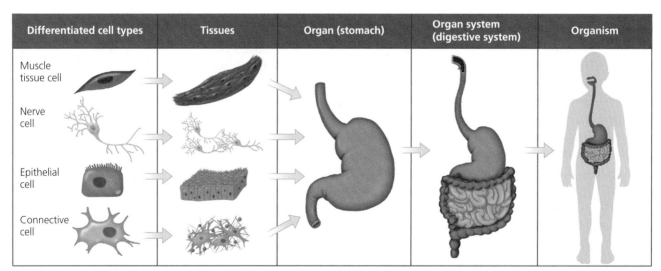

Differentiated cell types	Tissues	Organ (stomach)	Organ system (digestive system)	Organism
Muscle tissue cell				
Nerve cell				
Epithelial cell				
Connective cell				

FIGURE 5.3 In order to function smoothly, the body is organized in a structural hierarchy. Cells assemble into tissues, tissues make organs, organs form organ systems, and organ systems comprise an organism.

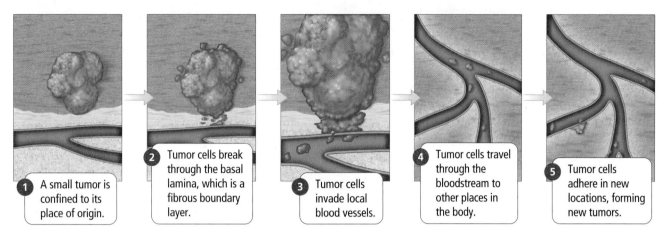

FIGURE 5.5 In a process known as metastasis, individual cancer cells break away from a tumor and enter the bloodstream, producing secondary tumors at new sites.

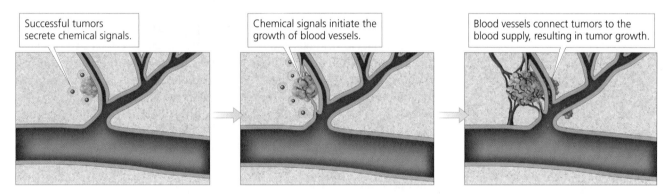

FIGURE 5.6 In angiogenesis, successful tumors secrete chemical signals that cause blood vessels to sprout from the circulatory system. These blood vessels connect tumors to the blood supply and the oxygen and food molecules it carries.

3. *Tumors can deprive other cells of oxygen.* In **angiogenesis** (**FIGURE 5.6**), successful tumors secrete chemical signals that cause blood vessels to sprout from the circulatory system—the organ system responsible for transporting nutrients and other materials throughout the body—grow toward the tumor, and connect the tumor to the blood supply of the body. This event results in the normal, healthy cells of the body receiving less oxygen and food molecules. Eventually, tissues and organs become physically stressed, seriously impairing normal function.

Cancer is a complex disease: it is really a set of diseases all unified by the characteristic of uncontrolled cell proliferation. In all, there are more than 100 types of cancer. At first glance, this number might provoke a sense of hopelessness, but the different types of cancer result from similar causes. All cancers involve some breakdown in the normal pathways of cell cycle regulation.

malignant :: cell growth that continues in an uncontrolled manner, resulting in cancer

metastasis :: a process by which individual cancer cells leave a tumor, enter the bloodstream, and produce secondary tumors at new sites

angiogenesis :: a process by which successful tumors secrete chemical signals that cause blood vessels to sprout from the circulatory system

5.2 How Do Cancer Cells Differ from Normal Cells?

In order to understand what goes wrong with cancer cells so that cell proliferation is no longer controlled, it makes sense to compare the behaviors and characteristics of normal cells to those of cancer cells. Cancer turns out to involve a breakdown in cell communication.

Normal cells communicate with one another using chemical signals that influence many cell behaviors: division, migration, differentiation or specialization, and even programmed cell death. Also, normal cells have specific morphologies, or shapes, and they attach to other cells in an orderly manner to form tissues and organs.

When grown outside the body in a culture dish, normal cells will reach a certain density and then stop dividing. When a cell encounters another in culture, both cease moving and dividing so that they don't pile up on one another. Also, normal cells in culture will divide only a certain number of times; the cells have a finite life span.

In contrast, cancer cells do not behave in an organized manner. Often, because they are damaged, they ignore the chemical signals produced by cells that tell them to stop dividing, to differentiate, or to die. In fact, some cancer cells act as if they are receiving signals telling them to divide, even when signal molecules are absent. Cancer cells also lose the characteristics unique to the tissue from which they originate. For example, a tumor initiated in the lung may be composed of cells that no longer look or function like lung cells.

In addition, cancer cells can lose their physical connections to a tissue or organ. They may then detach and secrete enzymes called proteases that break down other cells. These enzymes can burn a hole in tissue, exposing another tissue or organ to cancer, or permitting the cancer cells to escape and join the bloodstream. The cancer cells will circulate throughout the body, forming secondary tumors in new locations.

In culture conditions, cancer cells proliferate even after they reach a certain density. They continue to move around and divide even when they encounter other cells. As a consequence, cells pile up on top of each other in culture. Finally, for many cancer cells, there is no limit to the number of times they can divide. The cells are immortal.

5.3 What Is the Life Cycle of a Cell?

Even though there are more than 100 types of cancer, there are only a few kinds of defects in the cell communication mechanisms that control cell division. Signal molecules, receptors that bind these signals, and even genes can be abnormal. To understand exactly what goes wrong, consider first how cells *normally* regulate cell division.

The Molecules That Regulate Cell Division

First, many normal cells require signal molecules called **growth factors** in order to grow. They also respond to another type of signal molecule, a **growth inhibitor**, which prevents cell division. Cancer cells manage to grow in the absence of growth factors, they make a steady supply of growth factors themselves, or they may be insensitive to growth inhibitors.

Second, the growth factor, or signal that triggers cell division in a normal cell, does so by binding to a receptor present in the cell membrane. This binding of the signal molecule to the receptor activates the **signal transducer**, a molecule inside the cell that relays the information acquired when the signal molecule is bound to the receptor. This information passes from molecule to molecule, like a bucket brigade, until ultimately molecules called transcription factors are activated (**FIGURE 5.7**). This last step is essential because **transcription factors** regulate which genes are active in a cell. The proteins

growth factor :: a signal molecule that enables normal cells to divide

growth inhibitor :: a signal molecule that prevents cell division

signal transducer :: a molecule inside the cell that relays the information once a signal molecule binds to a receptor

transcription factor :: a molecule that regulates which genes are active in a cell

FIGURE 5.7 Cells communicate to regulate cell division. Different specific molecules act as signal transducers in the cell and relay the information along like a bucket brigade.

produced due to the activity of specific genes are directly responsible for triggering cell division.

In cancer cells, the receptors themselves can be permanently activated even when no signals are present. Similarly, the signal transducers can be stuck in an activated state even when there is no receptor at all. In addition, transcription factors may be produced and genes regulated even in the absence of all of the preceding steps of the communication pathway. Cancer cells have the capacity to bypass the regulatory steps of this control pathway. Different types of cancers do so in different ways.

Finally, normal cells eventually become damaged from age, wear and tear, or accidents. Sometimes the defects can be repaired. For example, cells have enzymes that can repair many types of DNA damage. However, when the defects are too severe to repair, especially when the DNA is seriously damaged, normal cells will undergo programmed cell death, or **apoptosis**. Cancer cells may lack the ability to repair DNA, as seen in *Xeroderma pigmentosum*, or they may ignore the signal to die no matter how abnormal their DNA might be (**FIGURE 5.8**). As a consequence, cancer cells continue to divide, accumulate more and more errors, and grow increasingly abnormal.

apoptosis :: cell death

The Cell Cycle

Cell division plays an important role in the lives of all organisms. Considering humans as an example, new cells are made regularly to replace worn-out blood cells, skin, and the lining of the intestines, and in males, to produce sperm. As we discussed in Chapter 3, cell division is also an essential component of embryogenesis, the growth of babies into adults. It is essential, too, to wound healing throughout one's life.

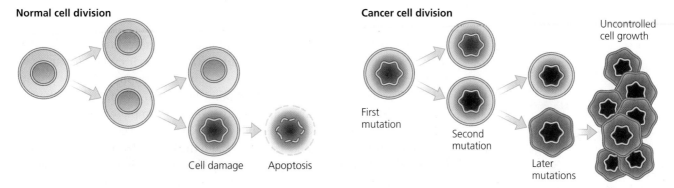

FIGURE 5.8 Normal cells eventually become damaged from age, wear and tear, or accidents. When the defects are too severe to repair, normal cells will undergo apoptosis, or programmed cell death, while cancer cells will continue to divide and grow increasingly abnormal.

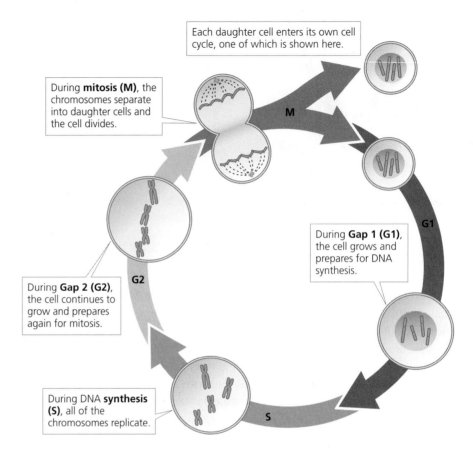

During **mitosis (M)**, the chromosomes separate into daughter cells and the cell divides.

Each daughter cell enters its own cell cycle, one of which is shown here.

During **Gap 1 (G1)**, the cell grows and prepares for DNA synthesis.

During **Gap 2 (G2)**, the cell continues to grow and prepares again for mitosis.

During DNA **synthesis (S)**, all of the chromosomes replicate.

FIGURE 5.9 **The cell cycle consists of four phases that prepare for and carry out cell division: mitosis (M), Gap 1 (G1), DNA synthesis (S), and Gap 2 (G2).**

cell cycle :: four distinct phases of cell life: mitosis (the chromosomes separate into daughter cells), Gap 1 (the cell grows), DNA synthesis (the chromosomes replicate), and Gap 2 (the cell grows and prepares again for mitosis)

FIGURE 5.10 **Under normal conditions, a series of checkpoints carefully regulate the timing and degree of cell proliferation.**

Clearly, cell division is a critical aspect of the entire life cycle. Learning how cells regulate proliferation is key to understanding what can go wrong when cells transform into cancer.

The **cell cycle** begins when the cell originates from the division of another cell and ends when it divides (**FIGURE 5.9**). The cycle comprises four distinct phases:

1. During mitosis (M), the chromosomes separate into the daughter cells and the cell divides (Chapter 3).
2. In Gap 1 (G1), the cell grows and prepares for the DNA synthesis.
3. During DNA synthesis (S), all of the chromosomes replicate (Chapter 4).
4. In Gap 2 (G2), the cell continues to grow and prepare again for mitosis.

G1, S, and G2 constitute interphase.

Under normal conditions, a series of "checkpoints" carefully regulate the timing and amount of cell proliferation within the cell cycle (**FIGURE 5.10**). Each of these checkpoints provides the cell with an opportunity for quality control. For example, there is a checkpoint late in G1 during which the cell assesses whether it has adequate raw materials to complete S. If it does not, cell division will be prevented. If the cell were to begin DNA replication and run out of the molecules needed to complete the job, the resulting cells would be severely defective. Similarly, another checkpoint in mitosis ensures that all chromosomes attach to the mitotic spindle before mitosis completes. Again, loss of a chromosome, such that daughter cells did not get the full set, would dramatically disable a cell.

Many cancer cells ignore the checkpoint controls. They go marching through the cell cycle even if conditions inside the cells are far from normal.

 # In What Ways Is Cancer a Genetic Disorder?

Compared to normal cells, all cancer cells have altered DNA, and as a result of cell division they pass the altered DNA on to their offspring. In this sense, cancer is certainly a genetic disease. The alteration or mutation of the DNA means that genes don't function normally. In some cases, the sequence of nucleotides in the DNA is changed so that a particular protein might be over-produced, made with an abnormal structure, or not produced at all. If the protein was one important in the communication pathway that regulates cell division, the mutation may be the first step toward transforming the cell into a cancerous one. Similarly, if the mutated gene is altered in some way in which its expression can no longer be properly controlled, and if the gene is part of the regulatory network controlling the cell cycle, the mutant cell may be on its way to becoming cancerous.

If the DNA damage is too severe, or if the genes that encode for the synthesis of DNA repair enzymes are themselves impaired, DNA repair won't occur. The mutated DNA will be passed on to daughter cells upon cell division. Eventually the inability to repair DNA damage will result in the accumulation of mutations and an increased likelihood of cancer. Mutations of certain types of genes make a cell vulnerable to transformation to the cancerous state.

Oncogenes and Tumor Suppressor Genes

There is a complex network of controls that regulate whether or not a cell divides. At each step along the pathways comprising the network, there are proteins that say "yes" or "no" to cell division. These proteins influence whether or not a cell will divide. One of the most amazing set of discoveries in the late 20th and early 21st centuries is that cells have genes that encode proteins that behave like a car accelerator and urge cells to go ahead and divide. These **proto-oncogenes** normally control cell division quite precisely. There are also **tumor suppressor** genes that act in a manner analogous to car brakes, telling the cell, "no, don't divide."

Proto-oncogenes can be altered to form **oncogenes** that are like an accelerator stuck "on" (**FIGURE 5.11**). Oncogenes tell the cell to divide. Similarly, tumor suppressor genes can be altered so they no longer function, disabling the "brakes" on cell division; this situation also encourages cell division. Cancer cells result from an accumulation of several mutations. In all cases, at least one tumor suppressor must be disabled and several oncogenes activated. For a discussion of experiments showing that genes may be inserted into normal cells to make them cancerous, see *How Do We Know? Cancer-Causing Genes from Malfunctioning Normal Genes*, p. 137.

The proto-oncogenes and tumor suppressor genes balance each other under normal circumstances but fail to do so in cancer cells. For example, in certain types of **leukemia**, a type of blood cell cancer, the tumor suppressor gene *NF-1* is mutated. The function of the normal *NF-1* gene is to inhibit

proto-oncogene :: a gene that encodes for a protein that regulates cell division

tumor suppressor :: a gene that encodes for a protein that tells a cell not to divide

oncogene :: a gene that tells the cell to divide in the absence of normal instructions to do so

leukemia :: a type of blood cell cancer in which the tumor suppressor gene is mutated

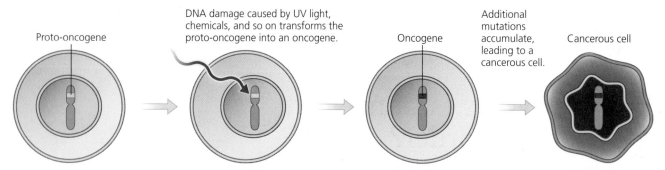

Proto-oncogene

DNA damage caused by UV light, chemicals, and so on transforms the proto-oncogene into an oncogene.

Oncogene

Additional mutations accumulate, leading to a cancerous cell.

Cancerous cell

FIGURE 5.11 Proto-oncogenes are genes that encode proteins to urge cells to divide. DNA damage can transform a proto-oncogene into an oncogene—one that tells cells to divide in the absence of normal instructions to do so, leading to cancerous cells.

Herceptin :: a drug that is used to treat a certain type of breast cancer by binding to the erbB-2 protein and stopping it from triggering cell division

a protein encoded by a proto-oncogene call *ras*. The normal *ras* gene activates cell division. When *NF-1* is mutated, it is unable to stop *ras* function. As a consequence, uncontrolled cell proliferation occurs.

There have been at least 100 oncogenes discovered so far, as well as a few dozen tumor suppressor genes. Certain types are found commonly in many cancers. Assessing the specific types of oncogenes present in tumors and determining which particular tumor suppressors are faulty provide valuable information for diagnosis, predicting disease outcomes, and in some cases treatment. For example, in approximately 30% of breast cancers, the oncogene *erbB-2* overproduces a growth factor receptor, the erbB-2 protein, on breast cell membranes. When too much erbB-2 protein is present, cell proliferation is out of control. The drug **Herceptin** binds to erbB-2 protein and stops it from triggering cell division. Herceptin is quite effective in regaining control of the division of breast cells. As we learn more about oncogenes and tumor suppressors, it is likely that we will develop new drugs and treatments in our arsenal to fight cancer.

Chromosomal Abnormalities

In addition to the relatively small genetic changes that occur when DNA mutates (Chapter 4), large-scale alterations in the inherited instructions can occur when chromosomes, or parts of chromosomes, break, rearrange their structures, stick to other chromosomes, or get lost (**FIGURE 5.12**). Just as was

FIGURE 5.12 Chromosome abnormalities may also lead to cancer. Pieces of chromosomes may be lost, inappropriately duplicated, turned upside down, or moved to another chromosome entirely.

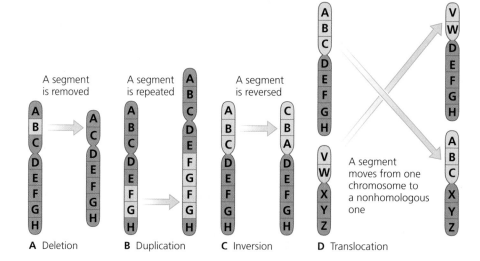

A segment is removed

A segment is repeated

A segment is reversed

A segment moves from one chromosome to a nonhomologous one

A Deletion   **B** Duplication   **C** Inversion   **D** Translocation

Cancer-Causing Genes from Malfunctioning Normal Genes

By the late 1970s, scientists had learned that certain viruses are tumorigenic, or capable of causing cancer. They also discovered that these viruses carried specific genes responsible for transforming the cells into a cancerous state and that these cancer genes could be found in normal cells too. Scientists wondered whether cancer was due to stray cellular genes that were introduced into cells when they were infected by tumorigenic viruses *or* if the cancers could be attributed to improperly functioning but previously normal genes.

Dr. Chiaho Shih and Dr. Robert Weinberg did an experiment to address this question. They took normal cells and cultured them in the laboratory. They treated these cells with a carcinogen, or cancer-causing substance, isolated from coal tar. As a consequence of the treatment, the cells transformed and became cancerous. Next Shih and Weinberg isolated the DNA from the cancer cells and inserted it into a second set of normal cells. They reasoned that if the cells became cancerous,

it would provide evidence that our cells contain genes that can become abnormal and cause cancer; no virus is needed. Indeed, that is exactly what happened. Some of the recipient cells became cancerous because of the added DNA.

Dr. Weinberg next wondered whether this result was an unusual event that would happen only if he used cells he had transformed in culture as the source of cancer genes. Would the experiment work if the source of potential cancer genes was an actual tumor isolated from a patient? To answer this question, Weinberg, along with his colleague Dr. Geoffrey Cooper, isolated DNA from a human bladder cancer. They inserted this DNA into normal cells in culture. As was the case in the Shih and Weinberg experiment, the added cancer DNA, this time from a human body, was able to transform cells in culture to a cancerous state. These experiments showed dramatically that cellular DNA can be altered in such a way that cancer can result. ⠒

Shih and Weinberg experiment

Shih and Weinberg took normal cells and cultured them in a laboratory.

They treated the cells with a carcinogen isolated from coal tar.

The cells transformed and became cancerous.

Shih and Weinberg then isolated the DNA from the cancer cells and inserted it into a second set of normal cells.

These normal cells became cancerous, indicating that viruses are not needed for a cell to become cancerous.

Weinberg and Cooper experiment

Weinberg and Cooper isolated DNA from human bladder cancer cells.

They inserted this DNA into normal cells.

The normal cells became cancerous, showing dramatically that cellular DNA can be altered in such a way that cancer results.

Dr. Chiaho Shih and Dr. Robert Weinberg conducted an experiment to address whether cancer was due to stray cellular genes introduced by a tumorigenic virus *or* whether it was due to previously normal genes functioning improperly. Then, Weinberg and Dr. Geoffrey Cooper isolated DNA from a human bladder to see whether the experiment would work if the tumor were isolated from a patient.

Before translocation

After translocation

FIGURE 5.13 The Philadelphia chromosome, which is associated with chronic myelogenous leukemia (CML), forms from a translocation event between two normal chromosomes. The chromosome rearrangement repositions two proto-oncogenes, *abl* and *bcr*, next to each other in such a way that both are permanently stuck "on."

the case for mutation, if chromosomal abnormalities turn proto-oncogenes into oncogenes or disable tumor suppressors, then the regulation of cell proliferation may be compromised and cancer can result.

The first case in which scientists were able to demonstrate that a specific chromosomal alteration was responsible for a particular cancer was the link revealed between the *Philadelphia chromosome* and *chronic myelogenous leukemia* (CML). In 1960, two scientists, Dr. Peter Nowell and Dr. David Hungerford, examined the chromosomes of CML patients and compared them to the chromosomes of healthy individuals. They discovered that all CML patients and no healthy patients had an unusual, small chromosome in their cells. Nowell and Hungerford named it the Philadelphia chromosome in honor of the city in which it was discovered.

In the 1970s, Dr. Janet Rowley examined the Philadelphia chromosome more closely, using methods that had not been available to Nowell and Hungerford when they did their study. She found that two chromosomes (numbers 9 and 22) exchanged pieces with each other. One of the products of this exchange was the Philadelphia chromosome.

Years after this discovery, in 1984, Dr. Gerard Grosveld used molecular biological techniques to show that the chromosome rearrangement that produced the Philadelphia chromosome actually repositioned two proto-oncogenes, *abl* and *bcr*, next to each other in such a way that both were permanently stuck "on" (**FIGURE 5.13**). As a result, cell division occurred without control.

The Philadelphia chromosome is not a unique incident. At least 200 specific chromosomal abnormalities have been consistently linked to particular types of cancer.

It is clear that all cancers share one common feature: DNA mutates in some way so that the network of cell division controls no longer operates effectively. How do these DNA changes originate? Are some people more susceptible to random mutation, or are there external causes of cancer that work by altering or damaging DNA?

5.5 What Risk Factors Are Associated with Cancer?

The causes of cancer are quite complex and sensitive to a person's environment and lifestyle. In fact, the majority of cancers, at least two-thirds, are influenced by factors we can control such as smoking, poor diet, excessive alcohol use, lack of exercise, and excessive exposure to UV radiation. We have less control (and sometimes no control) over other risk factors, including the infections we get, carcinogens that we encounter in the workplace, pollution, the genes we inherit from our parents, our age, and our wealth. All of these factors may affect cancer development: they damage or alter the DNA of the genes that regulate cell proliferation.

Smoking

Although the relationship between tobacco use and cancer was suspected for hundreds of years, it wasn't until 1948 that a young medical student, Ernest Wynder, decided to undertake a careful study of the subject. The idea came to Wynder after he had observed an autopsy of a man who had died of lung cancer. The man's widow had said that her husband had smoked two packs of cigarettes a day, and Wynder wondered whether there was a connection.

Because Wynder was a student, he did not have the resources to do a research project by himself. He therefore approached Dr. Evarts Graham, a physician and professor at Washington University's medical school. Dr. Graham, a heavy smoker himself, did not see the need for a study about smoking and lung cancer. Nevertheless, Graham was supportive of Wynder's project, and he approved the research. Wynder's well-designed and -implemented study showed that cigarette smokers were 40 times more likely to develop lung cancer than people who did not smoke. When Dr. Graham saw Wynder's data, he was so convinced about the link between smoking and cancer that he quit smoking. Sadly, Graham died of lung cancer six years later.

The risk of cancer and other diseases that result from cigarette smoking is quite high. Smoking is responsible for at least 30% of all cancer deaths. Approximately 80–90% of all lung cancers can be attributed to smoking (**FIGURE 5.14**). Even more striking, tobacco is responsible for as many as 4 million deaths per year worldwide. Of these tobacco-related deaths, 40% are due to lung cancer.

Even though public health education has helped to reduce cigarette smoking in the United States, tobacco use is still common in China, India, Eastern Europe, and Africa—all heavily populated regions of the world. Scientists estimate that unless smoking patterns change significantly, tobacco will kill 10 million people per year by 2030.

Diet and Exercise

Whereas smoking accounts for at least 30% of all cancer deaths in the United States, another 35% can be attributed to diet. In the United States, the problem comes from eating too much red meat, too much saturated fat, too few fruits and vegetables, and an inadequate amount of fiber. The link between diet and cancer is easy to see if one looks at the health of immigrants. For example, the incidence of breast cancer in American women is relatively high (12%), whereas in Japanese women (living in Japan), it is very low (less than half the incidence seen in American women). However, within one generation

FIGURE 5.14 Smoking is responsible for 80–90% of lung cancers. The risk of developing lung cancer increases with the amount smoked.

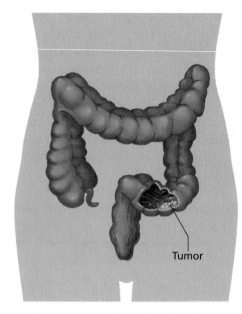

FIGURE 5.15 Consumption of meats and fats is linked to an increased risk of colon and rectal cancers. The graph relates the incidence of colon cancer in women to daily meat consumption.

of migrating to the United States and adopting an American diet, the incidence of breast cancer in Japanese women is just as high as in American women whose families have been in the United States for many generations.

Similarly, compared to countries in which diet is rich in fruits, vegetables, and fiber, the incidence of cancer of the colon and rectum are 10 times higher in areas where the consumption of meat and fats is significant: the United States, Western Europe, and New Zealand (**FIGURE 5.15**). Fortunately, studies have revealed that switching to a diet richer in fiber, fruits, and vegetables will lower cancer risk dramatically.

With respect to exercise, the most critical time period is in childhood. Excessive weight gain and the buildup of body fat increases the risk of eventually developing breast or prostate cancer. In fact, a lack of exercise and extra body fat in childhood are responsible for 5% of prostate and breast cancers in adults.

Excessive Alcohol Use

Although moderate alcohol use does not seem to increase cancer risk, there is a link between heavy alcohol consumption and certain cancers. For example, cancers of the mouth, throat, liver, larynx (voice box), and esophagus are associated with heavy drinking (four or more drinks per day). Excessive drinking may also be associated with breast cancer. In addition, the combination of excessive drinking and smoking increases cancer risk in a synergistic way. For example, the risk of esophageal cancer in a smoker is approximately 6 times that of a nonsmoker. Similarly, the risk of esophageal cancer in a drinker is approximately 6 times that of a nondrinker. But when drinking and smoking are combined, an individual has a 40 times greater likelihood of developing esophageal cancer than does a nondrinker/nonsmoker.

Radiation

radiation :: energy transmitted as waves or subatomic particles

X-ray :: a type of relatively high energy radiation

We are exposed to various types of **radiation** from the sun, medical procedures, communication devices, and even our own bodies, which themselves are a source. Although excessive exposure to high-energy radiation such as **X-rays** can indeed cause certain cancers, especially leukemia, the overall risk

FIGURE 5.16 Ultraviolet irradiation and X-rays are the most dangerous types of radiation with respect to cancer causation. Exposure to other forms of electromagnetic radiation—such as that emitted by electrical power lines, radios, televisions, cell phones, and microwave ovens—do not appear to increase cancer risk to a measurable degree.

is actually quite low. Exposure to other forms of electromagnetic radiation such as that emitted by radios, phones, cell phones, wireless systems, television, and electricity does not appear to increase cancer risk to a measurable degree (**FIGURE 5.16**). The most dangerous type of radiation with respect to cancer causation is ultraviolet irradiation.

Responsible for more than 90% of all skin cancers, the UV rays present in sunlight and tanning beds are certainly an avoidable risk. Unfortunately, the incidence of skin cancer is on the rise. In 1930, people started collecting data about the incidence of the most deadly form of skin cancer, melanoma. The number of cases of melanoma has risen dramatically since 1930. The current incidences, corrected for population size, are 10 times higher than those of 50 years ago. Melanoma is also a leading type of cancer among young adults. Melanoma, an aggressive, metastatic cancer, is diagnosed in 50,000 people each year. Of these, 7000 will die of the disease. In fact, one of Jeff Markway's brothers succumbed to melanoma.

Infection

The relative risk of cancer-causing infections varies a great deal throughout the world depending upon the relative degree of affluence and poverty. Although viruses are the most common type of infectious agents responsible for cancer, at least one type of bacteria (*Helicobacter pylori*) is associated with stomach cancer, and there are flatworms responsible for bladder and liver cancers.

Let's explore a specific example that demonstrates the relationship between infection and cancer—the **human papillomavirus (HPV)** (**FIGURE 5.17**). Responsible for most of the world's cancers of the genitals and anus, HPV is a sexually transmitted virus. The risk of being infected is directly related to the age at which a person becomes sexually active and also to the number of sexual partners. The earlier sexual activity begins and the greater number of partners, the greater the likelihood of infection. Sexually inexperienced individuals exhibit the lowest incidence of HPV infection.

The prevalence of HPV infection is higher in developing nations (15%) than in developed ones (7%). HPV is responsible for 80–90% of cervical cancers worldwide. Although rare in the United States, cervical cancer is the second most common cancer affecting women in the world. The relative rarity of cervical cancer in the United States is due to the wide use of the **Pap test**,

human papillomavirus (HPV) :: a sexually transmitted virus that is responsible for most of the world's cancers of the genitals and anus

Pap test :: a method used to detect cervical cancer by collecting and examining cervical cells

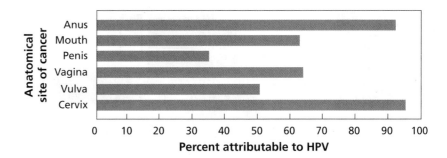

FIGURE 5.17 HPV is a sexually transmitted virus linked with several types of cancers, including cervical cancer and cancers of the genitals, anus, and mouth.

in which cervical cells are collected and examined to screen women for early signs of cervical cancer. In 2006, a vaccine, Gardasil, which prevents infection from certain types of HPV, was approved for use. (See *Scientist Spotlight: Peyton Rous* for an account of the discovery of tumor-forming viruses.)

Workplace Carcinogens and Pollution

The first description of workplace cancer dates back to the 1700s, when the physician Percival Pott noted the greater incidence of scrotal cancer among chimney sweeps compared to men and boys in other occupations. The World Health Organization's International Agency for Research in Cancer (IARC) has identified more than 40 specific materials in the workplace associated with increases in particular cancers (**TABLE 5.1**). Exposure to cancer-causing materials in the workplace is responsible for 5% of all cancer deaths in the United States. In 1950, before safety improvements were put into place, the cancer death incidence due to workplace exposures was 10%. Because these safety

Scientist Spotlight

Peyton Rous (1879–1970)

Although he did not appreciate it fully at the time, Peyton Rous ushered in an exciting era of cancer research with his discovery of tumor-forming viruses in 1909. Eventually he came to understand the importance of his work, as did everyone else; Rous was finally awarded a Nobel Prize in 1966.

Born in Texas in 1879, Peyton Rous grew up in Baltimore, where he earned his college and medical degrees from Johns Hopkins University. Rous decided to devote himself to medical research rather than clinical practice, and after a brief time at the University of Michigan, he joined the Rockefeller Institute in 1909. Rous remained at the Rockefeller until his death in 1970.

Rous's discovery of a tumorigenic virus came about in an unexpected way. In 1909, a farmer from Long Island made the journey into New York City in order to see Dr. Rous. The farmer brought his prize Plymouth hen with him, hoping that Rous could help: the chicken had a tumor in her right breast muscle. Rous knew he could not cure the chicken, but he somehow managed to convince the farmer to leave the bird for further study.

Peyton Rous ushered in an exciting era of cancer research with his discovery of tumor-forming viruses in 1909 and eventually received the Nobel Prize in 1966 at the age of 87.

Rous did two critical experiments with this bird. First, he removed the tumor and transplanted it to another chicken (from that same farmer's flock). The chicken that had received the transplant developed cancer. Evidently, cancer could be transferred from one bird to another. Second, Rous ground up the tumor tissue, collected the liquid that was squeezed out of the tumor, filtered the extract to make sure no cells or solid debris remained, and injected the filtrate into another chicken. Amazingly, the injected chicken developed cancer. Apparently some infectious agent, too small to be seen with the microscopes available in 1909–1911, caused cancer. We now know that this agent is a virus, named the Rous sarcoma virus in honor of its discoverer.

Rous abandoned his cancer work in 1915. He turned his attention to other biomedical questions, because he was unable to produce tumors in mammals by exposing them to tumorigenic viruses. As a consequence, Rous incorrectly thought the phenomenon he observed was specific to chickens

and therefore not of general significance. Although he returned to his cancer research in 1934, he did so in a scientific community that was not ready to believe that viruses could cause cancer.

Rous was 87 years old by the time he was awarded the Nobel Prize. Because the Nobel Prize is given only to living recipients and never posthumously, it is fortunate that Rous lived a long life. Still, it was a long wait for a well-deserved honor. It must have been an awkward but happy moment in the family when Rous's son-in-law, Alan Hodgkin, won a Nobel Prize before him, in 1963, for his discoveries about how nerve cells work. One can presume that when Rous won the 1966 prize, it brought great relief to everyone! ::

Table 5.1 :: A Sampling of Materials and Occupations Associated with Increases in Particular Cancers

Substance or Mixture	Occupation or Industry in Which the Substance Is Found	Site(s) of Cancer
Ionizing radiation and sources, including X-rays, γ rays, neutrons, and radon gas	Radiologists; technologists; nuclear workers; underground miners; plutonium workers; cleanup workers following nuclear accidents; aircraft crew	Bone, leukemia, lung, liver, and thyroid
Asbestos	Mining and milling; insulating; shipyard workers; sheet-metal workers; asbestos cement industry	Lung
Talc containing asbestiform fibers	Manufacture of pottery, paper, paint, and cosmetics	Lung
Wood dust	Logging and sawmill workers; pulp and paper and paperboard industry; woodworking trades	Nasal cavities and paranasal sinuses
Arsenic and arsenic compounds	Nonferrous metal smelting; use of arsenic-containing pesticides; wool fiber production; mining of ores containing arsenic	Skin, lung, and liver
Cadmium and cadmium compounds	Cadmium-smelter workers; battery production workers; dye and pigment production	Lung
Selected nickel compounds, including combinations of nickel oxides and sulfides in the nickel-refining industry	Nickel refining and smelting; welding	Lung and nasal cavity and sinuses
Benzene	Solvents in the shoe production industry; chemical, pharmaceutical, and rubber industries; printing industry; gasoline additive	Leukemia
Coal tars and pitches	Production of refined chemicals and coal tar products; coal gasification; aluminum production; road paving and construction	Skin
Mineral oils, untreated and mildly treated	Used as lubricant by metal workers, machinists, and engineers; printing industry; used in cosmetics and medicinal and pharmaceutical preparations	Skin, bladder, lung, and nasal sinuses
Soots	Chimney sweeps; heating-unit service personnel; brick masons; building demolition workers; insulators; firefighters; work involving burning of organic materials	Skin, lung, and esophagus

Source: Adapted from http://www.ncbi.nlm.nih.gov/pmc/articles/PMC1247606/table/t3-ehp0112-001447/. Environ Health Perspect. 2004 November; 112(15): 1447–1459. Published online 2004 July 15. doi: 10.1289/ehp.7047

FIGURE 5.18 Mutations in the *BRCA-1* and *BRCA-2* genes dramatically increase the chances of developing breast or ovarian cancer.

improvements are generally found in developed nations, the cancer incidence due to workplace exposure will continue to be high in developing nations and even grow because of the rapid industrialization that is presently under way.

As for the environment outside the workplace, scientists estimate that 2% of all cancers can be attributed to air, water, and/or soil pollution.

Inheritance

Most cancers, probably more than 90%, display no family history at all. Cancers appear sporadically in a family lineage. In approximately 5–10% of cancers, including some breast cancers, an inherited mutation might be the culprit.

As mentioned in Chapter 4, Paul Broca observed that many members of his wife's family had breast cancer. He mapped his wife's family on a pedigree chart and showed that out of 24 women, 10, including Broca's wife, had breast cancer. This cluster of cases in her family was strong evidence that an inherited factor was responsible.

More than 100 years later, scientists have discovered two genes that increase dramatically the chances of developing breast cancer. Mutations in these two genes, *BRCA-1* and *BRCA-2*, account for 90% of all hereditary breast cancers. A woman with mutant *BRCA* genes faces an 84% probability that she will develop breast cancer by the age of 70 (**FIGURE 5.18**). Scientists have also learned what the normal *BRCA* genes are supposed to do: they are responsible for DNA repair, and one type also acts as a tumor suppressor.

Age

Put simply, the likelihood of cancer increases as you age (**FIGURE 5.19**). Although there are childhood cancers, as well as some cancers that occur in young people, the majority of cancers appear later in life. Because cancer results from an accumulation of errors and damage to DNA, it takes time for enough mistakes to happen so that the cell passes a tipping point and becomes fully malignant and invasive.

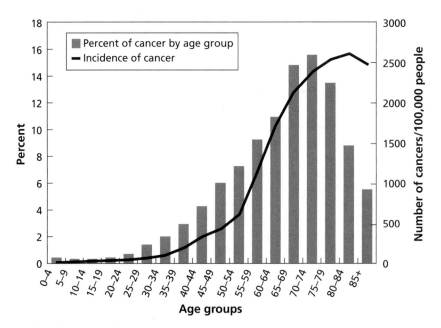

FIGURE 5.19 The likelihood of developing cancer increases with age. The bars show the percentages of all cancers in the population appearing in each age group, and the line shows the incidence of cancer in each age group.

Poverty

Living in poverty increases a person's risk of cancer. This is true for poor people living in wealthy countries and also those living in less prosperous ones. The incidence of all types of cancer except those of breast, prostate, colon, and rectum are higher in the poor.

The reasons for this association between poverty and cancer are many and complex. First, several factors that are linked with cancer are more prevalent in the poor: smoking, poor nutrition, exposure to carcinogens in the workplace, and exposure to infectious agents (sometimes due to lack of clean water). Second, poor people don't have easy access to health care. Cancer screening is not generally available. Problems are diagnosed too late, if at all, and medical treatment can be difficult to obtain and too expensive. This disparity between the health states of poor people and those who are more affluent is a serious concern that needs to be addressed for economic, political, and moral reasons. Similarly, workplace and environmental exposures to carcinogens are societal problems for which all stakeholders—individuals, businesses, society, and government—share a responsibility to act.

5.6 How Is Cancer Diagnosed?

Physicians have many effective tools for diagnosing cancer, but one of the most important sources of information is the patient. You are the most informed person about your body. You know if something is "different" or "does not feel right." Most of the time, the cough that seems to persist or the ache that won't go away is due to something that is not serious at all. Nevertheless, regular checkups and physician visits when you have a problem or concern are important ways of keeping healthy. **TABLE 5.2** includes the American Cancer Society's Guidelines for the early detection of cancer.

Some bodily events should prompt a trip to your doctor:

- A lump in your breast, testis, or elsewhere
- Changes in a wart or mole, especially in size or color
- Changes in bowel or bladder habits
- Persistent indigestion or heartburn; difficulty swallowing
- Unusual bleeding

Again, most of the time these symptoms don't indicate cancer, but sometimes they do.

Physicians use important tools to screen people for early signs of cancer. For example, women 40 years and older are encouraged to have yearly **mammograms**, a form of X-ray, to detect possible breast cancers (**FIGURE 5.20**). Women are also screened for cervical cancer by the Pap test to make sure they are normal. The relative rarity of cervical cancer in the United States is thanks to the wide use of the Pap test to screen women for early signs of cervical cancer. Both men and women can be screened for cancer of the colon or rectum using **colonoscopy**, a procedure wherein the interior of the large intestine, rectum, and anus are examined to determine whether any abnormal growths are present. Physicians also examine patients manually to see whether any lumps are present. For example, the prostate gland and testes in men are checked this way. If any of these screening methods reveals something that warrants closer attention, additional tests will be done.

FIGURE 5.20 Mammograms are a form of X-ray that can be used to detect breast cancers even at relatively early stages.

mammogram :: a form of X-ray used to detect possible breast cancer

colonoscopy :: a procedure used to examine the large intestine, rectum, and anus to screen men and women for colon or rectal cancer

Table 5.2 :: American Cancer Society 2013 Guidelines for the Early Detection of Cancer (Most Adults)

Cancer	Guidelines
Breast	• Yearly mammogram for women starting at age 40 • Clinical breast exam for women once every 3 years (ages 20–30) and annually (40 years and over) • Optional breast self-exam for women starting at age 20

Colorectal	• Screening should begin at age 50 for both men and women. Options include:
	  o Flexible sigmoidoscopy every 5 years
	  o Colonoscopy every 10 years
	  o Double-contrast barium enema every 5 years
	  o Virtual colonoscopy every 5 years
	  o Fecal occult blood test every year

Cervical	• Screening should begin at age 21 • Pap test once every 3 years (women ages 21–29) • Pap test plus an HPV test every 5 years (women ages 30–65)
Prostate	• Starting at about age 50, men should talk with a doctor about the pros and cons of testing to determine whether it is the right choice • If men decide to be tested, they should have the PSA blood test with or without a rectal exam—how often depends on their PSA level
Cancer-related checkups	• Women and men age 20 and older should have periodic health exams • Health exams should include health counseling and, depending on age and gender, exams for cancers of the thyroid, oral cavity, skin, lymph nodes, testes, and ovaries

Source: http://www.cancer.org/healthy/findcancerearly/cancerscreeningguidelines/american-cancer-society-guidelines-for-the-early-detection-of-cancer.

computerized tomography (CT) scan :: an advanced type of X-ray procedure that provides a more detailed look at internal structures by sending thin X-ray beams at different angles

magnetic resonance imagery (MRI) scan :: a medical imaging technique that uses a magnetic field and radio waves to visualize the inside of a body

The details of the next round of tests would be determined by the physician's specific concerns. Such tests could include blood sampling to look directly at the blood cells themselves or to measure specific proteins associated with particular cancers. For example, the amount of PSA, a protein found in high amounts in individuals with prostate cancer, could be determined. Also, visualization methods like ultrasound, **computerized tomography (CT) scans** (see *Technology Connection: Computerized Tomography [CT] Scans*), or **magnetic resonance imagery (MRI) scans** (**FIGURE 5.21**) provide a good look at the internal structures of your body, including the

FIGURE 5.21 Magnetic resonance imaging (MRI) (A) produces clear, detailed images of the internal structures of the body (B).

A

B

Computerized Tomography (CT) Scans

Using a standard X-ray machine, we can see inside the body because the rays can penetrate skin and other soft tissues. Other parts of the body are more opaque and do not permit the X-rays to pass through so easily; the X-rays are absorbed. These opaque tissues are structures that show up on the film as dark areas. For example, bones can be clearly viewed in this way. X-rays can also be used to view tumors, which in many cases are denser than the surrounding tissues and are able to absorb more X-rays. As a consequence, they will appear darker than the surrounding normal tissue.

Computerized tomography (CT) scanning is a more advanced type of X-ray procedure. CT scans provide a better and more detailed look at internal structures, including tumors, because rather than sending a single straight beam of X-rays through the tissue, CT scanners use a computer-controlled system to send thin X-ray beams at different angles, resulting in three-dimensional images. As with the

A **B**

CT scans use a computer-controlled system to send thin X-ray beams at different angles, resulting in three-dimensional images (A) that are better and more detailed than standard X-rays (B).

standard X-ray method, tumors will absorb more X-rays and appear darker than normal tissue. However, in this case, a tumor will appear in three dimensions rather than as a flat spot. CT scanning is one of several visualization techniques that have revolutionized medical diagnosis in general and for cancer in particular. ::

details of your organs. Finally, your physician may take a biopsy, a procedure in which cells are collected and examined to see whether they are cancerous (**FIGURE 5.22**).

Some cancer screens are designed so that *all* potential cancers will be detected. As a consequence, there will be some "false positives." For example, 10% of women who have mammograms will be called back for additional examination. The vast majority of the time, this second look reveals nothing worrisome. The mammogram is designed in this way so that there will be no "false negatives"—women who are told there is no cancer when there is.

In a needle biopsy, cells are removed with a syringe.

5.7 How Is Cancer Treated?

If the careful and extensive testing done by the physician reveals cancer, the next step is to make decisions about treatment. Three principal methods are used to treat cancer: surgery, radiation therapy, and chemotherapy. These treatments are often used in combination.

Surgery

Surgery is performed on more than 50% of all cancer patients. In some cases, it is relatively straightforward—the removal of a skin cancer, for example. In other situations, it is a daunting task, as in the removal of a cancerous lung. Surgery is also sometimes used to perform a biopsy.

Surgery is most successful when the tumor is localized and has not spread, or metastasized. The physician will remove the tumor and have

In a surgical biopsy, a larger piece of tissue is removed with a scalpel or tiny scissors.

FIGURE 5.22 In a biopsy, cells or tissue are removed and examined to see whether they are cancerous.

FIGURE 5.23 Radiation therapy is used to kill cancer cells that cannot be removed surgically or to shrink a tumor prior to surgery.

pathologist :: a doctor who specializes in the study of disease

radiation therapy :: radiation delivered either externally or internally to kill cancer cells

chemotherapy :: a treatment method using drugs that impair the ability of cells to replicate; these medications may be taken orally or injected directly into the bloodstream

immunotherapy :: treatment that directly targets a cancer or boosts the immune system to make other cancer therapies more effective

angiogenesis inhibitor :: a drug that stops the growth of blood vessels, inhibiting the ability of tumors to attach to the circulatory system and grow

pathologists, doctors who specialize in the study of disease, examine it in order to see how serious the cancer is by evaluating the degree to which the chromosomes and cellular structures appear abnormal. This information will provide guidance regarding other possible treatments.

Radiation Therapy

The objective of **radiation therapy** is to kill cancer cells that cannot be removed surgically or in some cases to shrink a tumor prior to surgery. Radiation is delivered to the cancer in one of two ways. For some cancers, the radiation exposure is external (**FIGURE 5.23**). After determining the appropriate dose and location, the tumor is bombarded with a focused beam of X-rays in order to kill the cells. In other situations, the radiation exposure projects from an internal source. In this circumstance, a localized source of radiation, namely, a tiny "seed" containing radioactive material, is placed near or directly in the tumor. Again, the objective is to kill the cancer cells.

Chemotherapy

Chemotherapy is generally used to treat cancers that have spread from the primary tumor. It is also sometimes used when there is a concern that the tumor may have metastasized even though the spreading is not clearly observable. Chemotherapy uses drugs that impair the ability of cells to replicate. These drugs may be taken orally or injected directly into the bloodstream (**FIGURE 5.24**). As with radiation therapy, the goal of chemotherapy is to kill cancer cells.

Chemotherapy sometimes produces challenging side effects. Because the drugs target all cells undergoing mitosis, populations of cells in the body that normally divide are especially vulnerable. For example, in a healthy person, blood cells, hair follicle cells, and the cells lining the intestines continually renew themselves. Chemotherapy drugs can impact these cell populations, causing anemia and a weakened immune system, hair loss, and nausea.

Cancer Treatments on the Horizon

Researchers continue to develop new potential treatments for cancer. Two promising avenues are immunotherapy and drugs to inhibit angiogenesis.

In **immunotherapy**, the patient's immune system is used to fight cancer. Some immunotherapies act to fight cancers directly, while others focus on boosting the immune system so that therapies can work better. In some cases, antibodies, or molecules produced by immune system cells, are made in the lab and administered to patients to target specific cancers, such as certain types of breast cancers.

Angiogenesis inhibitors are drugs that stop the growth of blood vessels. When used to treat cancer, they inhibit the ability of tumors to attach to the circulatory system and obtain the nutrients and oxygen necessary for growth. There are some angiogenesis inhibitors being used to treat a small number of specific cancers at the present time, with many more inhibitors undergoing clinical trials on patients.

Why Cancer Treatments Sometimes Fail

Even the most aggressive cancer treatments are not always successful. Cancer cells are not uniform or unchanging in their characteristics. In a given tumor, each new round of cell division can produce cells that are increasingly abnormal and ever more invasive.

Ironically, the process of natural selection can shift the cancer cell population so that the cells resistant to chemotherapy increase in number to the point where the drugs are no longer effective. When this occurs, the patient might be able to try a different battery of drugs, but eventually cancer cells may develop resistance to this new therapy. Ultimately, some cancers will recur and will be unresponsive to any treatment. At that point, the medical goal is to keep the patient pain free and to help with end-of-life issues.

FIGURE 5.24 Chemotherapy kills cancer cells by using drugs that impair the ability of cells to replicate.

5.8 How Can Cancer Be Prevented?

Jeff Markway suffered from a genetic disorder that dramatically increased his chances of developing melanoma. Except for avoiding UV light from the sun or other sources, he really had no other means of prevention. In contrast to Jeff's situation, many cancers are preventable for most people. As you have read, at least 65% of all cancers in the United States are due to smoking, a poor diet, and lack of exercise. Excessive exposure to sunlight is the principal cause of skin cancer. And in the United States, the most common way to become infected with a cancer-causing pathogen is through promiscuous sexual activity. Clearly there are ways that we can modify our behaviors to lower our individual risk for cancer.

Is there actually evidence that cancer can be prevented? After all, if a person has to make potentially difficult lifestyle changes, it would be comforting to know that the effort will pay off. Let's revisit smoking to see a case that demonstrates the effectiveness of behavior modification for the prevention of cancer.

The first publication detailing the link between cigarette smoking and lung cancer was published in the United States in 1950. A few months later, a paper detailed a similar study done in Great Britain. The British study verified the association between smoking and lung cancer that had been described in the U.S. publication by Wynder. In 1955, British men under the age of 55 had the highest lung cancer incidence in the world. After the reported relationship between lung cancer and smoking finally captured the attention of the government, things changed. First, because cigarette tar is a potent carcinogen, British cigarettes were redesigned to lower tar levels. Second, the government launched a public health education campaign that succeeded in reducing the number of people who smoked. Thanks to these efforts, and the willingness of people to change long-ingrained habits, the lung cancer incidence in British men under the age of 55 dropped by more than 65%. Prevention *does* work.

So what can you do to decrease your cancer risk? You know enough to figure it out, but **TABLE 5.3** provides some suggestions. To learn how the foods we eat can help prevent cancer, an idea referred to as **chemoprevention**, see *Life Application: Chemoprevention.*

chemoprevention :: the action of natural or manufactured materials used to prevent or halt cancer development

Table 5.3 :: Simple Steps to Reduce Your Chances of Cancer

- Don't smoke.
- Eat a diet rich in fruits, vegetables, and fiber, and lower in saturated fat.
- Avoid excessive exposure to UV light either from the sun or a tanning bed.
- Exercise and maintain a healthy weight.
- Don't engage in risky sexual behavior.
- Avoid excessive alcohol—especially if you smoke, because these two factors act synergistically to increase the incidence of certain cancers.
- Don't expose yourself to known carcinogens at home or work without protective clothing and proper precautions.
- Get regular checkups.

Life Application

Chemoprevention

It seems that new diets are constantly appearing in the media, including magazines and television. They have spawned a new wave of cookbooks describing different diets: Atkins, South Beach, anti-inflammatory, Mediterranean, and Paleolithic, to name a few. All of these food plans claim to provide health benefits. Is it possible that what we eat can indeed influence our health in a disease-specific manner?

We know that consumption of some foods increases our chances of developing cancer (note the correlation between animal fats and colon cancer), and that reducing or avoiding them will decrease our risk. But are there particular things we could eat that would decrease our risk even more? In the 1950s, some scientists began to examine this question. They wondered whether we could use natural or manufactured materials to prevent or halt cancer development. This idea is called chemoprevention, and research in this area has produced some promising results.

Studies have shown that a diet rich in fruits, vegetables, and fiber and low in saturated fat decreases the risk of developing certain cancers, such as those of the colon and rectum. However, scientists want to learn which phytochemicals, components derived mostly from plants, are responsible for this beneficial effect. One group of researchers studied dithiolethione, a phytochemical found in cruciferous vegetables such as broccoli, cabbage, and cauliflower. Controlled experimental studies have shown that dithiolethione inhibits the development of tumors in the lung, colon, mammary gland, and bladder when tested in laboratory animals. Other studies suggest that dithiolethione interferes with cancer formation because it activates liver enzymes that in turn detoxify carcinogens.

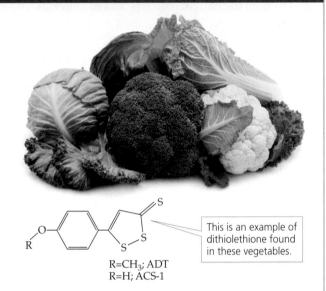

R=CH$_3$; ADT
R=H; ACS-1

This is an example of dithiolethione found in these vegetables.

Research has shown that dithiolethione, a phytochemical found in cruciferous vegetables such as broccoli, cabbage, and cauliflower, inhibits the development of tumors in the lung, colon, mammary gland, and bladder when tested in laboratory animals.

Chemoprevention research has blossomed in recent years. Many materials have been tested in laboratory experiments, and in some cases, human population studies are under way as well. It is likely that as we learn more, we may be able not only to decrease the likelihood of developing cancer, but also to take active steps to preserve health even more effectively. ::

Scientists estimate that if we paid more attention to cancer prevention instead of focusing mostly on treatment, at least 10 times more lives could be saved. Prevention of cancer also spares people fear, physical pain, and the financial burden of the disease. The realization that most cancers are preventable does not mean that someone who gets cancer is to blame. Instead, this knowledge should empower each of us to make good choices that will enhance our health.

Even if we were to eliminate all external causes of cancer, scientists estimate that we would still see 20–25% of the cancers that occur now. Some cancers are inevitable. This disease is complex and is the result of interactions between factors external to the body and events that occur in our cells. And chance plays an important role too.

 # Biology in Perspective

Cancer is a disease of uncontrolled cell proliferation. All cancers are the consequence of alterations or mutations of genes responsible for regulating cell division. Scientists have learned a great deal about how cancers form and how they might be prevented. There are several known factors that increase a person's risk of developing cancer. Some of them are avoidable, but not all, and some factors must be addressed by society at large, including the government.

Early diagnosis dramatically increases cancer survival rates. There are many methods for diagnosing cancer, including physical examination, certain blood tests, and visualization tools such as ultrasound, CT scanning, and MRI. Cancer treatments include surgery, radiation therapy, and chemotherapy, or a combination of these methods. As is true of cancer risk itself, poverty influences the chances that an individual will be able to get medical care for early detection and whether all possible tests and treatments will be available. Taking care of everyone in need of medical care requires more than what an individual can do for him- or herself.

Chapter Summary

5.1 How Does Cancer Make You Sick?

Use appropriate terminology to describe cancer and the reasons why cancer makes you ill.

- Malignant tumors can spread cancer; benign tumors are unlikely to be dangerous.
- Cells continue to divide uncontrolled in malignant tumors, but if cells stop uncontrolled cell division they form benign tumors.
- Tumors can press on other organs, nerves, and blood vessels; metastasize and form secondary tumors; and deprive other cells of oxygen through angiogenesis.

5.2 How Do Cancer Cells Differ from Normal Cells?

Contrast the behavior of normal and cancerous cells.

- Cells communicate to control cell division through chemical signals.
- Cancerous cells no longer respond properly to chemical signals and often lose their original function.

5.3 What Is the Life Cycle of a Cell?

Use a signal transduction pathway to explain how cells become cancerous.

- The cell cycle has four steps (M, G1, S, and G2). Chemical signals regulate cell division at checkpoints in the cell cycle.
- Chemical signals regulate gene activity through signal transcription pathways. Receptors sense chemical signals and transfer the signals to signal transducers that pass the signals to transcription factors that turn genes on or off.
- Mutations to any part of a signal transduction pathway can change cell division and prevent apoptosis.

5.4 In What Ways Is Cancer a Genetic Disorder?

Describe the role that mutations play in cancer development.

- Proto-oncogenes are normal genes that turn on cell division, while tumor suppressor genes stop cell division.
- Mutated proto-oncogenes or tumor suppressor genes can cause uncontrolled cell division and may ultimately become oncogenes that cause cancer.

5.5 What Risk Factors Are Associated with Cancer?

Identify actions you can take to reduce your risk of cancer.

- Any compound or activity that increases the risk of mutations has the potential to cause cancer.
- Risk due to infectious agents can be reduced through vaccines (e.g., Gardasil) and behavior.
- Some cancer risks, such as smoking and alcohol consumption, can work synergistically to create risk many times greater than either would cause on its own.
- Protective equipment, safety protocols, and manufacturing practices can reduce workplace exposure to carcinogens.
- Cancer risk increases with age, with certain inherited genes, and with poverty because of the likelihood of increased exposure to risk factors.

5.6 How Is Cancer Diagnosed?

Explain how cancers are diagnosed.

- Some bodily events should prompt a trip to a doctor, including the following: lumps in your body, changes in warts and moles, changes in bowel or bladder habits, persistent indigestion, difficulty swallowing, and unusual bleeding.
- Regular cancer screening can help catch cancers early.

- Once a cancer is suspected, further testing is usually required. Tests may include chemical methods (e.g., PSA) or visualization methods (e.g., ultrasound, CT scans, and MRI).
- Ultimately, doctors usually take a biopsy, which is sent to a pathologist to determine whether it is cancerous.
- Most cancer tests are sensitive to reduce false negatives but increase false positives.

5.7   How Is Cancer Treated?

Contrast the three methods used to treat cancer.

- Cancer is treated through three primary methods: surgery, radiation therapy, and chemotherapy; cancer treatment usually involves multiple methods.

- Cancers can become resistant to chemotherapy through the process of natural selection.
- Two promising avenues of cancer treatment on the horizon are immunotherapy and angiogenesis inhibitors.

5.8   How Can Cancer Be Prevented?

Enumerate steps to decrease the risk of cancer.

- You can take steps to reduce or eliminate cancer risk factors.
- Cancer prevention could reduce cancer-related deaths 10-fold.

Key Terms

angiogenesis 131	growth inhibitor 132	Pap test 141
angiogenesis inhibitor 148	Herceptin 136	pathologist 148
apoptosis 133	human papillomavirus (HPV) 141	proto-oncogene 135
benign 130	immunotherapy 148	radiation 140
cancer 128	leukemia 135	radiation therapy 148
cell cycle 134	magnetic resonance imagery (MRI)	signal transducer 132
chemoprevention 149	scan 146	transcription factor 132
chemotherapy 148	malignant 130	tumor 130
colonoscopy 145	mammogram 145	tumor suppressor 135
computerized tomography (CT) scan 146	melanoma 128	*Xeroderma pigmentosum* (XP) 128
cyst 128	metastasis 130	X-ray 140
growth factor 132	oncogene 135	

Review Questions

1. Explain the three primary ways that cancer makes someone sick.

2. Which of the following is NOT a reason why the production of extra cells creates serious health consequences for the body?
   a. Tumors can grow quite large.
   b. Tumors can be benign.
   c. Cancer can become invasive.
   d. Individual cancer cells can metastasize.
   e. Tumors can deprive other cells of oxygen.

3. Describe how cancer is a breakdown in cell communication and community.

4. Diagram the cell cycle. Where are potential checkpoints located?

5. What is the relationship between proto-oncogenes and oncogenes?

6. How do oncogenes contribute to tumor formation?
   a. Oncogenes tell the cell to divide; they are like an accelerator stuck in the "on" position.
   b. Oncogenes tell the cell to stop dividing; they act like a car brake.
   c. Oncogenes are similar to tumor suppressor genes; they tell the cells to stop dividing.
   d. Oncogenes encode proteins to urge cells to divide.
   e. Oncogenes do not contribute to tumor formation.

7. What is a tumor suppressor, and what role does it play in tumor formation?

8. The p53 tumor suppressor gene regulates whether the cell cycle can proceed between G1 and S phase, and G2 and M phase. It stops the cell cycle to allow for DNA repair, but if the damage is irreparable, it will trigger apoptosis. Why is it important to stop at those points in the cell cycle?

9. All of the following are risk factors associated with cancer EXCEPT which one?

   a. Lack of exercise

   b. Poor diet

   c. Smoking

   d. Young age

   e. Excessive exposure to UV radiation

10. Why is cancer more prevalent in the elderly and the poor?

11. Why is accurate DNA repair necessary to prevent cancer?

12. What cancer warning signs should signal that you need to see the doctor?

13. What are some of the routine tests that can be done to detect cancer?

14. Distinguish between false positive and false negative test results.

15. An oncologist might use _____ to reduce the metastasized cancer spreading throughout a person before removing or destroying the primary tumor through _____ or _____.

    a. radiation therapy; surgery; phytochemicals

    b. radiation therapy; surgery; colonoscopy

    c. surgery; chemotherapy; radiation therapy

    d. surgery; radiation therapy; chemoprevention

    e. radiation therapy; surgery; chemotherapy

16. What are some ways to prevent cancer?

The Thinking Citizen

1. Suppose you are responsible for developing cancer treatments. Explain how you would target your research based on what you know of signal transduction pathways related to cancer.

2. Why do people persist in dangerous behaviors that are known to cause cancer?

3. Explain how an individual eating all the right foods, exercising, and not exposed to carcinogens could develop cancer and another person who smoked a pack a day for 40 years might not have cancer.

4. What are specific choices you can make to reduce your chances of developing or dying from cancer?

5. Pick one of the occupations or workplaces listed in **TABLE 5.1** (on pg. 143). Describe safety precautions you would expect to be in place at that location. Explain how the precautions would help reduce the likelihood of cancer.

6. Do you think it is possible to achieve a cancer-free human population (0% incidence)? Why or why not?

7. An important area of research concerns whether angiogenesis inhibitors can stop tumor growth. Why would the inhibition of angiogenesis be a potential way to kill or limit the growth of tumors?

The Informed Citizen

1. Prevention is a proven method for reducing cancer incidence. How would you go about trying to communicate *effectively* with middle school and high school students about the choices they need to make for a more healthy life?

2. What is society's responsibility for individuals who develop cancer because of bad decisions they have made in their lives?

3. Many people seek alternative and complementary medical therapies for their cancer. Should insurance companies cover the costs of treatments that people *believe* to be effective but for which no evidence exists to support this belief? Why or why not?

4. Compare the likely cancer risks between two locations in your region with different socioeconomic levels. What are the sources of those different risks? How could you help decrease the cancer risks in each community? Why do you suppose people would move to a community that has a potential environmental source of carcinogens?

5. Suppose you were in charge of the part of the federal budget targeted to fund projects concerned with cancer. What proportions of the funds would you allocate for prevention and for treatment? How would you make your decision?

6. Despite all that people know about preventing harmful UV radiation and the availability of sunscreen, skin cancer is still on the rise. Why is that? What other factors might be responsible?

7. You have just been told that you have cancer based upon a preliminary test. The false positive rate is 10%, and the false negative rate is 1%. Further testing is painful but has a lower rate of false positives. The cancer treatment is painful, dangerous, and disfiguring. Do you get additional tests, go straight into treatment, or wait another year or two to get tested again? What additional information do you want? How would that information help you choose a course of action?

CHAPTER LEARNING OBJECTIVES

After reading this chapter, you should be able to answer the following questions:

6.1 How Do Males and Females Form?
Describe how chromosome and hormone instructions interact to determine sex.

6.2 What Happens If the Hormonal Signals Are Missing or Misread?
Predict the outcome of a change in hormone production or response.

6.3 How Do Men Produce Sperm?
Outline the steps of sperm production in males.

6.4 How Do Women Produce Eggs?
Outline the steps of egg production in females.

6.5 How Can Pregnancy Be Prevented?
Compare methods used to prevent pregnancy.

6.6 What Causes Infertility, and How Can It Be Treated?
Predict causes of infertility and propose potential treatments.

6.7 How Can We Tell If a Fetus or Baby Is Healthy?
Explain the purpose of and methods for testing fetal and newborn health.

6.8 What Tests Are on the Horizon?
Evaluate the ramifications of further refinements in genetic tests and treatments.

Reproduction is all about communication. It involves communication between two parents, such as this courtship display between two blue-footed boobies, but also communication between individual cells and between the embryo and the mother's body.

CHAPTER 6

Reproduction

What Kind of Baby Is It?

Becoming an Informed Citizen …

Sex and the urge to have offspring are biological drives. Understanding the biology of reproduction is, therefore, essential for understanding what makes us human and for preserving our natural resources.

*P*roducing offspring is the most important thing we do as a species. Although individuals may be childless, as a species we are diligently making babies—so many, in fact, that we risk overpopulating the world (Chapter 13). Understanding the biology of reproduction is thus essential for both responsible family planning and stewardship of our natural resources.

Reproduction is all about communication. Of course, human reproduction involves two people expressing their hopes and desires. But communication also takes place between individual cells. Embryos listen to chemical signals, too, to determine what sex organs to develop—male or female. Other communications regulate sperm and egg formation, and sperm and eggs must read each other's signals for successful fertilization to occur. The embryo and uterus communicate as well, so that the embryo can implant in the mother's body and stay there safely for an entire pregnancy. The fetus even signals the start of labor, with dramatic responses from the mother's body (Chapter 3).

Reproduction is at the core of who we are as individuals. One might define a man as a person capable of fathering a child and a woman as someone capable of giving birth. But is it really that simple? Do humans fall neatly into two categories? Does it even matter? It mattered a great deal to the fastest woman on Earth.

case study

The Fastest Woman on Earth

Running through the countryside of Limpopo on dirt roads, Caster Semenya (**FIGURE 6.1**) dreamed of being a champion. In 2008, Semenya took her first step onto the world stage of athletics, winning a gold medal at the Commonwealth Youth Games. She followed this victory with another gold medal at the 2008 African Junior Athletics Championship. In fact, Semenya's performance was so spectacular that she qualified for the 2009 World Championship in Berlin. Only 18 years old, Semenya appeared to be on top of her game—a world-class athlete with a dazzlingly bright future.

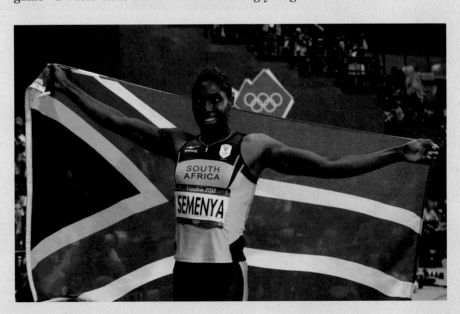

FIGURE 6.1 Caster Semenya was subjected to gender testing and accusations that she was not really a woman because she looked masculine and performed so much better than her competitors in track and field events.

But then the rumors and accusations began to surface. Some doubted that Semenya was actually a woman. She looked masculine, they contended, and performed so much better than her competitors that she *must* be male. Semenya was subjected to two rounds of gender testing, the first in Pretoria. The second was in Berlin, a day before her race. Here she was, a teenager from a tiny rural village in South Africa, being examined in ways she did not understand, without her family there for support. Her parents never even knew the tests were being done.

Successfully passing the tests did not stop the rumors. In many ways, the situation actually got worse. After Semenya ran successfully in the semifinal round, a TV reporter outside of the stadium hollered at her: "I hear you were born a man!" Although visibly distressed, she managed to keep her focus and win the world championship.

It turns out that the International Association of Athletics Foundations (IAAF) had broken confidentiality rules and leaked the rumor. Worse, Australia's *Daily Telegraph* claimed that test results showed that Semenya has external female genitals but no ovaries or uterus. The tabloid also claimed that she had undescended testes and testosterone levels three times higher than those of an average woman (but much lower than a man's). While we don't know if these reports are accurate (the tests are supposed to be private), we do know that Semenya was raised as a girl and sees herself as female. All of a sudden, at the age of 18, she was being told that she is not who she thinks she is.

Caster Semenya returned to the University of Pretoria, her athletic future uncertain. The IAAF allowed her to keep her gold medal because she did not commit fraud by competing as a woman. However, she did not run competitively with women in international track and field competitions for almost a year as she waited for the IAAF's decision regarding her eligibility. Finally, Semenya got some good news when, in 2010, she was declared free to compete professionally with other female athletes. She won a silver medal in the 800 meters at both the 2011 world championships and the 2012 Summer Olympics in London. Semenya also carried South Africa's flag in the opening ceremonies at those Olympics.

The IAAF had a huge problem. It wanted to answer a very difficult if not impossible question: what is the ultimate difference between males and females? Is it simply the ability to father a child or give birth once a person is sexually mature? If so, what about those who are unable to have children? Is there something that can be measured or tested that can show unequivocally that someone is a man or a woman? Or might there be still other outcomes to sexual development? Could someone be human and yet neither a man *nor* a woman?

6.1 How Do Males and Females Form?

When a baby is born, one of the first questions everyone asks is whether it is a boy or a girl. It is not always easy to answer. Some scientists estimate that as much as 1.7% of the population has a **disorder of sexual development** (**DSD**). For some, that means development is profoundly different from the standard definition of female and male. Others could go through life never knowing they *had* a DSD. Why is there so much variability? To answer, we need to know how males and females form.

disorder of sexual development (DSD) :: a condition sometimes noted at birth in which the genitalia are atypical

The Stages of Sex Determination

What determines sex? As we saw in Chapter 3, it all begins with *gametes*, the egg and sperm that unite to form a new organism, and it depends on the species (**FIGURE 6.2**). In some turtles and lizards, for example, the sex of the offspring is determined by the temperature at which eggs are incubated. Equal numbers of males and females will hatch at some temperatures, but if the environment is cooler or warmer, the sex ratio is skewed dramatically. In the marine worm *Bonellia*, newly hatched young move randomly, before settling on a solid surface. Exactly where the hatchling attaches determines its sex. If the young attaches to the front of a female body, it will develop as a male. If instead it attaches somewhere else, female development will be the outcome.

For some species of fish, sex is not even permanent. Born a male, the clownfish, *Amphiprion ocellaris*, changes into a female later on in life (**FIGURE 6.3**). In contrast, the black sea bass, *Centropristis striata*, starts its life as a male and can become female later. As these examples show, the development of male or female anatomy is not just complex: it may also have several stages, each with its own plan.

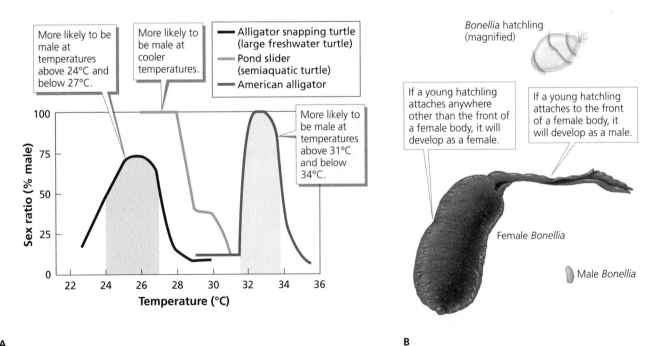

A   **B**

FIGURE 6.2 The mechanism of sexual determination differs by organism. In some reptiles, the sex of the offspring is determined by the temperature at which eggs are incubated (A), while in the marine worm *Bonellia,* the location at which a hatchling attaches determines its sex (B).

In humans and other mammals, the first set of instructions is found on the sex chromosomes (**FIGURE 6.4**). Humans have two types of sex chromosomes, X and Y, and normal cells have pairs of chromosomes. Males are XY and females XX. When gametes are produced for reproduction, males make sperm that have an X chromosome or a Y chromosome. Females make eggs that have an X chromosome. When the egg and sperm fuse together, the resulting zygote will get X from the mother and X or Y from the father.

So far this seems simple enough: XX means that the embryo will develop into a girl, and XY means that a boy will form. Girls have female sex organs, or **ovaries**. Boys have male sex organs, or **testes**. However, chromosomal sex is just the first piece of information in sexual development. It is responsible for **primary sex determination**: whether ovaries or testes form.

Secondary sex determination governs the development of other sex-related body characteristics (**FIGURE 6.5**). Men have external **genitals**, or sex organs. In addition to their external genitals (the vulva), women also have breasts, a **uterus** (or womb), and a vagina. Men and women also differ in the structure of the pelvic bone, voice tone, and the locations of body fat and hair. Chemical signals called **hormones** are responsible for these and other secondary sex characteristics. Let's take a look at the interplay of genes and hormones when sexual development goes "according to plan."

Chromosome Instructions

Until an embryo is six weeks old, it is not possible to tell whether it is male or female by its appearance; it is still bipotential for sex, meaning it has not yet made a physical commitment to be male or female. However, structures that will eventually become ovaries or testes are already present in the lower abdomen, awaiting instructions from specific genes (**FIGURE 6.6**).

Even these instructions are more than just XX in females versus XY in males. Ordinarily, the Y chromosome has the *SRY* gene, which is required for

FIGURE 6.3 For some species of fish, sex is not permanent. Born a male, the clownfish, *Amphiprion ocellaris*, changes into a female later on in life.

ovary :: female sex organ

testis :: male sex organ

primary sex determination :: the step in sexual development that determines whether ovaries or testes form

secondary sex determination :: the step in sexual development that governs the development of sex-related body characteristics such as external genitals, structure of pelvic bone, voice tone, and locations of body fat and hair

genital :: external sex organ

uterus :: the womb; the hollow, muscular organ located between the bladder and the rectum of females

hormone :: chemical signal responsible for physiological or developmental responses, including secondary sex characteristics

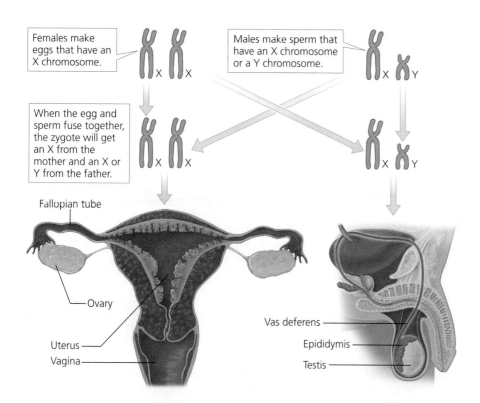

Females make eggs that have an X chromosome.

Males make sperm that have an X chromosome or a Y chromosome.

When the egg and sperm fuse together, the zygote will get an X from the mother and an X or Y from the father.

Fallopian tube

Ovary

Uterus

Vagina

Vas deferens

Epididymis

Testis

FIGURE 6.4 In humans and other mammals, primary sex determination comes from the sex chromosomes. They are responsible for whether ovaries or testes form.

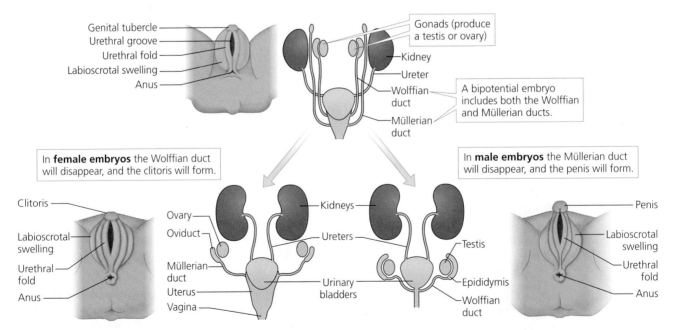

FIGURE 6.5 Secondary sex determination governs the development of other sex-related body characteristics.

the development of testes. If a mutation disables SRY, even though the embryo is XY, ovaries develop. If an XX embryo has the SRY gene because of a genetic error, testes develop. In normal females, the X chromosomes both have a gene called *DAX1*. An individual who is missing an X chromosome will be female but won't develop normal ovaries, because there is not enough DAX1.

Secondary sex characteristics develop in several stages—in the embryo, in fetal development, and at puberty. Until at least the seventh week of development, embryos, whether male or female, have the precursors of the internal features of both sexes. These features amount to a set of tubes and ducts—the Wolffian ducts in males and the Müllerian ducts in females. In XX embryos, the Wolffian ducts will disappear, while the Müllerian ducts will disappear in males.

How do genetic instructions handle all this internal sculpting? Cells in the testes and ovaries secrete the necessary hormones.

Hormone Instructions

In XY embryos, the testes secrete two critical hormones. Anti-Müllerian hormone triggers the demise of Müllerian ducts. **Testosterone** promotes the

testosterone :: a hormone that promotes the development of the Wolffian ducts, the prostate gland, and the penis and regulates the descent of testes into the scrotum

FIGURE 6.6 Until an embryo is six weeks old, it is bipotential for sex—it has not yet made a physical commitment to be male or female. However, internal structures that will eventually become ovaries or testes and external structures that will eventually become the clitoris or penis are already present.

development of Wolffian ducts, the prostate gland, and the penis. Testosterone also inhibits breast formation and regulates the descent of testes into a pouch called the **scrotum**. In XX embryos, ovaries secrete **estrogen**, which promotes the development of the Müllerian ducts, oviducts (the passageway from the outside of the body to the ovaries), the uterus, and the upper end of the vagina. Ovaries also secrete some testosterone, just not as much as testes do.

External genitals do not start out looking male or female either (Figure 6.6). At six weeks, the embryo has a small bud between its legs. Eventually, it will form either a penis or clitoris. By nine weeks, there are two swellings on either side of a urogenital groove. By 14 weeks, this groove has disappeared in male embryos, and the scrotum has formed from the swellings. In females, the groove becomes the opening of the vagina, and the swelling becomes other folds of the female genitals (called the labia majora and labia minora). Amazingly, all these changes are regulated by hormones. That is why most babies are easy to identify as male or female at birth.

The final changes come with puberty, thanks to a dramatic increase in sex hormones. These changes make the differences between males and females even more obvious—most of the time.

scrotum :: a pouch that holds the testes

estrogen :: a hormone secreted by the ovaries that promotes the development of the Müllerian ducts, oviducts, the uterus, and the upper end of the vagina

6.2 What Happens If the Hormonal Signals Are Missing or Misread?

In almost 2% of men and women, however, events play out differently. In fact, there are more than 40 different types of disorders of sexual development (DSD). Two dramatic examples are androgen-insensitive syndrome in women and pseudohermaphroditism in men.

Androgen-Insensitivity Syndrome

In **androgen-insensitivity syndrome (AIS)**, individuals who are XY do not develop as males because their cells are deaf to the instructions saying "make male parts." More technically, they lack receptors to bind to, or stick to, testosterone (**FIGURE 6.7**). Because AIS individuals are XY, the SRY gene functions correctly, and they develop testes. The testes make testosterone, which

androgen-insensitivity syndrome (AIS) :: a disorder of sexual development in which an individual who is XY does not develop as a male because the individual's cells lack receptors to bind to testosterone

AIS individuals develop internal testes and make testosterone, but they lack the right receptors, so no external genitals form.

SRY

gonad = testis

Sertoli cells line the seminiferous tubules and are involved in the production of sperm.

Leydig cells are found near the seminiferous tubules and secrete testosterone.

The sertoli cells secrete **MIS** (Müllerian-inhibiting substance).

Testosterone

Müllerian ducts degenerate.

Wolffian ducts degenerate.

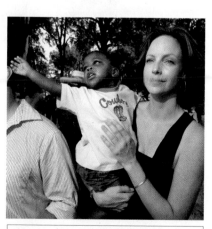

Eden Atwood is a jazz singer with AIS who lectures frequently about the disorder.

FIGURE 6.7 Individuals with androgen-insensitivity syndrome (AIS) lack the ability to respond to testosterone signals. Even though they are XY, they develop female physical traits.

Gender Testing in Sports

Athletic competition brings out the best in some people—and the worst in others. In the 1930s, Stella Walsh and Helen Stephens dominated women's track and field. During the 1936 Olympic Games, the press reported rumors that both women were actually men. Walsh herself accused Stephens of being male. The Olympic Committee examined Stephens and confirmed that she had female genitals.

Well into the 1960s, the press fueled claims about gender identity. Male athletes from Eastern Europe were said to have bound their genitals in order to compete as women.

In the 1930s, Stella Walsh (left) and Helen Stephens (right) dominated women's track and field. During the 1936 Olympic Games, the press reported rumors that both women were actually men.

The reports worked up a nationalistic fervor. *There is no way*, people cried, *that those Communists are going to get away with cheating!* To compete in the 1966 European Athletics Championship, women had to agree to a humiliating examination. They had to see a gynecologist and also stand naked in front of a committee to have their external genitals checked. Yet of 243 women examined, none had abnormalities.

Chromosome testing to verify gender began with the 1968 Olympics. Unfortunately, these tests introduced new problems, for not everyone with a Y chromosome develops as a male.

In 1991, testing for the SRY gene was the "new and improved" way to verify gender. Results were difficult to interpret. At the 1992 Barcelona Olympics, some 2000 women were evaluated, and 15 were positive for SRY. (Further information about these athletes has remained confidential.) In 1996, at the Atlanta Olympics, 3000 were tested, and eight were positive for SRY. Of these, seven had AIS and one had another DSD that prevented testosterone from being active. All eight were allowed to compete.

Gender testing of athletes has never revealed anyone deliberately misrepresenting her gender. However, many women have had their lives turned upside down. Maria Patino, a hurdler from Spain, went to the World University Games in Japan. Patino has AIS (she is XY) and forgot to bring the letter from her doctor verifying her as female. Patino was required to take a gender verification test, and the results leaked out. She was stripped of all past titles, her scholarship was taken away, and her boyfriend left her. Patino was eventually reinstated in 1988 because she was able to show that her body could not use testosterone.

There is no evidence that women athletes with DSDs have an edge. What drove the enthusiasm for gender testing, at least in part, was fear—fear that men would masquerade as women to win competitions. In 1999, the Olympics Committee finally abandoned compulsory gender testing. Now, just as it has always been for male athletes, female athletes will be spared the ordeal of proving who they are. ::

causes Müllerian ducts to degenerate. However, without the right testosterone receptors, no external male genitals form. Rather, the testes remain internal. And since the Müllerian ducts are destroyed, no internal female body parts form either.

AIS embryos are exposed to estrogen from the mother's ovaries and the embryo's internal testes, too. This estrogen is responsible for breast development and the female pattern of hair and fat. As a result, a person with AIS will look and be female, even though she is XY and has developed testes but no ovaries. In many cases, women who have AIS were completely unaware of it until they tried to find out why they did not menstruate or why they were unable to become pregnant.

Sometimes AIS can be partial. Whispers of testosterone signals get through. When this occurs, a woman may be XY and possess external female genitals, but she may also have some male characteristics—such as greater muscle mass, little breast development, and a less female-shaped pelvis. Could Caster Semenya have partial AIS? We don't know.

Pseudohermaphroditism

Sexual development can also take an unexpected path in males. In **pseudohermaphroditism**, for example, an XY individual develops testes but is missing an enzyme that enables testosterone to send the right signals. External genitals appear female, but the Müllerian duct degenerates, and no female internal ducts form.

Then, at puberty, the adrenal gland produces testosterone, and everyone is in for a big surprise. What seemed to be a clitoris enlarges to reveal itself as a penis. The voice deepens, and muscles enlarge. Breasts do not develop, and menstruation does not begin. Having started life as a girl, this individual develops into a man. Although male pseudohermaphroditism is rare in the general population, it is less unusual in an isolated village in the Dominican Republic. There, 2% of the population that is born looking female develop into males after puberty. People in this village call the condition *guevedoce*, or "penis at age 12."

So far, we have referred to individuals with DSD as male or female. But should we instead leave open how to designate anyone in that 2% of the population? Does it even make sense to say that someone changes sex at puberty? How else do men and women develop differently? See *Life Application: Gender Testing in Sports* to learn how the question of whether and how to designate male and female has consequences beyond the individual.

pseudohermaphroditism :: a disorder of sexual development in which an individual who is XY develops testes but is missing an enzyme that lets testosterone send the right signals until puberty; having started life as a girl, such an individual develops into a man

6.3 How Do Men Produce Sperm?

Males are sperm factories. They make sperm continuously, from the time they reach puberty practically until death. (Sperm quality degrades as a man ages.) The production of sperm takes about 9–10 weeks. It involves chemical signals and communication, cell division, and cell differentiation (Chapter 3).

The Testes

Sperm are produced in the testes. Measuring approximately 5 cm in length, the testes are located in the pouch called the scrotum (**FIGURE 6.8**). Small muscles hold the testes in place. These muscles also regulate the position of the testes relative to the body. If the air is cold, testes are brought closer. When it is warm, the muscles relax and the testes hang lower. This "yo-yo" keeps the sperm healthy. Sperm are very sensitive to temperature. If they get too warm or too cold, they can't function.

Testes are sensitive body parts. One good hit from a baseball, a foot, or anything else produces significant pain. Why, then, are testes external to the body, where they are vulnerable to injury or accident? As we'll see in

Spermatic cord

Blood vessels
and nerves

Vas deferens

Epididymis

Each testis contains 125 m
of seminiferous tubules—
longer than the length of
a football field!

Seminiferous tubules

Testis

Scrotum

FIGURE 6.8 Measuring
approximately 5 cm in length,
the testes are located in the pouch
called the scrotum.

Chapter 9, if the testes were inside the body, where the temperature is higher, they could not do their job.

The testes consist of around 30 wedge-shaped subunits. Each of these contains two or three tightly packed coils, called the seminiferous tubules. Astonishingly, each of the testes contains 125 m of seminiferous tubules— longer than the length of a football field! Lining the walls of the seminiferous tubules is a population of self-renewing cells, or **stem cells**. When these cells undergo mitosis (Chapter 3), on average one of the daughter cells replenishes the stem cell population. The other, called a primary spermatocyte, begins the long journey to becoming a sperm cell.

stem cell :: self-renewing cell

The primary spermatocyte undergoes meiosis I, producing secondary spermatocytes (Chapter 3). The secondary spermatocytes complete meiosis II. Finally, the spermatids made by this division differentiate into sperm (**FIGURE 6.9**).

Sperm and Semen

When sperm leave the testes, they are not completely mature. If these sperm were placed in a culture dish with a fertile egg, nothing would happen. The sperm don't have information yet about what to do. They get their next set of instructions when they move into another coiled tube, the **epididymis**. Cells

epididymis :: coiled tube sitting atop each testis, whose cells secrete chemical signals that help sperm to complete differentiation

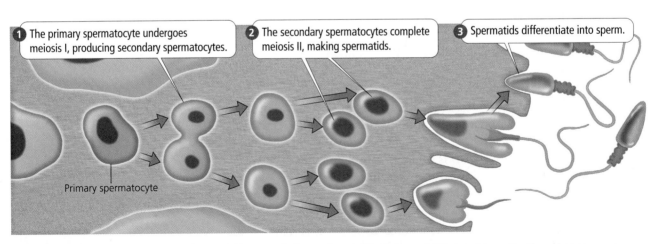

1 The primary spermatocyte undergoes meiosis I, producing secondary spermatocytes.

2 The secondary spermatocytes complete meiosis II, making spermatids.

3 Spermatids differentiate into sperm.

Primary spermatocyte

FIGURE 6.9 When stem cells lining the seminiferous tubules undergo mitosis, one of the daughter cells, the primary spermatocyte, begins the long journey to becoming a sperm cell.

of the epididymis secrete chemical signals that help sperm to complete differentiation. Sitting atop each testis, the epididymis would be approximately 6 m long if straightened out—another marvel of packaging. Sperm are stored in the last stretch of the epididymis and remain there until a male is sexually aroused. The sperm then speed into yet another tube (the vas deferens), past ejaculatory ducts, and into the urethra, which runs through the penis. Finally, in ejaculation, the sperm are propelled outside the body.

Sperm are swimming cells. They need some type of liquid to get anywhere. Several glands secrete materials into the mix of fluids that bathe the sperm. Besides serving as the passageway for sperm, the urethra also carries urine. Even before ejaculation, two bulbourethral glands secrete a clear fluid into the urethra to neutralize any urine residue that might be present. The **prostate gland** secretes a milky white fluid that neutralizes the acidity of the vagina so that sperm will survive. Finally, two seminal vesicles secrete the sugar fructose, which serves as an energy source for sperm, and **prostaglandin**, a chemical signal that causes muscle contractions in the female reproductive tract. These contractions help move sperm closer to the egg. The final combination of sperm and all of these secretions is called **semen**.

prostate gland :: secretes a milky white fluid that neutralizes the acidity of the vagina

prostaglandin :: a chemical signal that causes muscle contractions in the female reproductive tract

semen :: the final combination of sperm and secretions by the prostate gland, seminal vesicles, and bulbourethral glands

Hormones and Sperm Production

Sperm have only one job to do: fuse with an egg and together make a zygote. Inadequate sperm production could lead to infertility. Making too many sperm could lead to problems with sperm quality. Fortunately, the entire process is carefully regulated by hormones.

The hormone that starts the whole chain of events is secreted by a part of the brain called the **hypothalamus** (**FIGURE 6.10**). When gonadotropin-releasing hormone (GnRH) is released by the hypothalamus, it stimulates the release of other hormones by the **pituitary gland**, also located in the brain. The pituitary secretes follicle-stimulating hormone (FSH) and luteinizing hormone (LH). In females, FSH and LH play a role in egg formation. In males,

hypothalamus :: a part of the brain that secretes the gonadotropin-releasing hormone (GnRH)

pituitary gland :: a part of the brain that secretes follicle-stimulating hormone (FSH) and luteinizing hormone (LH)

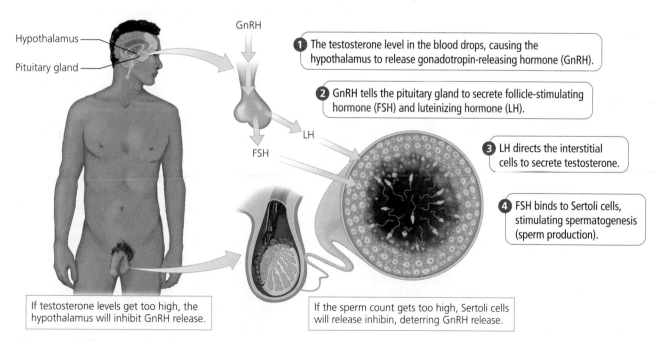

GnRH

Hypothalamus

Pituitary gland

LH

FSH

❶ The testosterone level in the blood drops, causing the hypothalamus to release gonadotropin-releasing hormone (GnRH).

❷ GnRH tells the pituitary gland to secrete follicle-stimulating hormone (FSH) and luteinizing hormone (LH).

❸ LH directs the interstitial cells to secrete testosterone.

❹ FSH binds to Sertoli cells, stimulating spermatogenesis (sperm production).

If testosterone levels get too high, the hypothalamus will inhibit GnRH release.

If the sperm count gets too high, Sertoli cells will release inhibin, deterring GnRH release.

FIGURE 6.10 Sperm production is regulated by a series of hormones. The hypothalamus releases GnRH and stimulates the pituitary gland, which in turn releases FSH and LH, stimulating testosterone production by the testes.

FSH stimulates sperm production, while LH triggers testosterone production by the interstitial cells of the testes. Interstitial cells are located in the tissues between seminiferous tubules.

The hypothalamus is the main player that monitors this hormone cocktail. If the testosterone level in the blood drops below a certain threshold, the hypothalamus secretes GnRH. Next, GnRH tells the pituitary to release LH and FSH. The LH directs the interstitial cells to secrete testosterone. FSH binds to Sertoli cells, which line the seminiferous tubules, stimulating spermatogenesis, the production of sperm cells.

If testosterone levels get too high, the hypothalamus can also shut the whole thing down by inhibiting GnRH release. And if the sperm count is too high, Sertoli cells release another signal, called inhibin, which deters GnRH release. Inhibin also acts directly on the pituitary to stop FSH release.

The careful dance between cells and hormonal signals makes the production of sperm sensitive to chemicals that resemble sex hormones, such as certain drugs or pollutants. Molecules that disrupt normal hormone signals can have serious consequences for sexual development, secondary sex characteristics, and fertility. Examples include anabolic steroids, which are sometimes abused for body building, and atrazine, a pesticide.

6.4 How Do Women Produce Eggs?

oogenesis :: the production of fertile eggs

Women have a much more involved task than men when it comes to gamete formation and reproducing. Even the production of fertile eggs, or **oogenesis**, is more complicated than the production of sperm. In addition to releasing eggs, a woman's body also needs to prepare for pregnancy. Whereas men develop sperm continuously, women are born with all the immature eggs, or **primary oocytes**, they will ever have.

primary oocyte :: immature egg

Men are able to ejaculate sperm multiple times per day, but women release one egg per month except during pregnancy. Behind that monthly pattern is the interaction between the brain and the female reproductive system—again orchestrated by hormones.

The Ovaries

Women have their gamete-making organs, or ovaries, tucked safely inside their bodies (**FIGURE 6.11**). In addition to maturing and releasing eggs, ovaries also produce hormones that are important for the development of secondary sex characteristics. The two ovaries alternate producing an egg each month. When an egg is propelled from the ovary, it enters a small tube called an **oviduct**. Each ovary is positioned next to its own oviduct. Motile hairlike structures line the inside of the oviduct and help to sweep the eggs into the uterus. Approximately the size and shape of an upside-down pear, the thick-walled, muscular uterus is the place where the fetus develops (Chapter 3). The **cervix** comprises the lower part of the uterus and the opening to the vagina.

oviduct :: a small tube down which an egg travels when it is propelled from the ovary

cervix :: the lower part of the uterus and the opening to the vagina

Oogenesis begins before a girl is even born. A female fetus develops approximately 2 million potential egg cells. By the time a baby girl is born, she has about 700,000, although by puberty only 350,000 remain. Fewer than 500 are ever released for possible fertilization.

Oogenesis, like sperm production, entails meiosis—but not at all continuously. Each oocyte starts meiosis I in the ovary and pauses, for years, until puberty. Think about it: a woman who is 35 years old has had oocytes stuck in

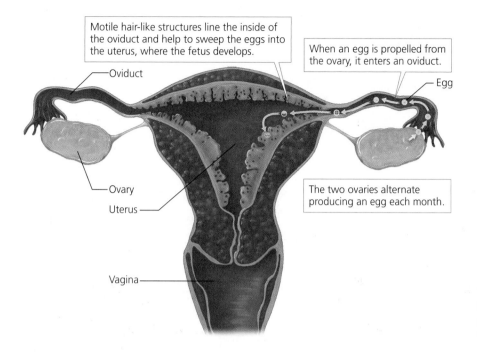

Motile hair-like structures line the inside of the oviduct and help to sweep the eggs into the uterus, where the fetus develops.

When an egg is propelled from the ovary, it enters an oviduct.

Oviduct

Egg

The two ovaries alternate producing an egg each month.

Ovary

Uterus

Vagina

FIGURE 6.11 Ovaries are the site of egg production in a woman's body. They mature and release eggs.

meiosis I for more than 35 years. No wonder the quality of eggs diminishes as women age.

Once puberty occurs, one oocyte completes meiosis I each month. One of these cells, the secondary oocyte, keeps most of the cytoplasm. The other tiny cell, called a polar body, does not develop further. The secondary oocyte starts meiosis II but stops and, about 12 hours later, bursts from the ovary. This event is **ovulation**. Some women feel ovulation as a sharp pain in the lower abdomen. If it fuses with a sperm, the secondary oocyte will complete meiosis II, producing a mature egg.

However, a woman's reproductive system is responsible for more than simply producing eggs. It must also support pregnancy should fertilization occur. Each month, an ovary releases an egg for possible fertilization, and the uterus prepares for pregnancy. These events are tightly coordinated by a sequence of hormones.

ovulation :: the release of an egg from the ovary

The Ovarian Cycle

At the start of an ovarian cycle, the hypothalamus secretes GnRH (**FIGURE 6.12**). As a consequence, the pituitary releases FSH and LH. (Remember that the hypothalamus, GnRH, LH, and FSH all play roles in forming sperm, too.) Each primary oocyte is surrounded by a single layer of cells; together, they are called a **follicle**. In women, FSH causes the follicle to grow. FSH and LH also stimulate estrogen secretion. Fluid builds up in the follicle, causing it to swell—until it bursts out of the ovary, releasing the egg.

After ovulation, the body has no idea whether the egg has been fertilized. It begins the preparation for pregnancy in any case. First, the ruptured follicle develops into a structure called the corpus luteum. The corpus luteum secretes estrogen and yet another hormone, **progesterone**, which plays a role in the development of the lining of the uterus needed for a potential pregnancy. In fact, the estrogen and progesterone levels in the blood surge so high that no more follicles ripen as long as the corpus luteum lasts. It's as if the body is saying, "Let's hold off getting another egg ready until we see what

follicle :: a primary oocyte surrounded by a single layer of cells

progesterone :: a hormone secreted by the corpus luteum that is used in the development of the endometrium or for preventing the development of a second follicle during one menstrual cycle

GnRH

Pituitary

FSH   LH

Hypothalamus

Pituitary gland

Follicle Egg

1 The hypothalamus secretes gonadotropin-releasing hormone (GnRH).

2 GnRH tells the pituitary gland to secrete follicle-stimulating hormone (FSH) and luteinizing hormone (LH).

3 LH and FSH stimulate estrogen secretion.

4 FSH causes the follicle to grow.

5 Fluid builds up in the follicle, causing it to swell.

6 The fluid bursts out of the ovary, releasing the egg.

7 The ruptured follicle develops into the corpus luteum.

8 The corpus luteum secretes estrogen and progesterone to maintain the endometrium.

9 If there is no pregnancy, the corpus luteum breaks down after 12 days, and the ovarian cycle starts anew.

Hormone levels in blood

LH
FSH

Mid-cycle peak of LH (triggers ovulation)

Ovulation

Corpus luteum

4 **5**   **6**   **7**   **8** **9**

Follicle growth →

Corpus luteum degeneration →

FIGURE 6.12 Egg production is regulated by a series of hormones. At the start of the ovarian cycle, the hypothalamus releases GnRH and stimulates the pituitary gland, which in turn releases FSH and LH, stimulating estrogen production and causing the follicle to grow until it bursts, releasing the egg.

menstruation :: the shedding of the lining of the uterus that supports fetal development

endometrium :: the lining of the uterus produced during the menstrual cycle to support the development of an embryo if a pregnancy occurs

happens to the one we just launched." But this holding pattern is temporary. If there is no pregnancy, the corpus luteum breaks down after 12 days, and the ovarian cycle starts anew.

The Uterine Cycle

If there is fertilization, it is up to the uterus to make sure that there is a place for the embryo to develop. Every month, if a woman is not already pregnant, the uterus prepares its inner surface for pregnancy. This uterine cycle (**FIGURE 6.13**) has three phases, starting with **menstruation**, the shedding of the lining of the uterus that supports fetal development. Once the old lining, called the **endometrium**, is gone, a new one is produced—stimulated by increasing estrogen levels. The body still has no clue whether a pregnancy or even intercourse is going to happen in the next couple of weeks. Yet the uterus gets ready just in case.

The second phase, ovulation, occurs roughly halfway through the 28-day cycle. The estrogen now prompts the cervix to secrete a thin, clear mucus that is easy for sperm to swim in.

You saw that the corpus luteum can survive around 12 days, even if there is no pregnancy. It also has an important job in the luteal phase—the third and final phase of the uterine cycle. It secretes estrogen and progesterone, which maintains the endometrium. If there is no pregnancy, then the corpus luteum degenerates, estrogen and progesterone levels plummet, and the endometrium breaks down. The menstrual fluid that flows from the body is a combination of blood and endometrial tissue.

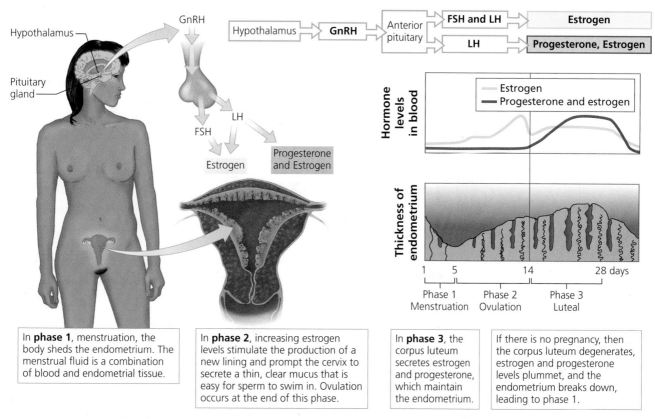

In **phase 1**, menstruation, the body sheds the endometrium. The menstrual fluid is a combination of blood and endometrial tissue.

In **phase 2**, increasing estrogen levels stimulate the production of a new lining and prompt the cervix to secrete a thin, clear mucus that is easy for sperm to swim in. Ovulation occurs at the end of this phase.

In **phase 3**, the corpus luteum secretes estrogen and progesterone, which maintain the endometrium.

If there is no pregnancy, then the corpus luteum degenerates, estrogen and progesterone levels plummet, and the endometrium breaks down, leading to phase 1.

FIGURE 6.13 Every month, if a woman is not already pregnant, the uterus prepares its inner surface for pregnancy. This uterine cycle has three phases.

Hormones and Pregnancy

Once a woman ovulates, the egg has only a brief time to encounter sperm. The journey from the ovary to the oviduct, uterus, vagina, and uterus takes about three days. However, fertilization leading to a successful pregnancy almost always occurs in the oviduct, where swimming sperm meet the egg.

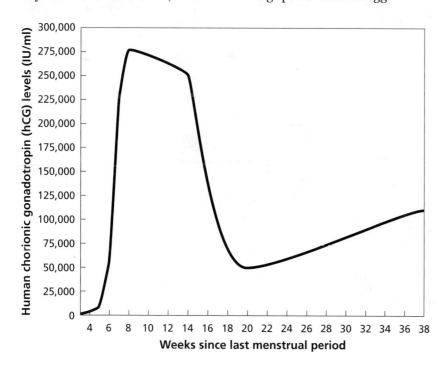

FIGURE 6.14 If fertilization occurs, the embryo travels down the oviduct and into the uterus, and the endometrium secretes human chorionic gonadotropin (hCG) hormone. During the early phases of pregnancy, hCG levels rise dramatically.

When fertilization does occur, the embryo travels down the oviduct and into the uterus. The embryo burrows into the wall of the uterus, a process called implantation. As a consequence, the endometrium does not break down and now secretes another hormone, human chorionic gonadotropin (hCG) (**FIGURE 6.14**). This hormone allows the corpus luteum to survive for another six to seven weeks. By the time the corpus luteum finally stops secreting estrogen and progesterone, the placenta makes enough progesterone to maintain the uterine lining on its own.

6.5 How Can Pregnancy Be Prevented?

The key to preventing pregnancy is keeping eggs and sperm apart. Of the roughly 300 million sperm ejaculated during sexual intercourse, only about 200 reach an oviduct. The female reproductive tract does its best to help the sperm succeed on their journey (see *How Do We Know? The Female Reproductive Tract Helps Sperm Find an Egg*). Even so, some of these sperm

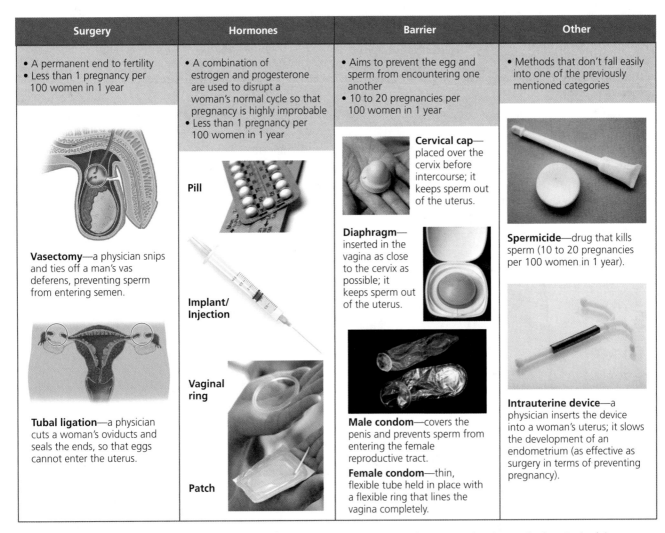

Surgery	Hormones	Barrier	Other
• A permanent end to fertility • Less than 1 pregnancy per 100 women in 1 year	• A combination of estrogen and progesterone are used to disrupt a woman's normal cycle so that pregnancy is highly improbable • Less than 1 pregnancy per 100 women in 1 year	• Aims to prevent the egg and sperm from encountering one another • 10 to 20 pregnancies per 100 women in 1 year	• Methods that don't fall easily into one of the previously mentioned categories

Pill

Implant/Injection

Vaginal ring

Patch

Vasectomy—a physician snips and ties off a man's vas deferens, preventing sperm from entering semen.

Tubal ligation—a physician cuts a woman's oviducts and seals the ends, so that eggs cannot enter the uterus.

Cervical cap—placed over the cervix before intercourse; it keeps sperm out of the uterus.

Diaphragm—inserted in the vagina as close to the cervix as possible; it keeps sperm out of the uterus.

Male condom—covers the penis and prevents sperm from entering the female reproductive tract.

Female condom—thin, flexible tube held in place with a flexible ring that lines the vagina completely.

Spermicide—drug that kills sperm (10 to 20 pregnancies per 100 women in 1 year).

Intrauterine device—a physician inserts the device into a woman's uterus; it slows the development of an endometrium (as effective as surgery in terms of preventing pregnancy).

FIGURE 6.15 Birth control methods can be divided into four categories: surgery, hormones, barrier, and other. Each of these methods has various rates of success in preventing pregnancy.

The Female Reproductive Tract Helps Sperm Find an Egg

Fertilization takes teamwork—and not just the kind you might be thinking of. Sperm face a long and difficult swim to find a fertile egg. They could not do it without assistance. Fortunately, the female reproductive system has ways to help.

After ejaculation into the vagina, sperm speed toward the cervix—and their first fresh challenge. Cervical mucus selects for sperm that swim well. Before ovulation, the molecules in the mucus are arranged like a mesh, capturing poor swimmers. When a woman has a fertile egg available, the same fibers line up side by side to make a smooth path for

Scientists have found that the sperm that stick to the oviduct are healthier and swim better than those that remained unattached.

sperm. Contractions in the muscles of the uterus help move the sperm along even more quickly.

Once sperm enter the oviduct, there are only a few thousand left from a population that started with hundreds of millions. Luckily for the survivors, the oviduct is ready and waiting. It contains a reservoir for sperm, where sperm can bind to the oviduct's inner surface.

Sperm that stick to the oviduct have been found to be healthier and swim better than those that remained unattached, according to scientists who cultured specimens from male volunteers with pieces of the inner surface of oviducts. Other studies in cows, pigs, and rabbits have confirmed that sperm collect in the oviduct, binding to specific molecules. When ovulation occurs, the sperm are set free—if only in small numbers—to locate the egg.

Dr. Joanna Ellington, a veterinarian and sperm physiologist, found an unusual way to demonstrate that a sperm reservoir exists in humans, too. After she and her husband decided they did not wish to have any more children, she scheduled a tubal ligation for herself. However, Ellington told her surgeon that she and her husband were going to have intercourse first. She asked to have pieces of her oviducts removed during surgery, so that she could see whether sperm were attached to the inner surface. They were.

Scientists have found that fertilization can occur as late as five days after intercourse. Sperm can "relax" in the oviduct all that time, waiting for ovulation to occur. Once an egg is released, the sperm are directed to it by chemical signals secreted by the egg. Sperm can also sense the higher temperature in its vicinity.

All this assistance is not just altruism on the part of the female reproductive tract. The path from the vagina to the egg is like an obstacle course, where each step helps select for healthy, strong sperm. They are the winners, and the egg is first prize. ::

enter the oviduct near the ovary that did *not* ovulate. They have no chance at finding a healthy, fertile egg. As discussed in Chapter 3, it is a wonder that babies are ever conceived. Yet they are—enough that many people want to plan whether and when to bring a new person into the world. The birth control, or **contraceptive**, industry earns billions of dollars worldwide each year. Birth control methods can be divided into four categories: surgery, hormones, barrier, and other (**FIGURE 6.15**).

contraceptive :: birth control

Surgery

Surgery can provide essentially a permanent end to fertility. Men can undergo a **vasectomy**. In this procedure, a physician snips and ties off the vas deferens,

vasectomy :: a surgical procedure in which the vas deferens is snipped and tied off to prevent sperm from entering semen

tubal ligation :: a surgical procedure in which a woman's oviducts are cut and the ends are sealed so that the eggs cannot get into the uterus

preventing sperm from entering semen. Women can have a **tubal ligation**. Here, a physician cuts the oviducts and seals the ends so that eggs cannot enter the uterus.

Hormones

Hormones can be used to disrupt a woman's normal cycle so that pregnancy is impossible either because follicle development is inhibited or because endometrium formation is prevented. A combination of estrogen and progesterone can be delivered in a pill (sometimes daily), an implant, an injection, a vaginal ring, or a patch.

Barrier

cervical cap :: a barrier contraceptive that keeps sperm out of the uterus

diaphragm :: a barrier contraceptive placed in the vagina as close to the cervix as possible; it bars the way to the uterus

condom :: a barrier contraceptive that covers the penis and prevents sperm from entering the female reproductive tract

After abstinence, barrier methods are the oldest form of birth control. Modern barriers aim to prevent the egg and sperm from encountering one another. Placed over the cervix before intercourse, the **cervical cap** keeps sperm out of the uterus. A **diaphragm**, inserted in the vagina as close to the cervix as possible, similarly bars the way to the uterus. Male **condoms** take a different strategy: by covering the penis, they prevent sperm from entering the female reproductive tract at all. Finally, female condoms—thin, flexible tubes held in place with a flexible ring—line the vagina completely. Besides preventing contact between eggs and sperm, condoms also help prevent the spread of sexually transmitted diseases, including HIV/AIDS.

Other

spermicide :: drug that kills sperm

intrauterine device (IUD) :: a contraceptive inserted into a woman's uterus to slow the development of an endometrium

Finally, there are some birth control methods that don't fall easily into one of the previously mentioned categories. **Spermicides** are drugs that kill sperm. They are most effective when used together with a barrier method. An **intrauterine device** (**IUD**), which is inserted into a woman's uterus by a physician, slows the development of an endometrium. If an embryo did form, it would not be able to implant in the uterus.

The world desperately needs reliable, affordable, culturally sensitive, and easily available contraceptives. Not only do many couples want to plan their families in a responsible manner, but the population is growing fast. One new approach, a vaccine against hCG, has been successful in limited clinical trials. Future vaccines for women could target sperm.

6.6 What Causes Infertility, and How Can It Be Treated?

According to the American Medical Association (AMA), approximately 15% of all couples are infertile. **Infertility** is defined as a failure to conceive and become pregnant after a year of regular, unprotected intercourse. In 40% of the cases, the cause can be traced to the man, in 40% to the woman, and in 20% to both.

Causes of Infertility

infertility :: failure to conceive and become pregnant after a year of regular, unprotected intercourse

A common reason for male infertility is a low sperm count. Sometimes the problem is not the overall number of sperm, but a shortage of healthy ones (**FIGURE 6.16**). Scientists do not completely understand why some men fail to make enough normal sperm. But there is a clear relationship between sperm counts and smoking, alcohol, or drug use. In addition, injury to the testes can

Normal

Double head

Multiple tails

Malformed head

FIGURE 6.16 One reason for male infertility is abnormal sperm.

damage the delicate network of ducts, making it difficult for sperm to enter the vas deferens.

Sperm counts are declining in men in developed, industrialized countries (**FIGURE 6.17**). Reasons proposed by scientists include a sedentary job or life-style (which elevates the temperature of the testes, and sperm cannot tolerate high temperatures), or exposure to harmful pollutants.

Infertility in women may also have several causes. Sometimes the hormonal cycles don't operate properly. Consequently, ovulation fails to take place or the uterine lining is not prepared properly. Another possibility is **endometriosis**, a painful condition in which uterine tissue grows elsewhere in the body—usually in the oviducts and in the ovaries. The tissue responds to hormones just like it would in the uterine lining. It grows and is shed with each cycle, but it is not easily eliminated from the body. This can lead to inflammation and disruption of ovulation. Sometimes female infertility is caused by blocked ovi-ducts due to **pelvic inflammatory disease (PID)**. PID can be the result of a sexually transmitted disease or even an infection caused by an IUD.

Disorders of sexual development (DSD) can cause infertility, too. For example, XY women who have complete androgen-insensitivity syndrome

endometriosis :: a painful condition in which uterine tissue grows elsewhere in the body, leading to inflammation and disruption of ovulation

pelvic inflammatory disease (PID) :: infertility caused by blocked oviducts that may be a consequence of a sexually transmitted disease or infection caused by an IUD

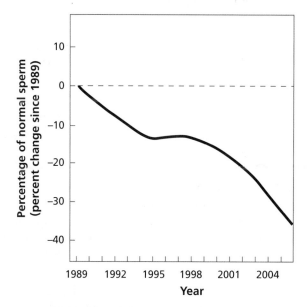

Source of Data: M. Rolland, J. Le Moal, V. Wagner, D. Royere, J. De Mouzon, Decline in semen concentration and morphology in a sample of 26,609 men close to general population between 1989 and 2005 in France. *Human Reproduction 2012*; **28**: 462-470.

FIGURE 6.17 Sperm counts and overall quality of sperm have been declining in men in developed, industrialized countries.
In France, for example, both the total number of sperm and the percentage of normal sperm have declined in 35-year-old men.

(AIS) are sterile; they lack ovaries and a uterus. This returns us to our earlier question: what defines male or female? Must a male have the potential to father a child? Must a female be able to give birth? If there are solutions to infertility, could they change who we are?

Infertility Treatments

The billions of dollars spent each year to prevent pregnancy are rivaled by the billions devoted to making pregnancies happen. For most of human history, an infertile couple who wished to stay together as a family had two choices—remain childless or adopt a child. In 1978, however, all that changed with the first "test tube" baby. Now a healthy woman in her 30s and a mother herself, Louise Brown led the vanguard of a revolution in reproductive medicine, **assisted reproductive technology (ART)**.

ART is used to bring healthy sperm and eggs together so that fertilization is possible. Its ultimate goals are the implantation of an embryo, a successful pregnancy, and the birth of a healthy baby.

assisted reproductive technology (ART) :: reproductive procedure that is used to bring healthy sperm and eggs together so that fertilization is possible

artificial insemination :: sperm from the male partner are placed in the female partner's vagina

intrauterine insemination :: sperm are placed directly into the uterus

in vitro fertilization (IVF) :: immature oocytes are extracted and placed in a petri dish in a solution that permits them to mature, and they are combined with sperm from a male partner and then transferred into the uterus

- With **artificial insemination**, sperm from the male partner are placed in the female partner's vagina. This method can be helpful for couples in which the man has a low sperm count, because sperm can be collected and concentrated. If the sperm concentration is still too low, sperm from another donor can be used instead.
- **Intrauterine insemination** is a variation of artificial insemination. Here sperm are placed directly into the uterus. The idea is that sperm with motility problems won't have to swim as far to meet an egg. Also, if the woman's vagina is a hostile environment for her partner's sperm, depositing sperm directly into the uterus can help.
- With *in vitro* **fertilization** (**IVF**) (**FIGURE 6.18**), the method used to conceive Louise Brown, a physician isolates one or more eggs from the female partner. The oocytes are placed in a petri dish in a nutrient-rich solution. Sperm, generally collected by the male partner, are then combined with the oocytes. In two to four days, a small number of embryos, each at approximately the eight-cell stage, are transferred into the uterus. One or more may implant, resulting in a pregnancy.
- Gamete intrafallopian transfer (GIFT) is a variation of IVF. The egg and sperm are placed into the oviduct together, or the one-celled zygote is placed there.
- Intracytoplasmic sperm injection (ICSI) is yet one more variation. With this method, a physician injects a single sperm into an egg. If an embryo results, it will be placed in the uterus no later than the eight-cell stage.

surrogacy :: an embryo is placed in the uterus of a woman who is not the biological mother in order to establish a healthy pregnancy that results in the birth of a baby

Sometimes ART doesn't work, and the couple opts for **surrogacy**. With surrogacy, an individual or a couple arranges for a woman to give birth to a baby. For example, artificial insemination with sperm from the male partner could initiate the pregnancy. The egg then comes from the surrogate. Alternatively, a couple's egg and sperm could make the embryo, and the surrogate then has the pregnancy. The baby then has no genetic relationship with the surrogate.

Finally, infertility is sometimes treatable by methods that do not involve manipulating eggs or sperm outside the body. Fertility drugs such as gonadotropic hormones can stimulate ovaries and help women who have problems with ovulation. Also, if infertility is due to a physical blockage of an oviduct, for example, surgery may help.

FIGURE 6.18 *In vitro* fertilization is one type of assisted reproductive technology used to treat infertility.

More than 3 million children worldwide have been born thanks to these medical breakthroughs. Starting with one IVF baby in 1978, the number of ART babies now exceeds 200,000 per year. The road to a biological child through ART can be arduous. The process is long and involved, expensive, potentially painful, and not guaranteed to succeed.

Happy families can come about in many ways. The love any of us has to give to our biological children can instead be showered on adopted children, extended family, and friends.

 # How Can We Tell If a Fetus or Baby Is Healthy?

In every welcome pregnancy, parents look to the future with great excitement. (See *Technology Connection: Home Pregnancy Tests* to learn how pregnancies are confirmed.) Whatever else their dreams for their child, the top priority for these parents is the baby's health. Fortunately, an array of tests can now detect possible medical problems, so that medical professionals can help and future parents can make appropriate decisions.

Blood and Urine Tests

Information about fetal health can be obtained from the mother's blood or urine. It is also gathered from measurement and observation of the fetus. In some cases, fetal cells are collected for analysis.

Technology Connection

Home Pregnancy Tests

Although it might not seem to rank up there with curing disease, home pregnancy tests were an important innovation. Before 1978, a woman wishing to know whether she was pregnant as early as possible had to see a physician. Some women would not see a doctor to find out. No doubt some did not want their doctor to know they were sexually active. But many simply wanted privacy: they wanted to learn whether they were pregnant, but not in front of someone else. And some women, worried they might be pregnant, did not have the courage to find out.

The need for a physician visit meant that more than 50% of pregnant women did not seek medical care until two or more months into the pregnancy. In other words, more than 50% of embryos and fetuses were at risk of inadequate prenatal care.

All of this changed once home pregnancy tests became available. Women can now purchase a kit at a pharmacy or supermarket. And they can take the test in the privacy of their own homes.

Home pregnancy tests all work the same way. An antibody detects hCG, a hormone present only when there is a pregnancy. Most tests rely on a stick, with a results window. Depending on the test, a woman either urinates on the stick or collects urine in a cup and then dips the stick in it. Then she waits for the results—often a symbol such as + or −. An indicator molecule causes the "pregnant" symbol to appear if the antibody binds to hCG.

The test is 99% accurate—especially if the first urine of the day is checked, when hCG levels are at their highest.

Negative pregnancy test

Positive pregnancy test

Home pregnancy tests use an antibody to detect hCG, a hormone present only when there is a pregnancy. Depending on the test, a woman either urinates on a stick or collects urine in a cup and then dips a stick in it; often a symbol, such as a double line or a plus sign, appears if the antibody binds to hCG.

The results are also more accurate if a woman waits at least a week after a missed period. Otherwise, because conception occurs days before implantation, a woman could get a negative result even if she is pregnant. Or she could get a positive result for a pregnancy that does not progress any further. Approximately 50% of all conceptions don't lead to successful pregnancy.

Regardless of the results, a woman should retest in a few days. And if the results remain positive, or if she would like to have negative results verified, the next thing to do is to see a physician. ::

screening test :: evaluation to see whether an embryo, fetus, baby, child, or adult is at risk for a particular problem

diagnostic test :: precise evaluation of risks to an embryo or fetus

amniocentesis :: a method for examining fetal cells that can be done between 14 and 17 weeks of pregnancy

chorionic villus sampling (CVS) :: procedure used to isolate fetal cells that can be done between 10 and 12 weeks of pregnancy

Blood samples from the mother can show whether she is in good health. Physicians can determine her blood type, whether she is anemic, and whether she has been exposed to infections that she could pass on to her newborn. Of particular concern are sexually transmitted diseases, especially HIV/AIDS, syphilis, and hepatitis. Most women will also be tested to see whether they have ever had rubella (German measles) or chicken pox. Screening is also done to determine if the mother carries defective genes that could be passed on. Depending on family history and ethnicity, other tests may be offered too.

Urine tests are routine, occurring with almost every visit to the obstetrician, to monitor the mother's health. Glucose levels, for example, can show whether diabetes is present. High protein levels sometimes mean the mother has preeclampsia, or high blood pressure that endangers some pregnancies.

Screens and Diagnostic Tests

Most women will be offered **screening tests** to see whether an embryo or fetus is at risk for a particular problem. If screens are positive, they may prompt **diagnostic tests**, which are more precise. As we have already seen, diagnostic tests are also important when family history or ethnic background indicates an increase in the risk of genetic disease (see Chapter 4). Other risk factors include problems in a previous pregnancy or the mother's ill health.

Several screens during the first and second trimesters of pregnancy can assess fetal health. Because the circulatory systems of the mother and fetus are connected, sampling the mother's blood provides a look at what is in fetal blood too. Some of the blood tests measure the levels of certain molecules that are known to vary when Down syndrome, spina bifida, or other disorders are present. Ultrasound is used to measure the fetus and to detect physical problems.

If the results of screening raise a concern, the next step is diagnostic testing. **Amniocentesis**, a method for examining fetal cells, can be done between 14 and 17 weeks of pregnancy (**FIGURE 6.19**). A long needle is inserted through the abdomen into the uterus, and fluid is withdrawn. This fluid contains fetal cells that can be isolated, cultured, and examined to see whether the chromosomes are normal. Also, chromosomes can be tested to see whether genes associated with hemophilia, muscular dystrophy, or cystic fibrosis are present.

Chorionic villus sampling (CVS) can be used to isolate fetal cells, too (**FIGURE 6.20**). This procedure, somewhat riskier than amniocentesis, can be done between 10 and 12 weeks of pregnancy. With CVS, the fetal cells are collected from the chorion, the fetal part of the placenta.

This toolbox of screens and diagnostic tests raises important questions. How should one decide which tests to have? If tests are done, what should one do with this information? Tests like these increase the chance of a healthy birth. However, in some cases, serious defects are detected for which there is no cure. Is there any point in having this information, or is it just a burden on already anxious parents? Even in the most dire of circumstances, treatments may still be available to lessen symptoms. In addition, parents may have time before birth to prepare for a child's special needs, or in other circumstances, parents may decide to terminate the pregnancy.

These tools may also force us to evaluate what we even mean by a "defect" or "normal." What characteristics in a child are acceptable to us? Once again, we must face our ideas of what constitutes a good life. With rigorous prenatal testing, many individuals with disorders of sexual development can be identified before birth. What might Caster Semenya's fate have been with prenatal screening? Could medical treatments have helped? Would different parents have chosen to terminate? Understanding the biology is not the hard part. Figuring out the right thing to do with the knowledge can be intensely difficult.

Newborn Tests

Finally, the nine months of waiting are over, labor and delivery are complete, and the obstetrician or midwife places the baby on the mother's chest. The baby, already alert, looks into the mother's eyes. It is a moment hard to describe (**FIGURE 6.21**). And then, after the hard work of being born, the baby is whisked away briefly so that tests can begin to see whether the baby is healthy.

At one and five minutes after birth, a healthcare provider assesses the baby using the **APGAR scale**. Developed by Dr. Virginia Apgar, this test records the baby's **a**ctivity, **p**ulse, **g**rimace response, **a**ppearance, and **r**espiration (see *Scientist Spotlight: Virginia Apgar*). Each category is scored 0, 1, or 2 (**FIGURE 6.22**). The totals alert doctors and nurses if some help is needed. The baby is also measured and weighed at this early stage.

All U.S. states require newborn screening tests and procedures, although specific requirements vary from state to state. Because newborns do not yet have all the molecules necessary for blood clotting, they receive an injection of

FIGURE 6.19 Amniocentesis is a method used to examine fetal cells between 14 and 17 weeks of pregnancy. A long needle is inserted through the abdomen into the uterus, and fluid is withdrawn.

FIGURE 6.20 Chorionic villus sampling (CVS) is used to isolate fetal cells between 10 and 12 weeks of pregnancy. The procedure is somewhat riskier than amniocentesis because the fetal cells are collected from the chorion—the fetal part of the placenta.

FIGURE 6.21 When labor and delivery are complete and the obstetrician or midwife places the baby on the mother's chest; it is a moment hard to describe.

APGAR scale :: an assessment of a newborn baby conducted at one and five minutes after birth to assess a baby's activity, pulse, grimace response, appearance, and respiration

	Score 0	Score 1	Score 2
Activity (muscle tone)	Absent	Arms and legs flexed	Active movement
Pulse	Absent	Below 100 beats per minute	Over 100 beats per minute
Grimace (reflex irritability)	Flaccid	Some flexion of extremities	Active motion (sneeze, cough, pull away)
Appearance (skin color)	Blue, pale	Body pink, extremities blue	Completely pink
Respiration	Absent	Slow, irregular	Vigorous cry
Score	0	1	2

Total score	
0–3	Severely depressed
4–6	Moderately depressed
7–10	Excellent condition

FIGURE 6.22 At one and five minutes after birth, a healthcare provider assesses a baby using the APGAR scale, which records the baby's activity, pulse, grimace response, appearance, and respiration. The totals alert doctors and nurses if some help is needed.

vitamin K. Babies also get eye drops to prevent infection from any microbes to which they may have been exposed on the trip from the uterus into the world. In many hospitals, newborns are also given hearing tests.

Since the 1960s, blood samples have also been taken. Early detection can give physicians the opportunity to suppress at least 30 diseases or lessen the symptoms. For example, all babies are screened for PKU, a disease that interferes with metabolism. Individuals with PKU are missing an enzyme that breaks down the amino acid phenylalanine. As phenylalanine builds up, it causes brain damage. Fortunately, if PKU is detected in a baby, a phenylalanine-free diet can prevent the disease entirely.

Scientist Spotlight

Virginia Apgar (1909–1974)

When women gave birth in a hospital in the early 20th century, many of them did so under anesthesia. And then the newborn baby went straight off to the hospital's nursery, without so much as an examination. For 90% of infants, this practice was just fine. However, 9% of babies need medical attention; 1% may not survive without it. As an anesthesiologist, Virginia Apgar had assisted in thousands of deliveries. She realized that there was a real need to identify which newborns need help. Although other physicians resisted at first, her test is now used all over the world.

Born in 1909, Apgar decided in high school to become a physician. She studied zoology at Mt. Holyoke and began medical school in 1929—a month before the stock market crash and the start of the Great Depression. Despite serious financial pressure, she graduated fourth in her class at Columbia University College of Physicians and Surgeons in 1933, where she won a highly competitive surgical internship.

In 1952, Dr. Virginia Apgar designed a test to determine whether a newborn needed to be resuscitated. Now known as the APGAR test, it has held up over decades of use and study.

For all Apgar's achievement, Dr. Allen Whipple, her mentor, discouraged her from a career in surgery. People simply would not go to a "woman surgeon." A talented surgeon himself, Whipple understood that surgery could not improve unless anesthesiology did. In 1937, however, anesthesia was generally handled by nurses. Whipple thought that Apgar was just the person to help transform the field. She took his advice.

After a year of training, Apgar returned to Columbia P&S as director of the anesthesia division. She was also the division's *only* member until the mid-1940s. Other physicians wanted no part of a low-prestige field. Finally, in 1946, anesthesiology became a medical specialty that required residency training. In 1949, Apgar was appointed full professor, the first woman ever at Columbia P&S to reach this rank.

Apgar studied the effects on newborns of anesthesia given to their mothers during labor. She originally designed her test in 1952 as a guide for whether a newborn needed to be resuscitated. Now known as the APGAR test, it has held up over decades of use and study. Data from more than 17,000 babies have demonstrated that Apgar scores, especially taken five minutes after birth, can predict with great accuracy both newborn survival and neurological development.

In 1959, Dr. Apgar decided to leave academic medicine so she could devote herself to the prevention of birth defects. She worked diligently to educate the public and to raise money for research. Perhaps we are fortunate that she did not become a surgeon after all. ::

6.8   What Tests Are on the Horizon?

Scientists continue to develop screening methods that are less invasive and more accurate. One new idea takes advantage of the discovery that fetal DNA, not enclosed in cells, circulates in the mother's bloodstream. Scientists can collect this DNA and make many copies of it (Chapter 7). The fetal DNA can then be examined for genetic disorders.

Scientists are also exploring whether **gene therapy**, the correction of a faulty gene, could be possible for embryos (Chapter 8). We can already screen for genetic disorders in embryos produced by IVF and related methods. Could gene therapy fix these defects so that the embryos will develop into healthy babies?

gene therapy :: the correction of a faulty gene

We can all agree on the goal of healthy babies. OK, but what does *healthy* mean? Biological and technological breakthroughs appear inevitable. But will everyone have access to these technologies? Would Caster Semenya be alive today if they did, and what sex would she be? And who pays the bill? We return to these and other issues in the next two chapters.

Biology in Perspective

Reproduction is the most important thing we do; it ensures the survival of our species. And one of the first questions asked upon hearing of the birth of a baby is whether it is a boy or girl. Perhaps we are so fascinated with sex and gender because reproduction matters so much.

Like any system in the body, the reproductive system usually develops in a fixed way. So do sexual characteristics. But not always. Individuals with disorders of sexual development (DSD) are often seen as profoundly different from others. A defect in the circulatory system or a heart problem does not often come with the same stigma. Why? What is it about sex and gender that we find so defining? Reproduction seems to be about more than just biology.

Biologically, the reproductive system's job is to produce offspring. Men produce sperm, women eggs, and if hormones and everything else line up, fertilization and pregnancy result. People spend a lot of effort and money to control pregnancy. Some try to prevent it. Infertile couples try to achieve it.

Once a welcome pregnancy has been established, we have a responsibility to do what we can to protect the health of the embryo and fetus. And when the baby is born, our responsibility expands to include this new member of the human family.

Chapter Summary

6.1 How Do Males and Females Form?

Describe how chromosome and hormone instructions interact to determine sex.

- Different species employ different methods of determining sex, including chromosomes, temperature, and chemical cues.
- Human primary sex characteristics are determined by sex chromosomes. Genes in female chromosomes (XX) result in ovaries; genes in male chromosomes (XY) result in testes.
- Other gender-specific characteristics (secondary sex determination) are the result of hormones produced by the ovaries (estrogen) and testes (testosterone).

6.2 What Happens If the Hormonal Signals Are Missing or Misread?

Predict the outcome of a change in hormone production or response.

- Defective hormonal signals and production, or signals that are misread, can result in disorders of sexual development (DSD).
- In AIS, receptors for testosterone are missing or too few and individuals develop testicles instead of ovaries but have female secondary sexual characteristics.
- In one form of pseudohermaphroditism, an enzyme that converts testosterone to DHT is defective. Pseudohermaphrodites with this condition initially develop female secondary sex characteristics but at puberty develop male secondary sexual characteristics because of increased testosterone production.

6.3 How Do Men Produce Sperm?

Outline the steps of sperm production in males.

- Stem cells in the seminiferous tubules of the testes produce sperm through meiosis.
- Immature sperm leave the testes and enter the epididymis, where they complete differentiation and are stored until sexual arousal.

- When sexually aroused, sperm are pushed into the vas deferens, where they encounter substances from different glands that stimulate swimming, neutralize acidity of the vagina, and stimulate contraction in the female reproductive tract.
- The final mix of fluid and sperm is called semen.
- In response to testosterone levels, the hypothalamus releases the hormone GnRH that stimulates the pituitary gland to produce FSH and LH.
- In males, FSH stimulates sperm production and LH stimulates testosterone production.

6.4 How Do Women Produce Eggs?

Outline the steps of egg production in females.

- Women are born with all the primary oocytes they will have.
- The two ovaries present in women alternate, producing an egg each month through a process called ovulation.
- In ovulation, an egg completes meiosis I, stops at meiosis II, and bursts from the ovary. The egg is swept into the oviduct (fallopian tubes) to the uterus.
- If an egg fuses with a sperm, it completes meiosis II, producing a mature egg.
- In the ovarian cycle, the hypothalamus secretes GnRH, stimulating the pituitary to release FSH and LH.
- In women, FSH causes estrogen secretion and a layer of cells surrounding an oocyte to grow and burst out of the ovary.
- The ruptured follicle develops into the corpus luteum, which secretes estrogen and progesterone, changing the lining of the uterus in preparation for pregnancy and preventing ovulation of other oocytes.
- Without pregnancy, the corpus luteum breaks down and restarts the ovarian cycle.
- The uterine cycle has three phases: menstruation, ovulation, and luteal.

6.5 How Can Pregnancy Be Prevented?
Compare methods used to prevent pregnancy.
- Keeping the egg and sperm apart prevents pregnancy.
- Contraceptive methods include surgery (blocks passage of the egg), hormones (inhibit follicle development or prevent endometrial formation), a physical barrier (blocks passage of sperm), or other methods (kill sperm or slow the development of the endometrium).

6.6 What Causes Infertility, and How Can It Be Treated?
Predict causes of infertility and propose potential treatments.
- Infertility is the failure to conceive after one year. Its cause can be traced to males and females equally.
- Male infertility results from low sperm count or few healthy sperm. It has been linked to smoking, alcohol, drug use, and damage to testes.
- Female infertility occurs when ovulation is disrupted, oviducts are blocked, or the endometrium does not develop properly.
- Assisted reproductive technology, including artificial insemination, intrauterine insemination, and *in vitro* fertilization can be used to help infertile couples conceive a child.
- Other solutions for infertility include surrogacy and fertility drugs.

6.7 How Can We Tell If a Fetus or Baby Is Healthy?
Explain the purpose of and methods for testing fetal and newborn health.
- Blood and urine tests can find conditions that may harm a fetus.
- Screening tests help determine whether a fetus is at risk, whereas diagnostic ones are more precise tests for specific conditions. Tests include ultrasound, amniocentesis, and chorionic villus sampling.
- Our ability to detect defects is greater than our ability to treat many defects.
- The APGAR scale represents a newborn's general health and awareness immediately following birth.

6.8 What Tests Are on the Horizon?
Evaluate the ramifications of further refinements in genetic tests and treatments.
- Fetal DNA may be sampled from fetal blood. In the future, this DNA could be tested to find genetic disorders.
- Gene therapy could potentially be used to correct genetic disorders early so that they never develop.

Key Terms

Review Questions

1. List three different methods for determining sex in species. Give an example for each.

2. Diagram the process of primary and secondary sex determination in human females. Be sure to include the hormones and organs involved.

3. Diagram the process of primary and secondary sex determination in human males. Be sure to include the hormones and organs involved.

4. What is primary sex determination?

   a. A step in sexual development in which genitalia are different with respect to the number of chromosomes associated with the male or female sex

   b. A step in sexual development that governs sex-related body characteristics

   c. A step in sexual development that determines whether ovaries or testes form

   d. A step in sexual development that determines whether external genitals form, the structure of the pelvic bone, voice tone, and locations of body hair

   e. A step in sexual development that initiates the production of hormones

5. What is androgen-insensitivity syndrome?

   a. A disorder of sexual development in which an individual who is XY does not develop as a male because their cells lack receptors to bind to testosterone

   b. A disorder of sexual development in which an individual starts life as a girl and then develops into a man at puberty

   c. A disorder of sexual development in which an individual develops both male and female genitalia

   d. A disorder of sexual development in which an individual who is XX develops as a male

   e. A disorder of sexual development in which an individual who is XX develops both male and female genitalia

6. How do the formation of eggs and sperm differ? How are the processes alike?

7. Complete the following table for these six sex hormones.

Hormone	Produced by which organ or tissue	Effect on males	Effect on females
GnRH			
FSH			
LH			
Testosterone			
Estrogen			
Progesterone			

8. How do hormones control how sperm form?

9. How do hormones control oogenesis and the uterine cycle?

10. What happens to the corpus luteum when there is a pregnancy? What happens to it when there is no pregnancy?

11. How does a bipotential embryo become male or female? (Consider both its internal and external structures.)

12. How can pregnancy be prevented? How effective is each method?

13. Which of the following is NOT a barrier method of birth control?

    a. Cervical cap
    b. Diaphragm
    c. Male condom
    d. Intrauterine device
    e. Female condom

14. What are possible causes of infertility in men? In women?

15. In which type of infertility treatment are sperm placed directly into the uterus?

    a. *In vitro* fertilization
    b. Intrauterine insemination
    c. Artificial insemination
    d. Gamete intrafallopian transfer
    e. Intracytoplasmic sperm injection

16. What prenatal tests are performed during pregnancy?

17. How is the health of a newborn evaluated?

The Thinking Citizen

1. If you were trying to develop a male birth control pill, what aspects of a man's reproductive biology might you target? Explain why.

2. What are the potential advantages for a fish species that changes sex over the course of its life cycle?

3. Some individuals are born with a defective enzyme that results in production of testosterone. If an individual is born XX with this defective enzyme, would this person have testicles or ovaries? What secondary sex characteristics would the individual have? Would the person be fertile or infertile? Be sure to explain your answers.

4. In the 1960s it came to light that some male children had XX chromosomes. An attempt was made to reassign their gender to female. These efforts were largely unsuccessful. Would you consider someone who looked male but had XX chromosomes male or female? What about individuals who are pseudohermaphrodites? Explain your reasoning for each.

5. BPA is a chemical used in plastics that mimics estrogen. How might BPA in baby bottles disrupt sperm formation? What effect might BPA have on preadolescent females?

6. A couple comes in to a clinic because they have been unable to conceive a child. What would you check on the man and on the woman? What options would you suggest if the problem was that sperm count was low and some sperm were deformed?

7. Why doesn't assisted reproductive technology (ART) always work? What are possible reasons for the failure of these methods to produce a healthy baby?

The Informed Citizen

1. Assisted reproductive technology (ART) such as *in vitro* fertilization generates embryos that will never be placed in a woman's body. What should be done with these embryos? Who should decide? Who "owns" them?

2. Is there a *right* to reproduce biological offspring? Why or why not?

3. Should health insurance companies be required to cover the costs of infertility treatments and ART? Why or why not?

4. Should parents be permitted to refuse all prenatal and newborn screening and testing? Why or why not?

5. What rights and responsibilities should sperm or egg donors or surrogates have for a child born as a result of their contribution?

6. Gender is an important category in our society. Not everyone fits neatly into "male" or "female." Who decides the gender of an individual? The person? Society? A medical professional? Is it always essential to assign gender? Why or why not?

7. Huntington's disease is a neurodegenerative genetic disorder that causes mental decline generally in middle age, along with other physical symptoms. Knowing that there is currently no treatment for Huntington's, would you test your fetus? Why or why not?

CHAPTER LEARNING OBJECTIVES

After reading this chapter, you should be able to answer the following questions:

7.1 Why Are Plants Such Good Sources of Food?
Explain how the bodies of plants are structured and specialized to store food molecules.

7.2 How Do Plants Make Food?
Summarize the light reactions and Calvin cycle phases of photosynthesis.

7.3 What Are the Goals of Genetic Engineering in Plants?
Describe the goals for genetic engineering of crop plants.

7.4 How Was Golden Rice Engineered?
Outline the steps used to produce a genetically modified organism.

7.5 How Else Is Genetic Engineering Being Used?
Describe how areas other than agriculture use genetic engineering.

7.6 What Are the Risks of Genetic Engineering?
Assess the risks of genetic engineering in terms of safety, the economy, and effectiveness.

7.7 How Ethical Is Genetic Engineering?
Evaluate the ethical considerations related to genetic engineering that confront scientists.

With modern genetic engineering, we can take genes from almost any species and insert them into almost any other species, but questions about the wisdom and safety of doing so persist. Bell peppers, *Capsicum annuum*, are just one of many plants species that have been genetically engineered.

CHAPTER 7

Plants, Agriculture, and Genetic Engineering

Can We Create Better Plants and Animals?

Becoming an Informed Citizen …

Domesticating plants and animals is a form of genetic engineering we have been using for over 10,000 years. Today, we can take genes from almost any species and insert them into almost any other species to create new features. But is genetic engineering safe, and is it wise?

enetic modification has been going on for billions of years—for as long as living organisms have been on Earth. Every day, each organism faces challenges to its survival and reproduction. Natural selection occurs because some individuals are more successful than others at meeting their daily challenges and passing on genes to the next generation. As a result, each generation differs genetically from the previous one, and members of each species adapt by becoming better at survival and reproduction. All species evolve, and evolution is a naturally occurring form of genetic engineering.

Domesticating plants and animals is a deliberate form of genetic modification that we have been using for over 10,000 years. Human intervention leads to new combinations of genes in organisms. With modern **genetic engineering**, however, we can accomplish far more in much less time. We can take genes from almost any species and insert them into almost any other species—at least in theory. With thousands of genes and millions of species to work with, we can create an almost unlimited number of new living features. This natural treasure trove of genetic information is one of the main reasons why many scientists want to protect biodiversity on our planet: if we lose species, we also lose the genetic information they carry. But can genetic engineering be done safely, and is it wise? Consider how two scientists used genetic engineering in the hope of feeding the planet.

genetic engineering :: the process of taking genes from one species and inserting them into another species

case study

Golden Rice

It all began in 1984 over beers at an international agricultural meeting in the Philippines. Gary Toenniessen of the Rockefeller Foundation asked a group of plant scientists this question: *how can we use the new technology of genetic engineering to improve rice?* Since rice is the principal food staple for billions of people, making it more nutritious would benefit a significant percentage of the world's population, including many in underdeveloped countries.

The scientists put their heads together and agreed on the answer: *engineer rice so that it can be a source of vitamin A.* This idea immediately generated great enthusiasm. Why? At least 750 million children in the world have diets that are deficient in vitamin A (**FIGURE 7.1**). Vitamin A deficiency (VAD) can lead to terrible consequences. According to the World Health Organization (WHO), 250,000 to 500,000 children go blind every year because VAD damages eyes. Half these children may die each year because VAD also increases susceptibility to many infectious diseases.

Vitamin A is found in many foods: carrots, tomatoes, meat, and milk, for example. But many economically disadvantaged people do not have access to balanced nutrition, and among these people VAD is a persistent problem. Most cases of VAD occur in parts of Asia, where rice is the major source of calories. Rice fortified with vitamin A could reduce malnourishment and help save lives.

FIGURE 7.1 **At least 750 million children in the world have diets that are deficient in vitamin A, but in countries with high GDP, this deficiency is not of public health significance.** Insufficient vitamin A in the diet can lead to serious health problems, including xerophthalmia; left untreated, this disorder may lead to blindness.

It's a long journey from an idea hatched over beers to a scientific payoff. At the time, the only way to get rice to make a biochemical precursor of vitamin A was to take genes from a species that can make it and insert them into rice—a formidable task. When eaten, this precursor can be converted into Vitamin A by the body. Toenniessen's question lay on the table for almost a decade. A chance meeting provided the spark needed for the answer.

Ingo Potrykus, a German scientist living in Switzerland, and Peter Beyer, a German scientist living in Germany, met in 1992 on a plane as they traveled to New York City to attend a conference organized by the Rockefeller Foundation. Potrykus was an expert at manipulating genes in rice, and Beyer was a specialist in the biochemistry of vitamin A production. They realized that their complementary specialties put them in a unique position to solve the problem of making rice a source of vitamin A.

Together they wrote an ambitious proposal to insert the genes for making vitamin A into rice seeds. The Rockefeller Foundation took a chance and funded them. And in 1999, *golden rice* was born.

Potrykus and Beyer pioneered many techniques and made discoveries that advanced the emerging field of genetic engineering. However, the genetic engineering used to develop golden rice and other crops has raised complex questions about safety, ethics, and other considerations. In this chapter, we will examine how genetic engineering works and how these techniques are being applied. We will also address an important social issue: what *should* we do with our new creative powers? But first we will take a look at how plants work to better understand why they make such good targets for genetic engineering.

Why Are Plants Such Good Sources of Food?

Plants come in a bewildering variety of sizes and shapes. Flowering plants are the largest group, with over 250,000 species (**FIGURE 7.2**). Nonflowering plants such as ferns, mosses, and evergreen trees like pine and fir provide us with many important resources. But most of the food we derive from plants comes from flowering ones, which is where we focus our discussion.

Plants are an excellent source of food because they are able to capture solar energy and turn it into food—energy-containing chemical compounds that our cells can use to do their work. Plants make food for their own purposes, not ours. Luckily plants and animals can use many of the same types of energy-containing compounds, so we are able to use these molecules almost as well as plants can. Photosynthesis is the process plants use to convert solar energy to food molecules; we will discuss this later in the chapter.

Plants also do a great job storing the food molecules they produce. They use some of them right away, but they store the rest in specific locations throughout their bodies—in roots, stems, leaves, or seeds, for example. This behavior provides us with specific packages of food, such as the starch in rice or potatoes, which we can harvest and eat without having to consume the entire plant. In this section, we will discuss how the bodies of plants are structured, and how different parts of their bodies are specialized to store food molecules.

Despite their astonishing diversity, the basic structure is similar for all flowering plants. They have four types of organs: roots, stems (including branches), leaves, and flowers that produce seeds and fruit (**FIGURE 7.3**). These organs are interconnected by two vascular, or internal transport, systems. The first system of tubes, xylem, transports water and minerals throughout the plant. The second, phloem, transports food and other chemicals like hormones. Each of the four organs can be used as a storage organ for vitamins and edible compounds such as starch and sugar.

Roots

root :: an organ that anchors a plant to its surface and enables it to absorb water and nutrients from the soil

root hair :: tiny lateral extension of a root's outer cells that absorbs water and minerals from the soil

The main functions of **roots** are to anchor the plant to its surface and to absorb water and nutrients from the soil. Typically (but not always) roots are below ground, and they grow downward and away from the light—the opposite of what the aboveground parts do.

The larger, older parts of roots do most of the anchoring. As roots grow into the soil, the younger growing tips produce a dense array of **root hairs**: tiny lateral extensions of the root's outer cells. Root hairs absorb water and

A B C D E F

FIGURE 7.2 There are more than 250,000 species of flowering plants. Some examples include: carrot (A), daisy (B), orchid (C), apple (D), tulip (E), and rose (F).

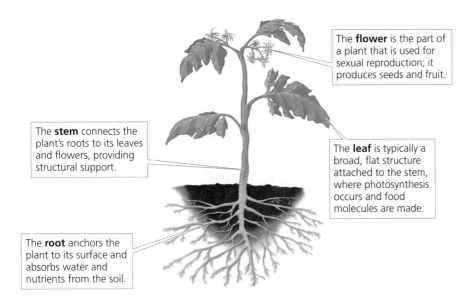

The **flower** is the part of a plant that is used for sexual reproduction; it produces seeds and fruit.

The **stem** connects the plant's roots to its leaves and flowers, providing structural support.

The **leaf** is typically a broad, flat structure attached to the stem, where photosynthesis occurs and food molecules are made.

The **root** anchors the plant to its surface and absorbs water and nutrients from the soil.

FIGURE 7.3 Flowering plants have four types of organs: roots, stems (including branches), leaves, and flowers.

minerals from the soil. Because there are so many of them, they have a very large surface area, with lots of soil contact. This allows root hairs to absorb as much water and as many minerals as possible (**FIGURE 7.4**). Rice plants, for example, have a fibrous root network that spreads out for maximum water and mineral absorption; more than 90% of the roots go no deeper than 20 cm (8 in.).

Many plants store food molecules, especially starches, in their roots. Since many plants lose their aboveground parts each winter, root storage is the only reliable location. From the plant's perspective, this food reserve supplies the energy to grow new aboveground parts in the spring, or perhaps to make flowers and seeds. From our perspective, root storage organs provide an important food source. We eat a lot of roots, including beets, carrots, parsnips, radishes, rutabagas, sweet potatoes, and turnips.

Roots are not always below ground; nor are they always recognizable as roots. Tropical strangler figs, for example, begin their lives as seeds germinating in the upper branches of other trees. They send aerial roots down to the soil, and as these roots grow, they surround the original trunk. Eventually the original tree may die and rot away, leaving a hollow cylinder of strangler fig roots as the "trunk." Also in rainforests, dirt and organic matter can accumulate on the upper branches of trees, forming what is essentially a mat of soil many meters above the forest floor. Many trees grow roots from their branches to penetrate these mats and extract rainwater and nutrients from them, just as they do from the forest floor.

FIGURE 7.4 Root hairs absorb water and minerals from the soil. Numerous fine root hairs on this radish seedling increase the absorptive surface area of the root.

Stems

Stems connect the plant's roots to its leaves and flowers. Stems provide structural support for the plant, holding all the parts together and defining the plant's overall shape. Stems allow plants to grow taller and compete for sunlight. The vascular system in stems ties the plant parts together physiologically. Their **xylem** transports water and minerals from roots to leaves and flowers, and the **phloem** carries food made in the leaves through the stems to the roots and other parts of the plant.

Although all plant stems share the same basic components, stem structure varies in different species. For example, the main stem of rice plants is round

stem :: connects a plant's roots to its leaves and flowers; it provides the structural support for the plant

xylem :: part of a stem that transports water and minerals from the roots to the leaves and flowers

phloem :: part of the stem that carries food made in the leaves through the stems to the roots and other parts of the plant

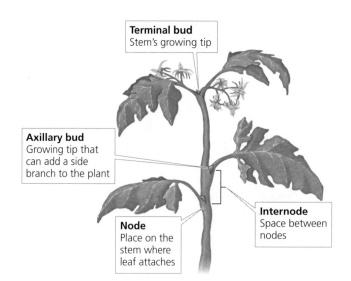

FIGURE 7.5 Plant stems connect plant roots to leaves and flowers and provide structural support. At various points along their length, stems have nodes.

FIGURE 7.6 Buds are the plant's insurance policy, a way to start up new growth if anything happens to the existing parts. If the terminal bud is damaged or does not exist, the axillary buds are released, and the plant may become bushier.

and hollow. At various points along their length, stems have nodes (**FIGURE 7.5**). Each node has a leaf and a structure called a bud. Some buds grow into leafy stems that become the plant's side branches, while other buds grow into flowers. There is also a bud at the tip of each stem, and as this bud grows, the stem itself gets longer.

The plant's shape is determined by which buds grow and by how long each stem becomes. In many plants, the bud at the stem's tip (called the terminal bud) "dominates" other buds on the stem, preventing them from growing. This allows the plant to elongate and grow taller to reach the sunlight. If the terminal bud is damaged—eaten by an herbivore or pinched off by a gardener, for example—the side buds are released from domination and the plant may become bushier (**FIGURE 7.6**). In a sense, buds are the plant's insurance policy, a way to start up new growth if anything happens to the existing parts.

Some stems are modified as storage organs, similar to what happens to some roots. Potatoes, which store starch, are the most familiar case. Potatoes are underground stems called tubers, and each "eye" (indentation in the potato) is actually a node from which a leaf and new stem can grow. Other stems are sources of nutrients, vitamins, fiber, and other useful products. Asparagus, bamboo shoots, cactus pads, and water chestnuts fall into this category. Sugar comes from the stems of sugarcane and maple syrup from the "stem" (trunk) of sugar maple trees. Aspirin, cinnamon, curare, and quinine come from the bark of tree stems. And although we can't eat it, wood comes from the stems of trees, and fibers from stems of flax and hemp are used to make textiles.

Like roots, stems can be modified in many ways by different species of plants. In some species, underground stems called rhizomes grow horizontally just below the surface to help plants spread out. Occasionally a rhizome sends up a vertical stem, which we see as the "plant." Irises and ginger have rhizomes. Other horizontal stems, called stolons, lie just above the ground surface, like the "runners" of strawberry plants. Stems can also be broad and flat, taking on many of the functions of leaves. Cactus pads are a good example.

Leaves

Leaves typically are broad, flat structures attached to stems by a stalk; in some species, such as rice, the leaves wrap around as a sheath. Each species has a characteristic leaf shape, and often leaf shape helps to identify plants.

leaf :: a broad or long, flat structure attached to the stems by a stalk; it is the place in the plant where photosynthesis occurs and food molecules are made

Leaves are the plant's biochemical factories, where photosynthesis occurs and food molecules are made. This is one of the reasons leaves are so nutritious.

Like roots and stems, leaves sometimes are modified to act as storage organs. When leaves serve as storage organs, they typically are located underground. **Bulbs** are short, underground stems with fleshy leaves that store nutrients and food. Onions and garlic are two familiar examples. The spicy chemicals they store in their bulbs prevent herbivores from eating them. Ironically, we eat them precisely because they have these spicy chemicals. In succulent plants, such as aloe, ice plant, and stone plants, found in arid places, the leaves are modified to store water.

In most cases, we eat regular aboveground leaves. We eat leaves mainly for their nutrients, vitamins, and fiber rather than for their food storage. The list of "leafy vegetables" is a long one and includes arugula, bok choy, cabbage, chard, cress, endive, lettuce, and spinach. Celery and rhubarb are actually the stalks of these plants' leaves (rhubarb leaves themselves are toxic). Many of the herbs we use to flavor our food are leaves as well.

Leaves are modified to perform many different functions (**FIGURE 7.7**). In cactus, the stems take on the photosynthetic function of leaves, and the leaves are modified as spines. In some climbing plants, all or part of the leaf is modified to produce a tendril that anchors the plant onto whatever it climbs. Some of the most dramatically modified leaves are found in butterworts, pitcher plants, sundews, and Venus flytraps. These plants are all carnivorous, and their leaves are adapted in various ways to capture and digest small insects and other prey.

Flowers

Flowers come in a delightful array of colors, sizes, shapes, and smells, and each species has its own unique blossom. Technically, flowers are not organs in their own right. Unlike other plant organs, flowers are used for only one purpose: sexual reproduction.

From an evolutionary perspective, reproduction is the goal of every living creature, and plants are no exception. In the case of plant flowers, reproductive energy is directed toward two purposes: providing offspring with a food supply to use when the offspring are very small (in seeds), and providing

bulb :: short, underground stem with fleshy leaves that store nutrients and food

flower :: part of a plant that is used for sexual reproduction; it produces seeds and fruit

In the cactus plant, the leaves are modified as spines, which protect the plant from being eaten. It is the stem that does photosynthesis.

In the Venus flytrap, leaves are adapted in various ways to capture and digest small insects and other prey.

In the pea plant, leaves are modified to produce tendrils that anchor the plant to whatever it climbs on.

A   B   C

FIGURE 7.7 Leaves are the plant's biochemical factories, where photosynthesis occurs and food molecules are made. Leaves come in many different shapes and sizes and are modified to perform different functions.

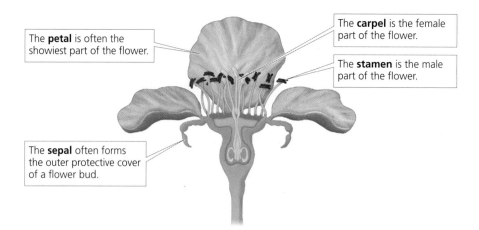

The **petal** is often the showiest part of the flower.

The **carpel** is the female part of the flower.

The **stamen** is the male part of the flower.

The **sepal** often forms the outer protective cover of a flower bud.

FIGURE 7.8 Flowers are the parts of the plant used for sexual reproduction. Flowers are derived evolutionarily from modified leaves and consist of four whorls: sepals, petals, stamens, and carpels.

sepal :: the outer protective cover of a flower bud

petal :: showiest part of the flower

stamen :: the male part of a flower

carpel :: the female part of a flower

seed :: a plant embryo surrounded by a protective coating

endosperm :: cells that surround a plant embryo and form the bulk of the seed

seed coat :: a protective cover over the seed that keeps it from drying out

animals with a food supply to entice them to disperse young plants to new homes (in fruits).

Flowers are derived evolutionarily from highly modified leaves, on a short stem. In fact, flowers consist of four sets, or whorls, of leaves, each modified for a different purpose (**FIGURE 7.8**). The outermost whorl makes the **sepals**, which often form the outer protective cover of a flower bud. The next whorl forms the **petals**, often the showiest part of the flower. The innermost whorls are, in order, the **stamens** and the **carpels**, which form the male and female parts of the flower.

While many flowers are edible (broccoli and cauliflower are examples), their main nutritional value comes from seeds, such as rice, and fruit. Their importance cannot be overemphasized: seeds and fruit form a large portion of the diet for virtually the entire human population.

Seeds A plant **seed** (**FIGURE 7.9**) contains a tiny embryonic plant, complete with a root, stem, and one or two leaves. Surrounding the embryo are cells called **endosperm**, which form the bulk of the seed. The **seed coat** forms a protective cover over the seed to prevent it from drying out.

When a plant makes a seed, it pumps the endosperm full of food molecules—sugar, starch, proteins and amino acids, oils, and other nutrients. The embryonic plant digests food in the endosperm when the seed germinates (begins to grow), and the endosperm provides energy until the young seedling is big enough to make its own food.

The endosperm contains everything a plant needs to begin its life. Luckily for us, we can use the endosperm for food as well. Corn kernels and beans are

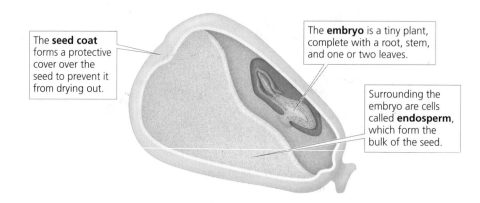

The **seed coat** forms a protective cover over the seed to prevent it from drying out.

The **embryo** is a tiny plant, complete with a root, stem, and one or two leaves.

Surrounding the embryo are cells called **endosperm**, which form the bulk of the seed.

FIGURE 7.9 A plant seed contains a tiny embryonic plant. Corn seeds house a small embryo and endosperm that is covered by a seed coat.

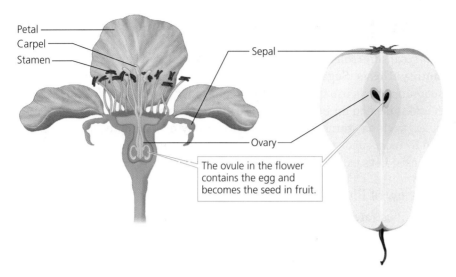

Petal
Carpel
Stamen
Sepal
Ovary

The ovule in the flower contains the egg and becomes the seed in fruit.

FIGURE 7.10 Seeds form inside the ovary of a flower, and as a seed ripens, it develops into a fruit. Botanically, any structure that develops from a flower's ovary is a fruit.

fruit :: any structure that develops from a flower's ovary

Dandelion fruits help seeds disperse by wind.

Coconuts are fruits that float around the ocean to disperse.

Touch-me-nots have fruits that explode and spray out seeds like tiny missiles.

Burdock fruits have tiny hooks that cling to passing mammals.

FIGURE 7.11 Fruits help seeds disperse.

actually seeds, and the endosperm of these two seeds alone comes close to providing a completely balanced diet. The nutrition we get from barley, beans, corn, peanuts, peas, rice, wheat, and many other "seed crops" also comes from the endosperm. We obtain many useful oils from the endosperm of corn, cotton, peanuts, safflower, and other plants. It is safe to say that a large proportion of the diet for the entire human race—and many other animal species—comes from plant endosperm.

Fruits Seeds form inside the female part of the flower, in a structure called the ovary (**FIGURE 7.10**). As the seeds ripen, the ovary develops into a **fruit**. Botanically, any structure that develops from a flower's ovary is a fruit, even things we consider to be vegetables and other structures we don't eat. Tomatoes are fruits, not vegetables, and the little tufts that help dandelion seeds blow in the wind are fruits as well.

While fruits come in many forms, they all have the same function: they help the seeds disperse (**FIGURE 7.11**). Some, like dandelion fruits, maple fruits, and cotton, help seeds disperse by wind. Coconuts are fruits that float around the ocean to disperse. Touch-me-nots have fruits that explode and spray out seeds like tiny missiles. Burdock fruits have tiny hooks that cling to passing mammals (on either fur or clothes), hitching a ride to disperse. Burdock fruits were the inspiration for Velcro.

But the structures we think of as fruits are the ones that are tasty and edible: apples, cherries, oranges, and the like—or maybe even pea pods. From the plant's perspective, the idea behind these types of fruits is simple: they are eaten by animals. The seeds pass through the digestive system unharmed, protected by their seed coat. Since plants are not mobile, they employ the mobility of animals to disperse their seeds. They reward these helpful animals with tasty and nutritious fruit to secure their services. By the time the seeds pass through, the animal has moved to a new location, effectively dispersing them. As an added bonus, the animal "plants" the seeds in a bit of fertilizer to help them get a good start in life. We have capitalized on naturally occurring fruits, selectively breeding plants to grow larger and more nutritious fruits—the ones we buy and eat today.

photosynthesis :: a process that plants use to capture solar energy and then use it to make sugar

7.2 How Do Plants Make Food?

Plants capture solar energy and use it to make sugar, a process called **photosynthesis**. Sugar has a lot of calories, a measure of the stored energy it contains. Cells use the energy stored in sugar to perform the biochemical reactions needed for life. Plants store the extra sugar they make in food molecules, such as starch in rice grains. Directly or indirectly, all the energy used by most species on the planet comes from sunlight captured by photosynthesis. Let's examine how photosynthesis works.

Overview of Photosynthesis

We can illustrate photosynthesis with this chemical "equation":

$$\text{water + carbon dioxide} \xrightarrow{\text{sunlight}} \text{sugar + oxygen}$$

Photosynthesis is a biochemical reaction that converts water and carbon dioxide into sugar. Sunlight provides the energy for the conversion. The result of photosynthesis is that the plant stores some of the solar energy as sugar, and molecules of oxygen are released into the atmosphere.

Sugar is a carbohydrate, which literally means carbon bonded to water. You can see this in the basic formula for a carbohydrate: CH_2O. In carbohydrates, the ratio of carbon to hydrogen to oxygen is 1:2:1. Different carbohydrates have different numbers of carbon atoms, so the formula scales up as needed. The sugar made in photosynthesis is glucose, which has six carbon atoms. Its formula is $C_6H_{12}O_6$ (note the 1:2:1 ratio among the atoms).

Using chemical formulas, we can make our equation more concise and more specific:

$$6H_2O + 6CO_2 \xrightarrow{\text{sunlight}} C_6H_{12}O_6 + 6O_2$$

Looking at the chemical bookkeeping, we see that it takes six molecules of water and six molecules of carbon dioxide to make one molecule of glucose, and six molecules of oxygen are released. Sunlight not only provides the energy for the reactions, but some of this solar energy is stored in the chemical bonds of glucose.

How efficient is photosynthesis? The answer may surprise you. As we'll see, plants are not able to use all sunlight, and only some of the energy they capture is successfully stored in sugar. Photosynthetic efficiency varies among plants and under different conditions, but a reasonable estimate is that the sugar that plants make stores 1–5% of the energy available in sunlight. Plants provide food for many species, but they do so using only a small fraction of the available solar energy.

Light Reactions and the Calvin Cycle

Photosynthesis occurs in two stages: light reactions and the Calvin cycle. Light reactions capture solar energy, and the Calvin cycle uses this energy to make sugar (**FIGURE 7.12**).

Light reactions

1 A molecule of chlorophyll absorbs sunlight, and an electron shoots out.

2 The electron eventually joins with another molecule to produce NADPH.

3 Each electron transfer releases energy, which is used to make ATP.

Light

e⁻

NADPH + ATP

Chloroplast

CO_2

Glucose

Calvin cycle

4 Carbon dioxide from the atmosphere attaches to another carbon-containing compound.

5 Enzymes use the energy from ATP and NADPH to take this new molecule through a series of chemical steps.

6 Eventually, the molecule is converted into energy-rich glucose.

FIGURE 7.12 Photosynthesis occurs in two stages: light reactions and the Calvin cycle. Light reactions capture solar energy, and the Calvin cycle uses this energy to make sugar.

Light reactions use a green pigment called **chlorophyll**, which is found in all the green parts of a plant. Incoming sunlight contains all the colors of the rainbow, as well as other colors that we can't see. Chlorophyll reflects (and therefore does not use) the green wavelength but can absorb both red and blue wavelengths. When chlorophyll absorbs this sunlight, it also absorbs some of the energy it contains.

When a molecule of chlorophyll absorbs sunlight, the solar energy is transferred to an electron, which shoots out of the chlorophyll. This energized electron is the key to capturing solar energy. The electron is transferred to a series of other compounds and eventually ends up in a molecule called NADPH. During each electron transfer, a little energy is released, which is used to make another molecule called ATP. Both NADPH and ATP are storehouses of chemical energy, and both are used in the Calvin cycle to make sugar.

But there is a problem: the chlorophyll molecules now have "holes" where their electrons used to be. Chemical reactions must be balanced, and photosynthesis would eventually stop if these holes were not filled. In one of the light reactions, an enzyme splits apart water molecules, separating the hydrogen from the oxygen. The oxygen is released, eventually becoming part of the atmosphere we breathe. When the water is split, free electrons are produced, which fill the holes in the chlorophyll molecules.

The **Calvin cycle** uses chemical energy produced by the light reactions: ATP and NADPH. As a result, the Calvin cycle does not require sunlight. In the Calvin cycle, carbon dioxide from the atmosphere attaches to another carbon-containing compound. This attachment is called **carbon fixation**. Now that the carbon dioxide has been trapped chemically, enzymes take this new molecule through a series of chemical steps. During these steps, energy is added to the developing molecule. This energy comes from the ATP and NADPH generated by the light reactions. Eventually, the molecule is converted into energy-rich glucose. Glucose lies at the center of many biochemical pathways in the cell. It can be used directly for energy, or it can be converted into many other compounds that cells need.

light reaction :: the first stage of photosynthesis in which solar energy is captured

chlorophyll :: a green pigment used by plants in photosynthesis

Calvin cycle :: the second stage of photosynthesis in which the solar energy captured from light reactions is used to make sugar

carbon fixation :: the attachment of carbon dioxide from the atmosphere to another carbon-containing compound

Chloroplasts have a **double membrane** that separates them from the rest of the cell.

Light reactions take place inside the **thylakoids**.

The Calvin cycle occurs in the **stroma**.

FIGURE 7.13 Photosynthesis occurs inside the chloroplasts.

chloroplast :: an organelle in a photosynthetic cell in which photosynthesis takes place

The Role of Cells in Photosynthesis

Each photosynthetic cell contains numerous organelles called **chloroplasts** (**FIGURE 7.13**). Chloroplasts have a double membrane separating them from the rest of the cell. Within the membrane are stacks of wafers, rather like stacks of poker chips. The individual wafers are called thylakoids, and the stacks are called grana (singular granum). The space inside the membrane surrounding the grana is filled with a fluid called stroma.

Photosynthesis occurs inside the chloroplasts. The thylakoids are green, and they are packed with chlorophyll molecules. The light reactions occur on the thylakoids. ATP and NADPH made by the light reactions move into the stroma. The reactions of the Calvin cycle take place in the stroma.

The Role of Leaves in Photosynthesis

Photosynthesis requires water, carbon dioxide, and sunlight. The main function of a leaf is to provide these commodities so leaf cells can photosynthesize.

The shape of a leaf promotes light capture. Most (though certainly not all) leaves are broad and thin or like rice, long and thin. Their width or length provides a large surface for capturing light. Most cells are close to the surface or just a few cells down, so all have access to sunlight and don't shade each other.

The internal structure of a leaf (**FIGURE 7.14**) helps water and carbon dioxide to circulate around the photosynthesizing cells. The inner cells are loosely packed and spongy, with space around them for gases to circulate. Water

The **phloem** carries food made in the leaves through the stems to the roots.

Xylem vessels transport water from the roots to the leaves.

The **waxy cuticle** makes the leaf waterproof and prevents water vapor from escaping inside the leaf.

The **inner cells** are loosely packed and spongy.

The **stomata** open to allow carbon dioxide to enter but also close when water is in short supply.

FIGURE 7.14 The internal structure of a leaf helps water and carbon dioxide circulate around the photosynthesizing cells.

enters the leaf through xylem vessels in its stalk. Water evaporates into the inner part of the leaf and diffuses to photosynthesizing cells. The humidity inside a leaf is close to 100% as long as water is available.

The outer layer of a leaf helps the plant conserve water and also controls the uptake of carbon dioxide. The surface cells are packed tightly together and secrete a layer of wax. This makes the leaf waterproof, and water vapor can't escape from inside the leaf. At various places on the leaf are small pores called stomata. These pores open to allow carbon dioxide to enter and diffuse throughout the inner spongy layer. When stomata are open, water vapor can be lost by evaporation, a necessary side effect of allowing carbon dioxide in. However, stomata can close when water is in short supply, and they often close at night when darkness prevents the light reactions of photosynthesis from occurring. Closed stomata shut down photosynthesis, but they do help conserve water.

7.3 What Are the Goals of Genetic Engineering in Plants?

Plants have the capacity to do something humans cannot—they can capture energy from the sun and transform it into food molecules. As a result, we depend upon plants for our survival. In early human history, people gathered and ate edible plants. With the development of agriculture, people learned to selectively breed plants to develop desirable traits. Genetic engineering promises the possibility of generating plants with traits that cannot be acquired through breeding.

Potrykus and Beyer had very clear social goals with their genetic engineering: they wanted to distribute golden rice to low-income farmers as a humanitarian gesture. They sought to design a **genetically modified organism** (**GMO**) that would be widely available for use. Any farmer anywhere in the world with an annual income less than $10,000 would have access to golden rice for just the cost of the seeds—no licensing fees. Working with a corporation named Syngenta, Potrykus and Beyer set up the Golden Rice Humanitarian Board to oversee the improvement, licensing, and distribution of golden rice.

genetically modified organism (GMO) :: an organism whose genetic material has been altered

On the other hand, many corporations see huge profits in genetically engineered crops. These crops are big business in the United States, Argentina, Brazil, and other countries, although they are banned in Europe and Japan because of popular opinion and fears about safety. In the United States, well over half of the corn, cotton, soybeans, and alfalfa grown are genetically engineered strains. Given the widespread use of corn syrup, cottonseed oil, and soybeans in processed foods, it is likely that many of the foods we eat come in part from genetically engineered plants.

What exactly is it that genetically engineered crops have to sell? The answer lies mainly in the biological goals that researchers have for crop improvement: pesticide production, herbicide resistance, and increased nutritional value.

Pesticide Production

Every year we lose billions of dollars' worth of crops to insect pests and spend billions to spray crops with pesticides, chemicals that kill insects and other pests. But what if our crops could make their own pesticides?

FIGURE 7.15 The bacterium *Bacillus thuringiensis* (*Bt*) produces toxins that kill several types of pests, including the European corn borer (shown), which causes billions of dollars of corn damage each year. The *Bt* toxin is a protein that binds to the intestinal cells of vulnerable insects and prevents them from feeding, leading to quick death from starvation.

A bacterium named *Bacillus thuringiensis* (*Bt*) produces toxins that kill several types of insect pests. The toxin is very specific and kills only certain caterpillars, beetles, and fly larvae, such as mosquitoes and black flies. This specificity is good because *Bt* does not appear to harm many beneficial insects like pollinators and predators, and it does not appear to harm us. However, the species it does kill are leaf eaters that can devastate entire fields of crops or other vegetation in short order. Insects that are vulnerable to *Bt* toxin include the cabbage worm, tomato and tobacco hornworms, European corn borer (which causes billions of dollars of corn damage annually), spruce and pine budworms, tent caterpillars, and the Colorado potato beetle (**FIGURE 7.15**).

The *Bt* toxin is a protein that binds to the intestinal cells of vulnerable insects and prevents them from feeding, leading to quick death from starvation. Because the toxin is a protein, there is a gene that codes for it. Actually, more than 50 genes for toxins have been discovered in *Bt*, carried by various forms of the bacterium.

A toxin gene provides the starting point for genetically engineering other plants. It is possible to extract the toxin gene from the bacterium and insert it into the cells of a young crop plant. The adult plant then has a working copy of the toxin gene and makes its own pesticide, with no additional work or expense on the farmer's part. The majority of corn and cotton grown in the United States has been genetically engineered to carry the gene for *Bt* toxin.

Herbicide Resistance

Weeds are an even bigger problem for crops than pests. One way we attack weeds is to spray crops with herbicides, chemicals that kill plants. But herbicides can inadvertently harm the crops—weeds and crops are both plants, after all. What if our crops were resistant to an herbicide that killed almost any other type of plant?

glyphosate :: a broad-spectrum herbicide; it is the active ingredient in Roundup

Glyphosate is the most widely used herbicide in the United States. Sold under the name Roundup, it is a wide-spectrum herbicide and kills virtually any plant (**FIGURE 7.16**). Once inside a plant, glyphosate interferes with a vital plant enzyme, one that makes certain amino acids. Without these amino acids, the plant can't make proteins and quickly dies.

Glyphosate is found in many different species, with slight variations. As it turns out, the enzyme in a bacterium named *Agrobacterium* is not affected by glyphosate. Genetic engineers have taken the gene for this resistant enzyme from *Agrobacterium* and inserted it into a variety of crop plants. Now these crops also make the enzyme that is not affected by glyphosate. These genetically engineered plants are marketed as Roundup Ready. If a farmer plants Roundup Ready crops, the fields can be sprayed with Roundup—which kills all the weeds and other plants but leaves the crops unharmed.

Increased Nutritional Value

Almost all the genetically engineered crops grown in the United States have been engineered for pesticide production, herbicide resistance, or both. As we will discuss in Chapter 11, we lose 25–50% of our global food crop to pests and weeds each year—a tremendous expense. If genetic engineering can reduce these losses, the economic and nutritional impact will be significant.

But there are other avenues of research for genetic engineering in plants, each with its own specific goals. One goal is to increase the nutritional value of the food we eat. Golden rice is a good example.

FIGURE 7.16 Glyphosate is the most widely used herbicide in the United States—capable of killing virtually any plant. Shown are weeds that have been treated with the herbicide.

In developing golden rice, Potrykus and Beyer managed to insert two genes that caused the seeds to produce a molecule called beta-carotene, which is the molecular precursor of vitamin A. Our bodies have an enzyme that converts beta-carotene into vitamin A. Beta-carotene is an orange pigment, which gives color to carrots and other fruits and vegetables. Beta-carotene gave the rice a yellowish color, which is how golden rice got its name (**FIGURE 7.17**).

Other plants have been engineered to keep humans healthier and better fed. Some of these crops are commercially available, while others are still in development:

- Soybeans and canola have been engineered to provide a source of omega-3 fatty acids, compounds that may help reduce the risk of heart attack and stroke.
- Genetically engineered virus-resistant squashes are commercially available, as is an engineered form of sugar cane that produces more sugar.
- Scientists have engineered a strain of potato that makes a more edible form of starch.
- Scientists are investigating techniques for engineering drought-resistant and freeze-resistant crops.

Some genetically engineered plants have achieved commercial success. But others have flopped—like the tomato engineered to ripen slowly, allowing it to be transported greater distances after harvesting. It simply didn't taste right. The list of genetically engineered crops gets longer every year.

FIGURE 7.17 In developing golden rice, Potrykus and Beyer inserted two genes that caused the seeds to produce beta-carotene, a molecule that contains an orange pigment and that gives the rice a yellowish color (top) compared with white rice (bottom). Our bodies have an enzyme that converts beta-carotene into vitamin A.

How Was Golden Rice Engineered?

Potrykus and Beyer started the golden rice project in 1992. They had considerable expertise and 20 years of scientific progress with gene manipulation to build on, but it still took seven years to make the first successful strains. What did Potrykus and Beyer have to do to make golden rice?

Each genetic engineering project is unique in some ways. Different species and genes are used, and a different set of biochemical techniques may be needed as a result. But each project seeks to insert working copies of genes from one organism into another, and most genetic engineering projects follow the same basic steps: define the problem, clone the genes, package the genes, transform the cells, and confirm the strain (**FIGURE 7.18**).

Define the Problem

Just like us (and most other organisms), plants need vitamin A, but most plants can make their own. Rice plants make vitamin A in their leaves and other organs, but *not* in the endosperm of their seeds—the part we eat.

Potrykus and Beyer had to figure out why vitamin A was not made in seeds. They discovered that the cells of the endosperm were missing two enzymes, called *psy* and *crt1*, which are needed to make vitamin A. Although rice has the genes for these enzymes, for some reason they are inactivated in

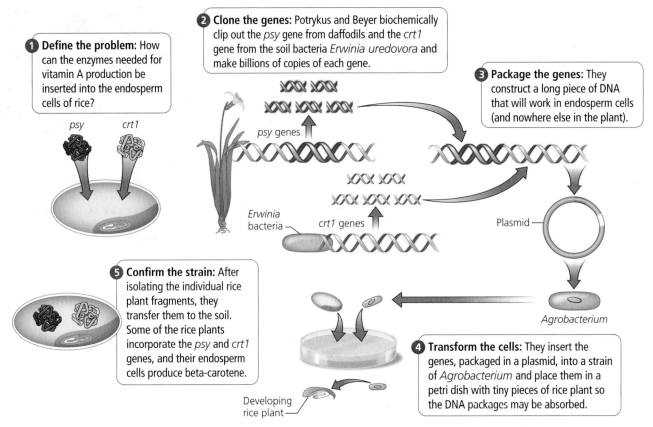

1 Define the problem: How can the enzymes needed for vitamin A production be inserted into the endosperm cells of rice?

psy   *crt1*

2 Clone the genes: Potrykus and Beyer biochemically clip out the *psy* gene from daffodils and the *crt1* gene from the soil bacteria *Erwinia uredovora* and make billions of copies of each gene.

psy genes

Erwinia bacteria   *crt1* genes

3 Package the genes: They construct a long piece of DNA that will work in endosperm cells (and nowhere else in the plant).

Plasmid

Agrobacterium

5 Confirm the strain: After isolating the individual rice plant fragments, they transfer them to the soil. Some of the rice plants incorporate the *psy* and *crt1* genes, and their endosperm cells produce beta-carotene.

Developing rice plant

4 Transform the cells: They insert the genes, packaged in a plasmid, into a strain of *Agrobacterium* and place them in a petri dish with tiny pieces of rice plant so the DNA packages may be absorbed.

FIGURE 7.18 Most genetic engineering projects follow the same basic steps that Potrykus and Beyer followed in the golden rice project: define the problem, clone the genes, package the genes, transform the cells, and confirm the strain.

endosperm cells. Armed with this information, Potrykus and Beyer were able to define their problem: they needed to insert working copies of the *psy* and *crt1* genes into endosperm cells of rice.

Clone the Genes

Since many organisms make vitamin A, there are a number of *psy* and *crt1* genes to choose from. The most obvious choice is to use the genes naturally found in rice. However, the mechanism that turns off these rice genes is effective, and Potrykus and Beyer were unable to turn them on. Instead, Potrykus and Beyer chose to get their *psy* gene from daffodils and their *crt1* gene from a soil bacterium named *Erwinia uredovora*.

Once the source species were selected, Potrykus and Beyer began a long and painstaking process. They had to biochemically clip out the relevant genes from the rest of the DNA in the daffodil and bacterial chromosomes, and they had to make billions of copies of each gene to have enough material to work with. This process of isolating and copying a gene is called **cloning**, and it is a central technique for a genetic engineering project on any type of organism (see *Scientist Spotlight: Kary Mullis*).

cloning :: the process of isolating and copying a gene

Package the Genes

If a gene's DNA is simply inserted into an organism, it is very unlikely to work. Enzymes inside the cell quickly break down the foreign DNA. Also, genes are

Scientist Spotlight

Kary Mullis (1944–)

Born in 1944 in North Carolina, Kary Mullis completed his bachelor's degree in chemistry at the Georgia Institute for Technology in 1966. He went on to graduate school at the University of California at Berkeley, earning his Ph.D. in biochemistry in 1973. Following two postdoctoral research positions, one at the University of Kansas Medical School and the other at the University of California San Francisco, Mullis joined Cetus Corporation in 1979 as a DNA chemist. It was there that he developed a method to produce copies of DNA quickly and efficiently.

Kary Mullis won a Nobel Prize in Chemistry for developing a method to produce copies of DNA quickly and efficiently.

Mullis was driving home after a party one night. His mind was active, and the winding roads with their double white lines got him to thinking about DNA. When a cell divides, each chromosome separates its double strands, and each of these single strands is copied. At this point, the cell has two complete sets of chromosomes—one for each daughter cell. Mullis's revelation was that you could use this process to make a copy of any *part* of a DNA strand: you don't need to copy the entire chromosome. And you could do it in a test tube. And you could make *lots* of copies.

First you define the part of the double-stranded DNA you want copied—perhaps a gene. You heat the DNA up almost to boiling to separate the two strands. Then on each strand, you attach a short piece of DNA just before the beginning and just after the end of the region you want to copy. These short DNA pieces are called primers, and they direct enzymes to where DNA copying should start and stop. The enzymes copy each DNA strand in between the start and stop markers, and the solution is cooled to allow the DNA strands to come together. Now there are two copies of the particular section of DNA you marked.

Mullis's main insight while driving that night was that you don't have to stop with just one round of copying. The enzymes don't get used up, so as long as you have enough raw materials to make DNA (you need the individual nucleotides A, T, G, and C), you can heat up the tube, copy, and cool down again and again. Each time, every DNA copy generates two copies—first two copies, then 4, 8, 16, and so on. Each copying session only takes 5–10 minutes. In half a day, you easily could go through 30 cycles—over a billion copies of the original DNA strand!

Mullis called his process the polymerase chain reaction (PCR), after DNA polymerase, the copying enzyme. PCR is the first step in every genetic engineering project. It allows you to clone the gene—make enough copies to work with. PCR is also the basis for forensic DNA analysis. Running PCR on two DNA samples makes enough DNA so the samples can be compared. This is used for paternity testing (comparing the possible father's DNA to the child's) and for determining whether DNA samples left at a crime scene match a suspect's DNA. PCR is a cornerstone of modern molecular biology. In 1993, Mullis and his colleague Michael Smith won the Nobel Prize in Chemistry for their work on PCR. ::

only one part of the biochemical network that makes an enzyme or other type of protein. For successful insertion, the gene copy must be protected and processed, and become part of a DNA package that the cell can use.

To make a gene package, extra DNA is added at the beginning and end of the gene (**FIGURE 7.19**). A section of DNA is added at the front of the gene to show where the gene starts and to provide a target where enzymes like RNA polymerase can bind to transcribe the DNA. Another section of DNA is added to the end of the gene to mark where the RNA polymerase should stop

FIGURE 7.19 To make a gene package, extra DNA is added to the beginning and end of the gene. The DNA "package" used to make golden rice includes the *psy* and *crt1* genes, which are essential for vitamin A synthesis.

DNA provides a target where RNA polymerase can bind to transcribe the DNA.

Other sequences ensure that the genes will work in the endosperm.

DNA indicates where the RNA polymerase should stop transcribing the gene.

crt1

psy

transcribing the gene. Other sections of DNA may be needed as well, depending on the details of how the gene is expressed and controlled.

Once they had the cloned genes, Potrykus and Beyer had to combine them with these other DNA elements. They had to construct a long piece of DNA that contained the *psy* and *crt1* genes, their start and stop sections, markers that would indicate successful insertion into a cell, and sections of DNA that would ensure that the genes would work in endosperm cells (and nowhere else in the plant).

Potrykus and Beyer constructed biochemical assembly lines that took DNA through a series of reactions, joining together the necessary sections of DNA. When the reactions were done, they had a complete DNA package that included everything needed for the genes to work.

Transform the Cells

After much work, Potrykus and Beyer had a large supply of packaged *psy* genes from daffodils and packaged *crt1* genes from *E. uredovora*. Now they had to get these genes inside rice cells and verify that they worked in the endosperm. When a cell takes in and uses DNA from a foreign source, the process is called **transformation** (see *Technology Connection: How to Transform Cells*). Potrykus and Beyer's next step was to transform rice cells using the foreign *psy* and *crt1* gene packages.

transformation :: the process by which a cell takes in and uses DNA from a foreign source

There are several ways to transform plant cells. Potrykus and Beyer settled on a method that uses *Agrobacterium* (the same bacterium that contributed to the Roundup Ready gene). *Agrobacterium* infects plants by inserting DNA into their cells. That is, *Agrobacterium* does the job that Potrykus and Beyer needed: it transforms plant cells.

Potrykus and Beyer first transformed a strain of *Agrobacterium* by inserting their *psy* and *crt1* gene packages. Then they cut up rice plants into tiny pieces and bathed each piece in a solution of their transformed *Agrobacterium*. Many (though not all) of these rice fragments became infected by *Agrobacterium*, and some of the cells successfully absorbed the DNA packages, completing the transformation.

Confirm the Strain

Potrykus and Beyer isolated the individual rice plant fragments and placed them on a growth medium. The individual fragments grew into whole rice plants, much like taking a cutting from a plant and growing a new one. Eventually these rice plants were transferred to soil, where they flowered and produced seed.

In many cases, the seed was regular rice seed—no beta-carotene, the precursor for vitamin A. In a few cases, however, the process was successful.

Technology Connection

How to Transform Cells

Transformation starts when a cell takes in foreign DNA from the environment. If transformation is successful, the new DNA becomes a working part of the cell's genetic material. The cell now has new genes that make functional proteins. When the cell divides, the new DNA is copied and passed to daughter cells just like the original DNA.

For DNA to get into a cell, it must cross the cell membrane, and possibly a cell wall. There are several ways to do this. Many species of bacteria transform naturally. When cells die in a bacterial habitat, they often split open, releasing their contents and their DNA. Other bacterial cells can take up this DNA, getting new genes for free. Many bacteria can also be transformed easily in the lab, just by exposing them to DNA and adjusting a few conditions like temperature. This makes bacteria good subjects for genetic engineering.

Transforming plant and animal cells is a little more involved because they don't take in foreign DNA naturally. Several techniques are used:

- *Agrobacterium.* This was the technique used to transform rice cells, and it is probably the easiest way. First *Agrobacterium* cells are transformed. Once they have the desired DNA, *Agrobacterium* can inject it directly into plant cells, almost like a virus.

- *Viruses.* It is possible to genetically engineer a virus to carry extra genes used to genetically engineer a plant. When the virus infects a plant cell, it injects its DNA—and the desired genes, too—much like what happens with *Agrobacterium.*

- *Electroporation.* Exposing cells to electric currents opens holes in the cell membrane and cell wall, allowing DNA to enter. This works on bacteria also.

- *Gene "guns."* Small particles of metal (often gold or tungsten) are coated with the transforming DNA. These particles are literally shot into cells using an air gun. The DNA dissolves off the particle, which is later expelled by the cell.

Once the DNA is successfully inside the cell, it must become functional. Sometimes the transforming DNA is in circular form and is called a plasmid. Bacterial cells often

One method for inserting DNA into plant and animal cells is a biolistic particle delivery system, or "gene gun." Small particles of metal are coated with the transforming DNA and literally shot into cells.

have many plasmids floating around inside, so once a bacterial cell is transformed with a plasmid, the transformation is complete.

Plant and animal cells don't typically have plasmids. In fact, these cells interpret foreign DNA as a sign of viral infection, and enzymes usually digest it. In these cells, the introduced DNA has to become part of a chromosome, and this has to happen before the digestive enzymes find it. Cells have enzymes that cut chromosomes, insert DNA, and chemically combine the parts into a seamless stretch of DNA. These enzymes are used to repair chromosomes, but they can also insert foreign DNA into a chromosome. When this happens, the foreign DNA becomes part of the cell's original DNA, and the cell is transformed. ::

These plants had incorporated the *psy* and *crt1* genes, and their endosperm cells produced beta-carotene. Potrykus and Beyer grew and interbred these seeds in the conventional way to produce a true-breeding strain of rice in which all plants produced vitamin A. Golden rice was born.

7.5 How Else Is Genetic Engineering Being Used?

Much of the world is malnourished, and we lose a large percentage of our annual crops to weeds and pests. That's certainly one reason why a lot of genetic engineering effort is directed toward crops. Even before the advent of molecular biology techniques, humans tried to improve crops by selecting for new combinations of genes in plants (see *Life Application: From Teosinte to Maize* to learn how modern corn was developed). But agriculture is only one of several enterprises that are investing heavily in genetic engineering.

Life Application

From Teosinte to Maize

Maybe your image of genetic engineers is people in white lab coats surrounded by test tubes and equipment in a modern genetics lab. But the first genetic engineers dressed much differently and had no equipment other than their hands and the power of observation. One "research team" lived 9000 years ago in what is now southern Mexico.

These people began a genetic engineering program that converted a native plant, teosinte, into maize (or "corn"). Teosinte and maize are closely related subspecies that do not look alike at all. Teosinte is a small, highly branched grass with many "ears." Each ear has only a dozen kernels in only two rows, and each kernel is covered with a hard case, which protects it but makes it hard to digest. When the ears ripen, the kernels break apart, dispersing the seeds. The maize you are most familiar with consists of a single stalk with a single ear, although some types of sweet corn can have two to four ears per stalk. The ear contains many rows of kernels, which are not covered by a case. The kernels stay on the ears when they ripen.

If you think these are big differences, you aren't alone. For decades, botanists debated where maize came from. But genetic analysis along with archeological evidence now gives us a clear picture. Maize developed from a particular type of teosinte that grew in the Central Balsas River Valley in southern Mexico. Despite the fact that teosinte and maize look so different, most of the major differences are due to only five genes (though there are probably hundreds of minor genetic differences among these five genes).

Those ancient people were true genetic engineers: they altered the genetics of one species to produce a new subspecies that was more nutritious and easier to grow. The procedure they used was selective breeding. They selected plants that had useful traits, ate some of them, and crossed the rest with other plants with useful features. The frequency of useful features increased over time, and these engineers guided the evolution of teosinte to maize.

We can make some educated guesses about how the domestication progressed. Harvesters would naturally select for plants that kept their kernels on longer. They would have preferred plants that had the most kernels (more food) and probably the fewest ears (easier to harvest). And they certainly would have preferred kernels with the softest cases. If seed harvesters selected plants according to these preferences, over time teosinte kernels would remain fixed to the ears, fewer and larger ears would be produced per plant, and the hard kernel cases would become softer and perhaps reduced in size. Just the traits we see in modern maize.

Genetic evidence supports these possibilities. One of the five major genetic differences, for example, is a gene that makes the glume: the hard case that covers teosinte kernels. In maize, this gene produces a soft, small glume. It's the part that gets stuck in your teeth when you eat corn on the cob. These early genetic engineers did their jobs well. Today maize accounts for over 20% of human nutrition globally. ::

Teosinte   Modern Corn

Genetic engineering has existed for thousands of years. Over 9000 years ago, ancient people used selective breeding to produce corn that was more nutritious and easier to grow than the existing native plant teosinte.

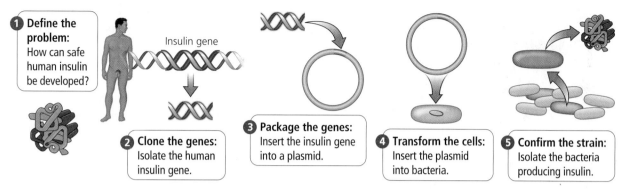

1 Define the problem: How can safe human insulin be developed?

Insulin gene

2 Clone the genes: Isolate the human insulin gene.

3 Package the genes: Insert the insulin gene into a plasmid.

4 Transform the cells: Insert the plasmid into bacteria.

5 Confirm the strain: Isolate the bacteria producing insulin.

FIGURE 7.20 Genetic engineering methods like those used to produce golden rice can also produce medically important products like insulin. Once bacteria-producing insulin are isolated, additional steps must be undertaken to purify it for safe use.

Medicine, industry, and research, among other areas, are looking toward genetic engineering to solve many of our biggest biological challenges.

Medicine

Diabetics used to be treated with insulin from pigs or cattle—human insulin was too expensive and was not available in large quantities. In 1978 Herbert Boyer successfully inserted the human insulin gene into a bacterium, which produced large quantities of human insulin. Today, most of the insulin used is human insulin produced by bacteria, yeast, or other organisms (**FIGURE 7.20**).

Insulin was the first drug produced by a genetically engineered organism. After several decades of research and development, genetically engineered organisms now produce a number of medically and commercially important drugs. The list includes human growth factor, a hormone used to treat dwarfism; blood clotting factors to treat hemophilia; erythropoietin to treat anemia; and vaccines like Gardasil that help prevent cervical cancer.

Bacteria and yeast are used to make most genetically engineered drugs because they are relatively easy to transform and grow quickly. The first genetically engineered drug from an animal was approved in 2009. Scientists inserted the gene for human antithrombin, which reduces blood clotting, into a goat. The protein forms in its milk, and after purification it is injected into patients at risk for life-threatening blood clots.

Industry

Although genetically engineered organisms are often designed to make things, they also can be used to break chemicals down. This occurs during **bioremediation**—the biological cleanup of pollution and contamination. Some bacteria have been engineered to break down oil spills. These organisms were used to clean up the 1989 *Exxon Valdez* spill; the Chinese government released 20 tons of genetically engineered bacteria to help clean up a massive oil spill in the Yellow Sea in 2010; and these bacteria were called upon again to help with the BP Deepwater Horizon spill in the Gulf of Mexico in 2010 (**FIGURE 7.21**).

Bacteria often live in harsh environments—hot, highly acidic, very salty, or low in oxygen. As result, they can be right at home inside many industrial processes, including deep within mines. Some bacteria have been genetically engineered to produce enzymes that dissolve metals from their ores, making the metal easier to extract. This is called **bioleaching**, and it provides an

A

B

FIGURE 7.21 Certain types of naturally occurring and genetically engineered bacteria can break down oil. Oil-eating bacteria were used in the clean up efforts of the 2010 massive oil spills in the Yellow Sea (A) and the Gulf of Mexico (B).

bioremediation :: the biological cleanup of pollution and contamination

bioleaching :: the action of bacteria that have been genetically engineered to produce enzymes that dissolve metals from their ores, making the metal easier to extract

FIGURE 7.22 Strains of mice, known as "knockouts," have had specific genes removed by genetic engineering so that scientists can better study what genes do. The mouse on the left has had the gene affecting hair growth "knocked" out.

A

B

C

FIGURE 7.23 Novelties produced by genetic engineering methods include GloFish (A), blue roses (B), and fluorescent bacteria (C) that can be grown to create pictures in petri dishes.

genome :: all of an organism's DNA and every gene required to make it all work

environmentally safer alternative to some traditional mining techniques. These bacteria are also useful for bioremediation projects involving heavy metal cleanup.

Genetically engineered bacteria make compounds used in the production of plastics, textiles, and cheese. Several companies are working to develop bacteria that digest waste material like woodchips and wheat straw and excrete compounds very similar to crude oil. These bacteria already exist in small quantities in the laboratory, but companies still face challenges producing them on a commercial scale.

Research

The ability to manipulate genes gives researchers many ways to study genes and genetic diseases. For example, there are strains of mice that have been genetically engineered to be highly susceptible to cancer. These mice provide researchers with a large population of subjects on which to test treatments that may ultimately help human cancer patients.

Other strains of mice, known as "knockouts," have had specific genes *removed* by genetic engineering (**FIGURE 7.22**). Scientists try to correlate missing biological functions with the missing genes, which helps them discover what the genes do.

Novelty

Some genetically engineered organisms were developed simply for their novelty (**FIGURE 7.23**). GloFish are genetically modified zebra fish that give off a green, red, or orange glow. Originally, glowing zebra fish were developed for research, with the hope that the glow would help scientists see internal structures better. But now they have become a popular addition to home aquariums—the first genetically engineered animal you can buy in a pet store.

In 2004, after 13 years of research, two companies working together succeeded in inserting a pigment gene from a petunia into a rose. They created a blue rose, a color not found in any rose in nature.

Bioart is a new form of expression that "paints" with living organisms that have been genetically modified to glow or to produce specific colored pigments. One form of bioart uses bacteria that have had genes for fluorescent proteins inserted, similar to GloFish. These bacteria can be grown in patterns on petri dishes to produce colorful displays.

Construction of an Organism's Genome

What would it take to create an entirely new form of life? One approach would be to design and construct the organism's **genome**—all of an organism's DNA and every gene required to make it all work. Then DNA could be removed from a cell, and the entire synthetic genome could be inserted. Would this cell be a new form of life?

We are not yet able to design genes from scratch. But scientists at the J. Craig Venter Institute have taken the first steps in this process. In May 2010, the scientists announced that they had synthesized the entire genome of one species of bacteria. They made an exact copy of the genome in a test tube using genetic engineering procedures. They removed the DNA from cells of a different bacterial species and inserted the synthetic genome. These transformed cells grew and reproduced normally. And they began to take on the characteristics of the species that provided the synthetic genome. Interestingly, these cells were alive but did not have a living parent.

This is not quite the same thing as creating a new form of life. The research team did not design new genes; they copied existing ones. But they did construct the genes themselves, and they demonstrated that a synthetic genome could be inserted successfully into a recipient cell. In the foreseeable future, scientists may be able to design a variety of synthetic genomes by cobbling together genes from many different species. While none of the genes may be new, the cells containing these unique combinations of genes would be different from all existing species. In many ways, these cells could be considered new, artificially created forms of life.

7.6 What Are the Risks of Genetic Engineering?

Ultimately, genetic engineering could produce changes in society similar in scope to the Industrial Revolution. Custom-designed life forms could change the way we handle agriculture, medicine, manufacturing, and many other enterprises. Changes as sweeping as these promise large benefits, but they also involve risks. This situation produces debates, often quite impassioned, about whether the benefits are worth the risks (**FIGURE 7.24**). Golden rice provides a good example of how these debates play out.

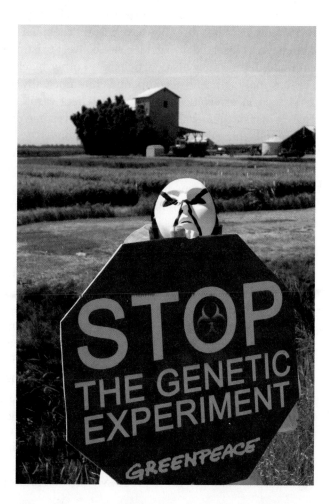

FIGURE 7.24 Many worry that genetically engineered crops pose unacceptable risks to the food supply and the environment. A Greenpeace activist protests the location of a rice crop genetically engineered to produce human proteins for drug production.

The golden rice project ran into opposition right from the start. Many people, including a number of scientists, were concerned about the safety of genetically engineered crops. Are they safe for human consumption, and are they safe for the environment? Others were concerned that only huge businesses can produce genetically engineered plants, which gives them control over this part of the agricultural markets. Still other opponents argued that it would be better and cheaper to find ways to provide people in developing countries with a more balanced diet that automatically would provide vitamin A and other vitamins and nutrients as well.

Scientifically, golden rice is a success, but the scientific success has not paid off yet in terms of the commercial or humanitarian success Potrykus and Beyer had intended to achieve. Critics of golden rice have challenged its usefulness on three grounds: safety, economic considerations, and whether it effectively solves the social problems associated with vitamin A deficiency.

Safety

The beta-carotene produced in golden rice seeds is chemically identical to beta-carotene found naturally in other plants. Are there reasons to be concerned about safety when the beta-carotene is produced by genetic engineering? Some potential risks from consuming genetically modified food may include an unanticipated harmful reaction caused by the inserted gene, a product that converts to an unanticipated molecule, or a product or one of the reactants that triggers an allergic response.

There are biochemical differences in genetically engineered cells. A great many biochemical reactions occur in every cell, and these reactions are linked in complex ways because they have certain chemical compounds in common. The enzymes inserted into golden rice convert existing compounds into beta-carotene. But do they also participate in other biochemical reactions? If so, are they making harmful compounds that normally would not be produced? A related concern comes from the excess beta-carotene now found in the cell. Does it just sit there waiting to be eaten, or is some of it converted into other compounds that could have adverse effects on us?

Genes make proteins, and proteins can trigger allergic reactions. Genetic engineering combines genes from several species into one, which also combines potential allergens from several species into one. Every species makes tens of thousands of proteins, and genetic engineering uses genes for only a couple of them, so there is a good chance that the introduced proteins don't cause an allergic reaction. Still, some people are allergic to daffodils (or at least certain compounds in daffodils). Would these people also be allergic to golden rice? Allergies appear to be on the rise. It can be very difficult to diagnose the causes of an allergy, but many consider the possibility that the increase in genetically engineered food may be partly responsible.

Some consumers want to see the word "natural" on the label of foods they buy, and organically produced foods have gained great popularity. Critics of genetically engineered crops often refer to them as "Frankenfoods," after the cobbled-together body parts of Frankenstein's monster in Mary Shelley's novel. Some consumers are hesitant to eat genetically engineered foods and are concerned that labels currently are not required to say whether any of the ingredients come from genetically engineered sources. Another source of concern is that many of the crops we've chosen to engineer are used mainly in processed and fast foods—high-fructose syrup from corn and protein from soybeans, for example. Since processed and fast foods are so prevalent in our

How Do We Know?

Evaluating the Safety of Genetically Engineered Products

People question the safety of genetically engineered products on several levels, including the effects on humans, effects on animals, and effects on the environment. Some of the arguments for and against these products are biological, while others involve the aesthetic value of using them. Here we consider some of the guidelines used to make decisions.

The most fundamental guideline is substantial equivalence, a principle defined in 1993. The principle states that a genetically engineered product is safe provided it is "substantially equivalent" to the nonengineered product. For genetically engineered drugs, this means that the engineered drug has the same or better effectiveness and the same or more positive effects (including side effects) than the nonengineered drug.

For crops, substantial equivalence means that the nutritional content of the food and the levels of allergens and toxins fall within the range of variation for nonengineered foods. The food developer is responsible for performing tests and presenting findings to the Food and Drug Administration for approval. Approval is often granted if the engineered product is found to be substantially equivalent. If any differences are found, additional rounds of testing are required, or the developer may decide not to continue development. The same rules apply for crops intended for human or animal consumption.

Critics of genetically engineered products argue that substantial equivalence is not a strong enough criterion. Foods can be shown to be substantially equivalent on the basis of biochemical tests, for example, without doing tests to see what happens when humans eat them. There is also concern that crop developers, who intend to market the crops, are the ones responsible for testing their safety.

Proponents of genetically engineered products argue that they are tested much more thoroughly than conventionally produced products. Many foods naturally contain toxins and allergens. Consider how many people are allergic to peanuts or the gluten in wheat, but these products don't require additional testing or regulation. Also, genetic engineering involves transferring only small numbers of genes between species, and under controlled conditions. Conventional crossbreeding transfers thousands of genes, with no hope of testing the effects of all of them.

The question of safety does not yet have a clear answer. At least three major interests are active in the debate: health and safety, humanitarian efforts to increase food production and nutritional value, and the profit motives of major biotech corporations. Until consensus is reached on how to deal with genetically engineered products, our best bet as individuals is to become as knowledgeable as we can, so that we can make decisions that are right for us. ::

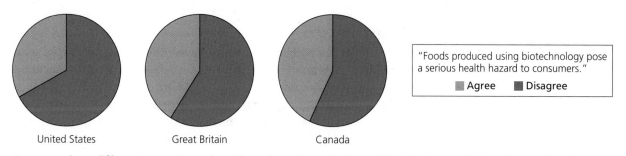

United States   Great Britain   Canada

"Foods produced using biotechnology pose a serious health hazard to consumers."
■ Agree   ■ Disagree

Consumers have different perceptions about the safety of genetically modified food. In one 2005 poll, respondents from the United States were less likely than respondents from Canada and Great Britain to believe that foods produced using biotechnology posed a serious health hazard to consumers.

diets, these engineered ingredients show up in a wide variety of foods, making a "natural" diet harder to put together.

The scientists and researchers who work on golden rice believe they have tested golden rice thoroughly and that it is safe to eat. Critics argue that the tests may have measured the wrong things, or that the results have been interpreted incorrectly and that safety issues remain a concern (see *How Do We Know? Evaluating the Safety of Genetically Engineered Products*).

Economic Considerations

It takes a large corporation with good financial, biotechnical, and legal resources to produce a genetically engineered crop. When a corporation makes such an investment, it expects a profit in return. Some of the concern about golden rice comes from worry over how these economic realities will affect the ability of farmers in developing regions to afford the seed.

Corporations working on golden rice obtained over 70 patents for new techniques they developed. Syngenta, the corporation that partnered with Potrykus and Beyer, held most of these patents, but over 30 corporations patented some parts of the development process. Potrykus, Beyer, and Syngenta worked with the Golden Rice Humanitarian Board to persuade each of these corporations to forgo licensing fees on their patents for low-income farmers. Agricultural programs in developing countries and farmers making less than $10,000 annually would be able to purchase golden rice seed for the same price as other rice seed—there are no additional licensing fees. Farmers also can keep seed from one year to grow the next with no additional charge.

Perhaps Syngenta and the other corporations expected to make a profit by selling golden rice to wealthier farmers and agricultural conglomerates in developed countries. However, vitamin A deficiency is simply not a problem in these countries, so there is no market for golden rice there. Syngenta no longer has a commercial interest in golden rice, but they continue to work with the Humanitarian Board on improving the crop and developing plans for its eventual distribution.

However, the role of large corporations may not be so benign in other cases. The patents held on techniques used to make golden rice may affect the price of other genetically engineered crops that use these patented techniques. In general, it takes a large investment to produce a genetically engineered crop, and corporations protect their patents aggressively. A number of biotech corporations are suing each other over patent infringement, and some corporations have even sued individual farmers.

Effectiveness

A final concern is whether golden rice is a solution, or even a partial solution, to the problem of vitamin A deficiency in developing Asian countries and elsewhere. Is relying on a fortified rice to provide a source for the body to make vitamin A the best approach to combating malnourishment and vitamin A deficiency? Would it be better to invest money in finding ways to provide a more balanced diet of foods that would contain vitamin A and a host of other vitamins and nutrients?

Critics argue that golden rice will simply encourage low-income people to continue eating a diet consisting largely of rice. Proponents counter that golden rice is not intended as a substitute for a balanced diet, but rather as a more nutritious component of a balanced diet, like our vitamin-fortified milk and cereal.

Without a doubt, the development of golden rice is a significant scientific achievement. But after more than a decade of testing and debate, golden rice is still not commercially available, and no consensus has been reached on its potential benefits or risks. The debate over golden rice rages on, both sides are impassioned, and both do their best to make a strong case for their position.

How Ethical Is Genetic Engineering?

What happens when a genetically engineered organism is released into the environment, or if the genes it carries "escape"? Can we trust ourselves to use genetic engineering wisely? What happens when we turn genetic engineering on ourselves? These are just some of the ethical considerations confronting scientists and others as the scientific techniques of genetic engineering progress.

Genetic Engineering and Our Environment

Normally genes in nature are passed from parent to offspring, often referred to as **vertical gene transfer**. Genetic engineering uses **horizontal gene transfer**: transferring genes between individuals of the same generation, and often between individuals of different species. Horizontal gene transfer occurs naturally as well. Many species of bacteria participate in horizontal gene transfer as a regular part of their life cycle. As with genetic engineering, this provides the bacterial cell with a combination of new genes from a variety of species. If a genetically engineered bacterium were released into the environment, it is quite possible that its engineered genes could find their way into other bacteria.

Plants also engage in horizontal gene transfer, though not to the same level as bacteria. Certain viruses that infect plants can transfer genes horizontally. Pollen from a genetically engineered plant carries foreign genes. This enables engineered genes to spread beyond a farmer's field. Many plants can hybridize: different species can mate to produce hybrid offspring. Suppose that Roundup Ready corn accidentally hybridized with a weedy grass species. If the grass became Roundup Ready, then the whole strategy of using Roundup to control weeds would be in trouble. In fact, Roundup-resistant weeds have been reported in Iowa and are spreading throughout the state due to the spread of pollen and seeds into new sites. Resistance to the herbicide was essentially nonexistent before the introduction of Roundup.

Many of the concerns over genetically engineered organisms arise from the possibility of accidental or unintended consequences. A darker specter is the possibility of deliberate harm, perhaps from genetically engineered weapons. Imagine the havoc that would be wreaked if a genetically engineered virus produced a toxin as deadly as anthrax, and was as virulent as Ebola and as contagious as the flu. Genetic engineering may provide many benefits, but we must also guard against its potential misuse.

Genetic Engineering and Human Life

Genetic engineering gives us extraordinary powers. Someday we may be able to create life forms to our own specifications. This raises several ethical concerns. Do we have the wisdom to use these powers in a safe and beneficial way? Do we have the right to genetically alter naturally occurring species to suit our needs? Do we have a responsibility to do all we can to use genetic engineering to improve our food supply and agricultural systems, potentially saving millions of people from malnutrition or worse? These are difficult questions, and clear-cut answers have yet to emerge.

Perhaps the greatest ethical dilemma is the question of whether we should genetically engineer ourselves, far-fetched though this may seem. There are

vertical gene transfer :: the transference of genes from parents to offspring

horizontal gene transfer :: the transference of genes between individuals of the same generation, and often between individuals of different species no technical reasons why genetic engineering procedures won't work on us, other than that we are a different species so the procedures need to be modified accordingly. We could use genetic engineering to extend life, improve health, and cure diseases, as we will discuss in Chapter 8. We could try to engineer for greater intelligence or athletic ability, or different aspects of beauty.

The things we could do with genetic engineering cover a spectrum from beneficial to scary. It is important that we become knowledgeable about the benefits and risks, and that we carefully consider the ethical issues. We are a long way from being able to do most of these things. But the basic techniques have already been worked out, and our technical skills increase every year.

 # Biology in Perspective

Can we create better plants and animals? The short answer appears to be *yes*. Because they photosynthesize and store food conveniently in their organs, plants have always been a good source of nutrition. Through thousands of years of selective breeding, we have produced many new varieties of plants and animals, more nutritious or more useful in other ways.

Genetic engineering allows us to speed up this process and make it dramatically more precise. By identifying and cloning genes and using them to transform cells, we can recombine genes (and genetic traits) from almost any species with those of any other species. The new combinations of traits we can produce are limited only by the range of traits found in nature and our technical skill. Already we have engineered vitamin-fortified crops, as well as crops that make their own pesticide or are resistant to herbicides, drought, freezing, and other hazards. Genetically engineered organisms make medicine and play important roles in industry and environmental cleanup.

While it provides potential benefits, however, genetic engineering raises concerns about economics, the environment, and safety. It also creates ethical concerns, particularly in the context of genetically engineering humans, as we will discuss in the next chapter.

Chapter Summary

7.1 Why Are Plants Such Good Sources of Food?
Explain how the bodies of plants are structured and specialized to store food molecules.

- Roots anchor plants, absorb water and nutrients, and often store resources for use later.
- Stems connect the roots to the leaves, transporting nutrients and water.

- Leaves are plants' biochemical factories but are often modified to store food, protect plants, help plants attach to surfaces, and even to catch and digest insects.
- Flowers are modified leaves that produce the egg and pollen used for reproduction.
- Seeds contain embryonic plants.
- Fruits are any structure developed from the ovary of the flower; they help disperse seeds.

7.2 How Do Plants Make Food?

Summarize the light reactions and Calvin cycle phases of photosynthesis.

- Plants capture energy and store it as chemical energy through photosynthesis.
- Photosynthesis occurs in organelles called chloroplasts and converts 1–5% of available sunlight energy into energy stored in sugar.
- Photosynthesis is divided into two phases: light reactions and the Calvin cycle.
  - o In light reactions, chlorophyll captures light energy and transfers that energy to an electron, which is used to generate ATP and NADPH, which are passed to the Calvin cycle.
  - o The Calvin cycle converts the energy captured by the light phase of photosynthesis into the chemical bonds of sugar.

7.3 What Are the Goals of Genetic Engineering in Plants?

Describe the goals for genetic engineering of crop plants.

- The primary goals of genetic engineering in crops include protection from insect pests, herbicide resistance, and increased nutritional value.
- Other goals include disease resistance and drought and freeze resistance.
- Many crop plants have been engineered to include a gene for a toxin produced by a bacterium (*Bt* toxin) that kills many insect plant pests.
- Golden rice was developed to produce beta-carotene in rice and relieve vitamin A deficiency in underdeveloped regions.

7.4 How Was Golden Rice Engineered?

Outline the steps used to produce a genetically modified organism.

- Define the problem: identify the genes that need to be altered.
- Clone the genes: isolate the gene to fix the problem and make lots of copies.

- Package the genes: add control regions of DNA to the cloned gene to make sure that the gene turns on properly.
- Transform the cells: put the packaged DNA into the DNA of the target organism.
- Confirm the strain: grow whole plants from the plant tissue and select transformed individuals with the new trait.

7.5 How Else Is Genetic Engineering Being Used?

Describe how areas other than agriculture use genetic engineering.

- Genetic engineering is used to produce medicine, to clean up pollution, to produce products such as plastics, to produce art and novelty organisms, and to help us understand how cells work through research.

7.6 What Are the Risks of Genetic Engineering?

Assess the risks of genetic engineering in terms of safety, the economy, and effectiveness.

- Some people have concerns about the safety of genetically modified organisms (GMOs), particularly as they relate to food allergies.
- Most genetically engineered crops have patented genes and do not allow farmers to save seed to plant the next year. This can be a significant problem for farmers in developing countries.
- Some critics claim that GMOs do not help the problems they are designed to solve.

7.7 How Ethical Is Genetic Engineering?

Evaluate the ethical considerations related to genetic engineering that confront scientists.

- Many concerns over genetically engineered organisms arise from the possibility of accidental or unintended consequences.
- Some ethical issues include the following: misuse, genetically engineered biological weapons, inadvertent horizontal transfer of herbicide resistance to weeds, and genetic engineering of humans.

Key Terms

bioleaching 205
bioremediation 205
bulb 191
Calvin cycle 195
carbon fixation 195
carpel 192
chlorophyll 195
chloroplast 196
cloning 200
endosperm 192
flower 191

fruit 193
genetic engineering 186
genetically modified organism
  (GMO) 197
genome 206
glyphosate 198
horizontal gene transfer 211
leaf 190
light reaction 195
petal 192
phloem 189

photosynthesis 194
root 188
root hair 188
seed 192
seed coat 192
sepal 192
stamen 192
stem 189
transformation 202
vertical gene transfer 211
xylem 189

Review Questions

1. What are the four types of plant organs?

   a. Roots, stems, leaves, and seeds

   b. Roots, stems, leaves, and fruit

   c. Roots, stems, flowers, and seeds

   d. Roots, stems, flowers, and fruit

   e. Roots, stems, leaves, and flowers

2. List the functions of each plant organ and explain how each plant organ provides food.

3. What are the four "whorls" that make up a flower?

   a. Sepals, petals, stamens, and carpels

   b. Sepals, petals, xylem, and carpels

   c. Sepals, petals, phloem, and stamens

   d. Sepals, petals, phloem, and carpels

   e. Sepals, petals, bulbs, and stamens

4. Describe the reproductive role of seeds and fruits, explaining how they are similar and how they are different.

5. What occurs during light reactions and during the Calvin cycle of photosynthesis?

6. What are several differences and similarities between selective breeding that led to the domestication of crops and "modern" genetic engineering?

7. Why did Potrykus and Beyer decide to work with rice as a way to reduce vitamin A deficiency worldwide?

8. What are the biological goals that researchers have for crop improvement?

   a. Pesticide reduction, herbicide resistance, and increased nutritional value

   b. Pesticide production, herbicide enhancement, and increased nutritional value

   c. Pesticide production, herbicide resistance, and decreased nutritional value

   d. Pesticide production, herbicide resistance, and increased nutritional value

   e. Pesticide reduction, herbicide enhancement, and decreased nutritional value

9. How do farmers benefit from crops that are resistant to herbicides like Roundup?

10. Describe two examples of genetic engineering used in medicine and suggest an organism used to produce the product.

11. Describe one example in which genetic engineering has been used "for fun": to produce an entertaining novelty.

12. What are three reasons why genetically engineered plants that produce a vitamin may not be safe to eat even if the vitamin is chemically identical to "natural" vitamins?

13. What is horizontal gene transfer? How does it compare to vertical gene transfer?

14. What are the five general steps involved in designing and carrying out a genetic engineering project?

The Thinking Citizen

1. What do the endosperm of seeds, egg yolks, and milk and other dairy products all have in common? Why are they so nutritious?

2. Some plants have a different version of photosynthesis. They open their stomates and take in CO_2 only at night. They store CO_2 internally until daylight, when they close their stomates and begin to photosynthesize. What are the benefits of this type of photosynthesis? What sort of habitat might these plants live in?

3. Many genetically engineered crops have a gene for a toxin produced by *Bacillus thuringiensis* (*Bt*). Many bacteria and other microorganisms produce toxins, pesticides, and antibiotics. Penicillin comes from a soil fungus, for example. What benefit might these organisms get from producing these compounds? Does this suggest that other useful genes might be found in nature?

4. Human growth hormone is a pituitary protein that can be used to treat dwarfism. Suppose you wanted to genetically engineer a strain of bacteria (*E. coli*) to produce human growth hormone. Outline the general steps you would take. (You don't have to know the detailed biochemical procedure.)

5. At first it may seem surprising that genes are largely interchangeable: genes that work a certain way in one species work the same way if they are inserted into a different species. This suggests that despite all their obvious differences, biochemically species are quite similar. Do you think this result provides evidence that species are related evolutionarily, all descending from a common ancestor? Why or why not?

The Informed Citizen

1. The licensing agreement for golden rice states that the crop cannot be released in a country lacking governmental or other regulations for ensuring that it does no environmental damage. However, the targets for golden rice adoption are developing countries, which can lack these types of regulatory agencies. Does this licensing requirement protect the environment, or does it present an obstacle to adopting golden rice?

2. In 1996, Monsanto tested a genetically engineered strain of soybean. They had inserted a gene from Brazil nuts that made a protein particularly rich in certain amino acids, which they hoped would make the soybean more nutritious. This strain of soybean was intended for animal consumption, not human. However, because some people are allergic to Brazil nuts and there was a chance that the soybeans could enter the human food chain, they tested whether the strain could cause an allergic reaction in humans. Some people were allergic to the soybean, and Monsanto discontinued work on the project. Is this an example of a testing and regulatory system that works well to prevent unsafe genetically engineered products from coming to the market? Or does it show that the dangers of allergies to genetically engineered crops are real and that we should not support the crops' development?

3. How do you feel about the potential to genetically engineer humans? Would you ban all research for this subject? Are there some genetic therapies or modifications you would support but others you would prohibit? What guidelines would you use to decide whether a particular proposal for human genetic engineering was ethical?

CHAPTER LEARNING OBJECTIVES

After reading this chapter, you should be able to answer the following questions:

8.1 Do Complex Human Characteristics Have a Genetic Basis?
Assess the legitimacy and risks of genetic determinism to society.

8.2 What Is Gene Therapy?
Distinguish between somatic and germ-line gene therapies.

8.3 What Are the Benefits and Risks of Genetically Altering Humans?
Contrast the benefits and risks of somatic and germ-line gene therapies.

8.4 How Can Stem Cells and Cloning Be Used to Alter People?
Compare the different methods and uses of stem cells and cloning.

8.5 What Are the Benefits and Risks of Stem Cell Research?
Use information on risks and benefits to evaluate the ethics of stem cell research.

8.6 What Other Challenges Result from Advances in Medical Technology?
Assess the ethical challenges we face as a result of advances in genetic technology.

Thanks to the creativity, curiosity, and hard work of a great number of scientists, we now know things about inheritance, how genes work, and how to manipulate genes that would have been the stuff of science fiction 40–50 years ago.

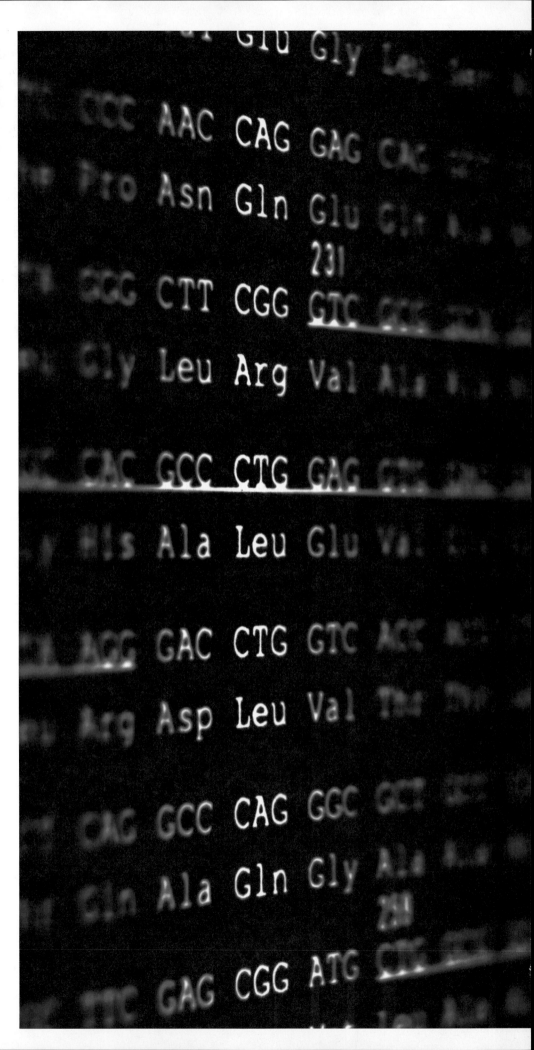

CHAPTER 8

Health Care and the Human Genome

How Will We Use Our New Medical and Genetic Skills?

Becoming an Informed Citizen …

Although we live in an exciting time and have the capacity to do wonderful things with the genetic and medical technologies that exist and are being developed, we also have the capacity to do harm. How can we ensure that we maximize the positive consequences of our actions while minimizing the negative ones?

We are at a profoundly important junction in our history. Thanks to the creativity, curiosity, and hard work of a great number of scientists, we now know things about inheritance, how genes work, and how to manipulate genes that would have been the stuff of science fiction 40–50 years ago. Even 20 years ago, some of today's routine procedures would not have seemed possible. We look now toward a horizon where potential preventions or cures for many diseases may become realities. The challenges we face are not the scientific or technical obstacles, but rather how to define the ethical boundaries for our actions.

This is not the first time society has faced the collision between what we know (or think we know), what we *can* do, and what we *should* do. Let's look at a similar scenario from the early part of the 20th century to see if it holds any helpful lessons.

Carrie Buck and the American Eugenics Movement

FIGURE 8.1 When her widowed mother, Emma (right), was committed to the Colony for Epileptics and Feebleminded in 1920, Carrie Buck (left), only 14 years old, was sent to live with the Dobbses. Carrie's life with the Dobbses was clouded by her mother's reputation, and she was treated poorly.

John and Alice Dobbs seemed to be good-hearted, decent people. After all, they took in Carrie Buck (**FIGURE 8.1**) when her widowed mother, Emma, was committed to the Colony for Epileptics and Feebleminded in 1920. Carrie was only 14 years old at the time and had no other family to care for her. But although it appeared that Carrie was quite fortunate to be taken in by the Dobbses, she had actually just traded one problem for another.

Emma Buck had not been a well-respected, contributing member of society. Poor and uneducated, she went downhill after her husband died. People in her rural Virginia community thought Emma was a "worthless" person, and so she was brought before the Commission on Feeblemindedness. These local commissions were found in cities and towns across the state. Once the commission learned that Emma had been arrested for prostitution and that she suffered from syphilis, a sexually transmitted disease, they declared her "notoriously untruthful" and "suspected of being feebleminded or epileptic." While we have no way of evaluating the claims of her untruthfulness, we do know that Emma was neither feebleminded nor epileptic. Nevertheless, once sentenced, she spent the rest of her life in the colony.

Carrie's life with the Dobbses was clouded by her mother's reputation. She never felt like she was part of the family. The Dobbses expected Carrie to do chores around the house, and she did them skillfully. She also did well in school. However, once Carrie was promoted to the sixth grade, the Dobbses removed her from school and gave her more housework to do. They also "loaned" her to other families to do housework for them. Still, although life might not have been great for Carrie, it was tolerable.

Everything changed in 1923. Carrie, only 17 years old, became pregnant. She said she was raped. The Dobbses did not believe her, and we will never know what actually happened. However, the truth of how she became pregnant hardly mattered, because in small-town Virginia in this era, *any* pregnancy outside of marriage was deeply shameful. The Dobbses wanted Carrie out of the house—immediately.

Mr. Dobbs filed papers to have Carrie committed, and Justice Shackleford, who had sentenced Emma Buck, scheduled a hearing. The Dobbses claimed that Carrie was prone to outbursts and hallucinations. They said she behaved oddly. Shackleford declared Carrie feebleminded and sentenced her to the same colony to which he had committed her mother; she was to report to the colony after the baby was born. On March 28, 1924, Carrie gave birth to a daughter, Vivien, and the Dobbses took Vivien in.

In the time between Emma's and Carrie's commitments, the state of Virginia passed a eugenics law. **Eugenics** is the idea that the way to improve "human stock" (analogous to breeding better cows, for example) is to prevent biologically defective people from having children and encourage "worthy" people to have many. It is an idea based on a flawed understanding of inheritance and evolution. Virginia's eugenics law stated that a person could be ordered to be sterilized for "the good of society." Carrie was the test case. She had been declared flawed, as had her mother before her. Moreover, feeblemindedness was deemed to be passed from parent to offspring according to the laws of heredity. So was Vivien feebleminded too?

Seven-month-old Vivien (**FIGURE 8.2**) was examined by a social worker who reported, "There is a look about it that is not quite normal, but what it is I can't tell." This testimony was enough to have Vivien labeled feebleminded. The constitutionality of the Virginia law was immediately challenged in a higher court. The case, *Buck v. Bell*, finally reached the U.S. Supreme Court, where, in 1927, Justice Oliver Wendell Holmes wrote the Court's opinion upholding the Virginia law: "It is better for all the world, if instead of waiting to execute degenerate offspring for crime, or to let them starve for their imbecility, society can prevent those who are manifestly unfit from continuing their kind. The principle that sustains compulsory vaccination is broad enough to cover cutting the Fallopian tubes. Three generations of imbeciles are enough."

Carrie Buck was sterilized on October 19, 1927. She was only 21 years old. And what happened to the "imbecile" Vivien? Sadly, she died at the age of eight from an infection—but not before going to school and making the honor roll.

Why did so many people think that society's ills were biologically inherited and could be cured by controlling who was permitted to be born? We know more about genetics and inheritance now than people did during the early part of the 20th century. Do we now have the ability, dreamed of by eugenicists, to eliminate all genetic "defects"? Should we? Is eugenics simply a shameful chapter in our history, or is its goal of improving "human stock" still alive with us today in a modern form?

SOURCES: Edwin Black, *War Against the Weak: Eugenics and America's Campaign to Create a Master Race*, Dialog Press, 2012.

eugenics :: the notion that the way to improve human stock is to prevent biologically defective people from having children and encourage worthy people to have many

FIGURE 8.2 Vivien Buck (shown here with Mrs. Dobbs) was labeled feebleminded at the age of seven months after being examined by a social worker.

Do Complex Human Characteristics Have a Genetic Basis?

According to the eugenics sterilization laws that were prevalent in the United States in the first half of the 20th century, a board of experts could order someone sterilized if the person was judged to be "insane, idiotic, imbecilic, or moronic." Convicted criminals could also be ordered to be sterilized. In some states, offenses that could lead to sterilization included alcoholism, drug addiction, blindness, deafness, homelessness, or pauperism. The crux of the argument was that these defects were heritable, and therefore sterilization was the only way to stop these defects for the good of society (**FIGURE 8.3**). Were eugenicists correct in thinking that a lot of what is wrong with us is encoded in our genes? Are there "bad" genes?

Genetic Determinism

A 1998 article in *Life* magazine proclaimed: "Were you born that way? Personality, temperament, even life choices. New studies show it's mostly in our genes." This article captured the overly simplistic and largely incorrect ideas of genetic essentialism and genetic determinism.

genetic essentialism :: the notion that being human means having a human genome

genome :: the term for all genes in an organism

genetic determinism :: the idea that our genes determine, direct, or cause everything about us

Genetic essentialism is the notion that being human means having a human **genome**, the term for all genes in an organism. Humans do indeed have a human genome. In fact, each of the trillions of cells in the body contains a human genome. But no one would think of each individual cell as being a human person. Having a human genome is certainly not the only thing that makes us human. **Genetic determinism** is the idea that our genes determine, direct, or cause *everything* about us: our physical appearance, emotions, and behaviors. According to genetic determinism, our genes are in charge: the environment in which we live has little effect on how we grow and develop.

Common sense and a great deal of empirical evidence demonstrate that genetic determinism is simplistic and largely incorrect. Consider identical

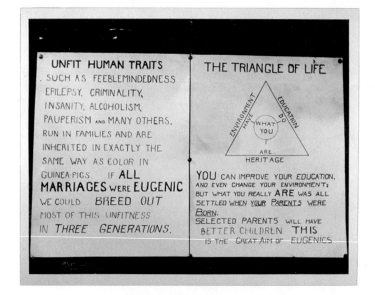

FIGURE 8.3 Eugenicists believed that many "unfit human traits" were inherited, and therefore sterilization was the only way to eliminate these defects for the good of society.

twins. They have identical genomes, but anyone who has known twins has observed that they are very different people with distinct personalities, likes and dislikes, and behaviors (**FIGURE 8.4**). As you read in Chapter 3, even conjoined twins, who share not only the same genes but nearly identical environments, are unique. Each of us inherits a genome that defines a biological potential. We are going to be humans, not giraffes or dandelions. However, the genome is responsive to the environment even while we are *in utero*, as well as after we are born. Did my mother drink excessively while she was pregnant? Did I obtain adequate nutrition, so my brain developed well during childhood? Neither of these conditions is "in our genes."

FIGURE 8.4 Identical twins can have dramatic differences. One sister (right) smoked for 16 years before quitting and spent a lot of time in the sun, while the other sister did neither.

It is fortunate for us that our genomes are not completely deterministic. If they were, we would all be dead. For example, most mammals can make vitamin C, an essential nutrient necessary for survival. Humans can't. In fact, 100% of people have a mutation that leads to a deficiency in an enzyme needed in the pathway for vitamin C synthesis (see Chapter 9). Luckily, humans have a way to survive these faulty genetic instructions: we get vitamin C from our environment by eating foods, such as citrus fruits, that are rich in this nutrient. Problem solved.

This returns us to our initial question: do complex human characteristics have a genetic basis? The answer is *not really*, or at least not in the way eugenicists or genetic determinists think. Despite reports trumpeting these latest "findings," no specific genes have been identified for aggressive behavior, altruism, bipolar psychosis, criminality, drumming, figure skating, homosexuality, mathematical ability, schizophrenia, shyness, or television viewing. We are biological beings, so of course genes play a role in many of our characteristics, but this does not mean they determine outcome.

Defining Normal

The experts who committed Emma and Carrie Buck, sterilized Carrie, and labeled Vivien feebleminded thought they had a good understanding of what it means to be normal. The actions that followed their judgment that the Bucks were defective were undeniably terrible, but were they correct that Emma, Carrie, and Vivien were not normal?

The definition of "normal" is quite difficult to articulate. It can refer to a statistical average, like the average height in a population. In this context, normal does not generally suggest a value judgment. Normal can also be used to mean how similar something is to an ideal. For instance, we might have an ideal for beauty and evaluate some people as normal or abnormal accordingly. When normal is used in this way, it is clear that normal is considered superior.

The challenge of defining normal is that in biological beings like us, there is no ideal form. Variation in both how we look and how we function *is* normal. Variation gives us the flexibility to respond to different and changing environments. Natural selection selects traits that happen to work, not traits that are perfect. *Most* individuals in a population will deviate from the average.

Even what humans view as normal varies. What we consider to be beautiful or admirable behaviors have changed over time. Women working as professionals outside the home and men taking on the role of stay-at-home parents would have been seen as seriously abnormal in 1900. Men without facial hair and thin, muscular women would have also seemed odd.

We can shift the question from what is normal or abnormal to what constitutes a diseased or disabled state. In some cases, it might be easy to answer. For example, breast cancer is a disease, not simply a variation in a trait.

But consider dwarfs who suffer no health problems but are simply short in stature. Is this a disability that needs to be prevented or remedied?

We are not any better at defining normal now than people were in Carrie Buck's day. And like them, many of us act as if much of disease and abnormality is genetic at its roots. In Carrie Buck's world, this meant that so-called experts could label someone as shiftless or apply another vague diagnosis. That individual would then be sterilized, stopping those "shiftless genes" from being passed on. Today we have genetic screening, prenatal diagnosis, and gene therapy to hunt down the "bad genes"—and perhaps try to fix them. Is this modern approach any more rational and effective than were eugenics laws? Have good outcomes resulted, or have we merely created new concerns?

8.2 What Is Gene Therapy?

Thousands of diseases and disorders can be attributed to mutations in specific genes (**TABLE 8.1**). Examples include cystic fibrosis and sickle cell disease (Chapter 4), androgen-insensitivity syndrome (Chapter 6), heart defects, cleft palate, certain types of deafness, lack of eyes, extra fingers or toes, male sex reversal, and Huntington's disease (see *Scientist Spotlight: Nancy Wexler* to learn about the search for the Huntington's gene). Serious mutations resulting in spontaneous miscarriage are not at all unusual; approximately one-half to two-thirds of all conceptions result in spontaneous miscarriage.

Prenatal screening tries to identify embryos or fetuses that have genetic or developmental abnormalities. Parents can then use this information to plan and prepare for a child with special needs or to decide to terminate the pregnancy because they do not want to pass on an unwanted trait. One of the challenges we face as a society is that prenatal screening permits us to decide which traits are acceptable to us and which should be avoided or eliminated.

Even with perfect prenatal screening, there will always be individuals born with genetic syndromes. Not all of these are predictable. For example, although the trait for achondroplasia (**FIGURE 8.5**), a type of dwarfism, is passed from parent to child, seven out of eight cases of achondroplasia are due to new mutations. No one else in the family is a dwarf.

FIGURE 8.5 Achondroplasia, a form of dwarfism, can be inherited from parents. However, most cases are due to new mutations (meaning that no one else in the family suffers from the disorder).

Table 8.1 :: Examples of Diseases or Disorders That Can Be Attributed to Specific Gene Mutations

Disease or Disorder	Chromosome Location
Huntington's disease	4p
Juvenile epilepsy	6p
Cystic fibrosis	7q
Wilms tumor-1	11p
Retinoblastoma	13q
Polycystic kidney disease	16p
Cataract	16q
Malignant hyperthermia	19q
Alzheimer's disease	21q
Retinitis pigmentosa	Xp

Scientist Spotlight

Nancy Wexler (1945–)

A small village in Venezuela called Lake Maracaibo holds the key to understanding Huntington's disease (HD). Nancy Wexler began studying the people in this village in 1979 because it had been reported that they have an extremely high incidence of this neurological disorder. She has spent her life trying to solve the problem of HD and other genetic syndromes, and her interest is personal. Wexler's mother was diagnosed with HD in 1968. Wexler's maternal grandfather and two uncles died of the disease. And she and her sister are at risk for developing it.

Nancy Wexler was born in 1945 to Milton Wexler, a psychoanalyst, and Leonore Sabin Wexler, a geneticist. Wexler went to Radcliffe College and graduated with a bachelor's degree in social relations and English. However, her mother's diagnosis shifted Wexler's career plans. She began studying HD in graduate school. Her father founded the Hereditary Disease Foundation in 1968. The Wexler family made fighting HD their number one priority.

Her work in Venezuela turned out to be pivotal. Wexler collected medical and family history information from more than 18,000 people. She also collected 4000 blood samples and sent them to her collaborator James Gusella at the Massachusetts General Hospital. He used the blood samples to map the location of the HD gene to the tip of chromosome 4. In 1993, an unprecedented collaboration among a half dozen research laboratories led to the identification of the actual HD gene. The willingness of this Huntington's Disease Collaborative Research Group to work together instead of competing hastened this discovery. This work has also led to the development of a test that can detect HD long before any symptoms appear. However, since there is no cure for HD, not all people who are at risk opt to take it. In fact, Wexler

Nancy Wexler has spent her life trying to solve the problem of Huntington's disease (HD) and other genetic syndromes. Her work in Venezuela led to the identification of the HD gene.

chose not to have the test done or to have children. She recognizes how difficult the decision can be and has wondered whether she might have been happier had she learned that she would not develop the disease. However, she understands firsthand that knowledge of a disease's certainty may be more than a person wants to live with.

Wexler continues her work on HD and other hereditary diseases. She is a professor at Columbia University and the president of the Hereditary Disease Foundation. She also returns to Lake Maracaibo every year to work with HD patients. Wexler's research has revealed that all of the people there who suffer from HD are related to one woman who died of the disease. Her work has also helped to map some other disease genes, including those for familial Alzheimer's and a certain type of kidney cancer. ::

Some people argue that the emphasis on prenatal screening to prevent certain births shows a lack of understanding of disability. They argue that the problems experienced by the disabled are due not to the physical challenge itself, but rather to the response of others to the disability. While this may be true in some situations, it would be hard to make the case that suffering from a disease like cystic fibrosis or muscular dystrophy occurs principally because of the way that others act. Most people think it would be great to have a way to cure such diseases. Gene therapy is designed to do just this. There are two types of gene therapy: somatic gene therapy and germ-line gene therapy.

Somatic Gene Therapy

The idea of **somatic gene therapy** is pretty simple: cure a genetic disorder by inserting a normally functioning gene into the patient. The actual implementation of the idea has been difficult, however.

somatic gene therapy :: the insertion of a normally functioning gene into a patient to cure a genetic disorder

The first successful gene therapy was done in 1990 to help children who suffered from severe combined immunodeficiency (SCID) (**FIGURE 8.6**). These children are missing an enzyme called adenosine deaminase (ADA) that is needed for normal lymphocyte function. **Lymphocytes** are a type of white blood cell essential for the body's **immune response**, or ability to fight off infection. Children with SCID are vulnerable to infection from *everything* and must therefore live in a sterile environment to survive.

Scientists took immature lymphocytes and lymphocyte stem cells from SCID patients and cultured them to grow a population of cells in the lab. Lymphocyte stem cells divided to make new lymphocytes. The scientists isolated the ADA genes from healthy cells and inserted them into the genomes of harmless viruses. They then infected the cultured cells with the viruses. Once the cells were infected, the ADA genes popped out of the viruses and inserted themselves into the cells' genomes. The cells that had incorporated the ADA gene were injected back into the patient (**FIGURE 8.7**). In successful cases, the new gene produces the missing enzyme, treating the disease.

Somatic gene therapy may indeed hold the promise of curing some diseases besides SCID. But as with any new medical treatment, there are risks both technical and ethical. We will discuss these risks later.

Germ-Line Gene Therapy

Somatic gene therapy is similar to other types of medical treatments: something is not working properly in a patient, and physicians try to repair it. **Germ-line gene therapy** aims to correct genetic problems in the germ line—eggs and sperm—so that harmful mutations can never be passed on. This technique opens a new world that some think is exciting and wonderful and others consider perilous and wrong. Imagine being able to eliminate the gene for cystic fibrosis, for example, from a family so that no descendants would ever inherit this gene or develop the disease. It's as if the authorities could

FIGURE 8.6 Individuals with severe combined immunodeficiency (SCID) lack normal immune function. They must be protected from exposure to anything that could cause an infection.

lymphocyte :: a type of white blood cell

immune response :: the body's ability to fight off infection

germ-line gene therapy :: the correction of genetic problems in the eggs and sperm so that harmful mutations can never be passed on

① Scientists took immature lymphocytes and stem cells from SCID patients and cultured them in the lab.

② The ADA genes were isolated from healthy cells and inserted into the genomes of harmless viruses.

③ The cultured cells were infected with the viruses.

④ The ADA genes were released from the viruses and inserted into the cells' genomes.

⑤ Scientists injected the cells that had incorporated the ADA gene back into the patient.

In successful cases, the new gene produced the missing enzyme, treating the disease.

FIGURE 8.7 Somatic gene therapy involves the insertion of a normally functioning gene into a patient and has been used to cure patients with severe combined immunodeficiency (SCID). These children are missing an enzyme called adenosine deaminase (ADA) that is needed for normal lymphocyte function.

have eliminated Emma Buck's "feeblemindedness gene" (if such a thing actually existed) before it doomed both Carrie and Vivien.

So far, no one has tried to do germ-line gene therapy on humans. It has been done in other mammals, however, and is actually a routine research technique in laboratory mice. Also, the rhesus monkey ANDi (the backward abbreviation for "*i*nserted *DNA*") has been altered so that all of its cells have the DNA and RNA encoding for the molecule GFP, green fluorescent protein. However, for some as yet unknown reason, the GFP RNA is not translated. As a result, the GFP protein itself is not present. Nevertheless, all of ANDi's descendants will have the GFP gene in their genomes. Strangely, two other monkeys that also had the GFP gene and RNA in every cell, but which were stillborn, did glow green, meaning they did make the GFP protein (**FIGURE 8.8**). Scientists are not sure why ANDi and these two monkeys were different in this regard.

The goal of this research was not to make "glow monkeys." Scientists wanted to develop methods to insert human disease genes into primates so that the disease could be studied in organisms more similar to humans than mice or rats. Some people argue that it is unethical to genetically engineer animals in this way for this purpose. Even though the scientists who produced ANDi do not support the extension of their work to humans, many are concerned that it is just a matter of time until we engineer our own germ lines.

Besides the ethical questions, significant technical and scientific ones remain. The process of genetically engineering the germ line is extremely inefficient. The experiment that produced ANDi began with more than 200 fertilized eggs. From these, only 40 embryos were produced. The 40 embryos resulted in five pregnancies. And only three live births occurred: ANDi and two monkeys that did not incorporate the GFP gene into their genomes. If this research were to be done in humans, and at this time no one is trying to do so, it would entail research with human embryos and similar inefficiency, at least at first.

For an idea of what would need to be done to alter human germ lines, let's look at how ANDi was actually produced (**FIGURE 8.9**). Scientists inserted the

FIGURE 8.8 Germ-line gene therapy aims to correct genetic problems in the eggs and sperm so that harmful mutations can never be passed on. The rhesus monkey ANDi was altered so that all of its cells encode for the molecule GFP.

1 Scientists inserted the GFP genes into harmless viruses.

Egg cell

Pipette

2 The viruses were injected into the space around the unfertilized rhesus monkey eggs.

3 The eggs were then fertilized in petri dishes with rhesus monkey sperm.

4 The healthiest embryos were implanted into surrogate rhesus mothers.

Three of the monkeys born—ANDi and two stillborn males—had incorporated the GFP into their genomes.

FIGURE 8.9 The germ-line gene therapy procedure used to produce ANDi was not efficient. Out of more than 200 attempts, only one live monkey was born with the GFP gene incorporated into its genome.

GFP genes into harmless viruses and injected the viruses into unfertilized rhesus monkey eggs. The eggs were then fertilized with rhesus monkey sperm. The embryos that divided twice to produce four cells were next implanted into surrogate rhesus mothers. The surrogates had undergone hormone treatments so that they would accept a pregnancy. A tiny number of surrogates carried their babies to term. Three of the monkeys that were born—ANDi and two stillborn males—had incorporated the GFP into their genomes.

Even though the technical issues remain formidable, many scientists and bioethicists are concerned that we will eventually be able and willing to alter human germ lines. Are there benefits that might result, and what are the risks?

8.3 What Are the Benefits and Risks of Genetically Altering Humans?

When Carrie Buck was sterilized in 1927, the experts truly thought they were doing something that was both scientifically sound and good for society. These were mainstream ideas. The geneticists who were convinced that characteristics like being shiftless or loving the sea were heritable were not on the extreme edge of their profession. Are we more knowledgeable now? Are we sufficiently knowledgeable to alter genes in people in a way that always results in a beneficial outcome?

Scientists and the general public view somatic gene therapy and germ-line gene therapy as quite different from each other. Somatic gene therapy aims to fix or replace a defective gene in *an individual*. This alteration is not passed on to subsequent generations. In contrast, germ-line gene therapy is designed to make genetic changes that will be inherited by *all* descendants. The ethical challenges for these two forms of gene therapy are not the same. Neither are the potential benefits and risks.

Somatic Gene Therapy

Somatic gene therapy holds the promise of curing genetic disorders, some of which are quite terrible. For example, Lesch-Nyhan syndrome is associated with a mutation in a gene that encodes for an enzyme necessary to break down uric acid. Individuals with this disorder are missing this enzyme, and the waste product builds up in the body. Symptoms include muscle problems, kidney malfunction, mental retardation, and self-mutilating behaviors. Lesch-Nyhan syndrome is inevitably fatal, usually owing to kidney failure, generally sometime in childhood. If the technical details were worked out so that it was safe to replace the mutant gene associated with this disease with a normally functioning one, why hesitate?

Among arguments made against somatic gene therapy are the following: it is not safe, it might not be effective, it is expensive, and we should not "play God." All of these arguments have been made every time a new medical therapy—whether blood transfusions, vaccinations, or even surgery—has been developed. New medical treatments are usually risky at first, and this is true for somatic gene therapy as well.

As for effectiveness, it will be predictably variable. Diseases in which the mutant gene causes a "loss of function" are much easier to cure. For example,

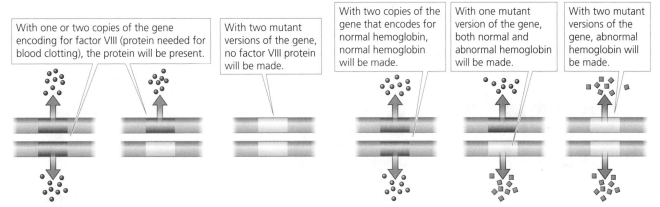

With one or two copies of the gene encoding for factor VIII (protein needed for blood clotting), the protein will be present.

With two mutant versions of the gene, no factor VIII protein will be made.

With two copies of the gene that encodes for normal hemoglobin, normal hemoglobin will be made.

With one mutant version of the gene, both normal and abnormal hemoglobin will be made.

With two mutant versions of the gene, abnormal hemoglobin will be made.

A With a **loss-of-function disease** such as hemophilia, it is necessary to ensure only that the normal protein is made because the mutant allele makes no protein.

B With a **gain-of-function disease** such as sickle cell, the production of bad proteins must be stopped in addition to ensuring that the correct protein is made.

FIGURE 8.10 Loss-of-function diseases (A) are much easier to cure with somatic gene-line therapy than gain-of-function diseases (B).

the gene associated with **hemophilia** fails to direct the production of a clotting factor. As a result, an individual with hemophilia cannot form blood clots in the event of an injury. If a patient with hemophilia received somatic gene therapy to insert the functional gene, he or she would be able to make the clotting factor and would be cured (**FIGURE 8.10**). In contrast, mutant genes that produce a "gain of function" will be very difficult—maybe impossible—to treat. Sickle cell disease, for instance, is the result of a mutant gene that produces abnormal hemoglobin. To correct this problem with somatic gene therapy, it would be necessary to insert the functional gene *and* to disable or eliminate the mutant genes in all blood cells.

Somatic gene therapy is expensive. But so are all new medical technologies, especially when they are first developed and implemented. This is true for heart surgery, organ transplantation, chemotherapy, and many other treatments that are routine in modern hospitals today. As for "playing God," one could argue that everything we do to heal, treat, or overcome illness, disability, or accident is playing God. Before the advent of open heart surgery to bypass clogged arteries, many people died at a much younger age than they do today.

The most successful somatic gene therapy so far has been the treatment of severe combined immunodeficiency (SCID). But scientists do not have complete control over where an introduced gene inserts, and the consequences are potentially lethal.

Even though a great deal of scientific work still needs to be done, it seems likely that development of somatic gene therapy will continue. It is a medical intervention designed to help. And who wouldn't like a world in which people are free from the suffering caused by cancer, cystic fibrosis, muscular dystrophy, and a host of other diseases?

Germ-Line Gene Therapy

What happened to Carrie Buck and the tens of thousands of individuals sterilized under the provisions of the eugenics laws was the ultimate attack on germ lines. Because of prejudices and beliefs about what could be inherited, the experts and legal authorities simply put an end to the germ line altogether. To ensure that no "bad" genes were passed on, they prevented the passing of *all* genes into the next generation.

hemophilia :: a disease in which an individual cannot form blood clots in the event of an injury

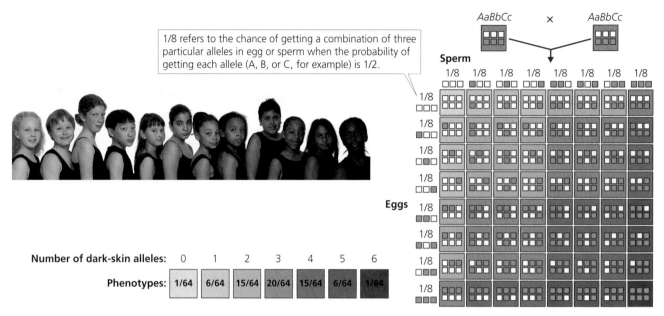

FIGURE 8.11 Genes do not act alone as individual entities. Some traits, such as skin color, are influenced by multiple genes.

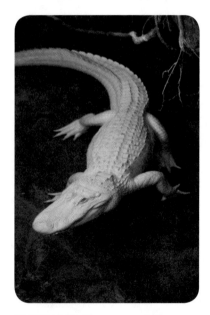

FIGURE 8.12 Many genes influence more than one trait. Albino organisms are unable to make the pigment melanin because of a single mutation, which affects the traits of coat color, eye color, and vision, among other characteristics.

genetic counseling :: an approach used to inform couples about their relative risk of passing on certain specific genes associated with a disease

Germ-line gene therapy is not nearly as crude. Nevertheless, it does share two things in common with eugenics: (1) confidence about the deterministic genetic basis of characteristics and (2) confidence about defining desirable and undesirable traits. Are these confidences well grounded?

The relationships between individual genes and specific characteristics, physical or otherwise, are not straightforward. Genes don't act alone as individual entities. Some traits—such as height, weight, skin color, eye color, and susceptibility to cancer—are influenced by multiple genes (**FIGURE 8.11**). Also, many individual genes influence more than one trait (**FIGURE 8.12**). For instance, the globin gene affects both how well red blood cells carry oxygen and the shape of the cells themselves. And the environment influences the way many traits develop. For example, genes that affect how the brain and face form during embryogenesis and fetal development are sensitive to whether a mother drinks alcohol during pregnancy. Do we have adequate knowledge to alter genes for all of the descendants of an individual? Even if all of the technical details were worked out and we could alter germ lines safely, is it ethical to do so?

Many people argue that we have safe and effective alternatives to germ-line gene therapy. **Genetic counseling** can inform couples about their relative risk of passing on certain specific genes associated with a disease. Prenatal diagnosis of embryos can identify whether any of these harmful, disease-causing mutations are present. If the embryos are the result of *in vitro* fertilization (IVF), which we studied in Chapter 6, healthy embryos can be selected for implantation. A couple could opt for sperm or egg donation from individuals who are not at risk for passing on these particular defective genes. And adoption is always an option.

Many people consider the following ethical issues surrounding germ-line gene therapy to be so serious that efforts to develop this technology for people should stop:

- First, safety is by no means certain, and a mistake is permanent, affecting more than one person. The insertion of a gene can unexpectedly disrupt the functioning of another gene, sometimes producing grave consequences.

- Second, we are not sure which genes are "good" and which are "bad." In some cases, this depends on the environment. For example, the sickle cell mutation has the potential to produce sickle cell disease in individuals who get two copies of the gene. However, a person with only one copy enjoys protection from malaria. Whether or not having a copy of the sickle cell gene is beneficial or harmful depends on whether a person lives in an area where malaria is rampant.

- Third, as with somatic gene therapy, we need to distinguish between medical treatment and enhancement. Is it appropriate for parents to make specific choices about their children's genomes? Is it acceptable to "design" children according to the parents' tastes or needs? And what if "genetic fashion" were to change and as a society we stopped favoring a trait that we had once valued? (See *Life Application: Sex Selection* to learn about choices some parents are making about their children.)

- Finally, this type of therapy has the potential to alter how we feel about the sick and disabled in society. If we were able to design people, would we judge a person's family harshly because they hadn't prevented this affront to perfection? What about poor people who could not afford access to this technology? Would we see a growing gap between the rich and poor based on their biology?

Life Application

Sex Selection

Who would have predicted that the availability of inexpensive ultrasound machines would result in increased crime, sex trafficking, and social unrest? Yet that is exactly what is happening in several places in the world, especially China and India, where some people engage in sex selection—choosing whether to allow a female embryo or fetus to come to term and be born.

Male children are preferred in many cultures for varying reasons. In China, for example, sons are important for the care and protection of the elderly. They are also considered essential for continuing the family line. In India, sons are seen as a financial asset to the family. It costs a family money, in the form of a dowry, to "marry off" a daughter, and when she does marry, she joins the husband's family.

The shift in sex ratio is dramatic in both of these countries. In some Chinese provinces, 120–130 boys are born for every 100 girls. A study in India evaluated 1.1 million households and collected data for 133,738 births. They found that for families that already had one girl, the ratio of boys to girls for the second child was 1000 to 759. If the first child was a boy, the sex ratio for the second child was 1000 boys to 1102 girls. As long as they had one boy, girls were acceptable.

This pattern of sex selection is not confined to China and India. The 2000 U.S. census showed a similar bias in U.S.-born children of Chinese, Korean, and Asian Indian parents. The bias in favor of sons was especially dramatic in families in which the first two children had been girls: the number of boys born in this third position outnumbered girls by 50%.

This shift in gender ratio worldwide is due to the selective abortion of females. A 2002 study showed that 300 out of 820 women in a central Chinese village had abortions, and 70% of the aborted fetuses were female. A 1985 survey in Bombay (now Mumbai), India, reported that of the female fetuses aborted, 96% occurred after the sex had been determined by amniocentesis.

The gender gap created by these practices is very serious. The 2000 census in China showed that there were 19 million more boys than girls in the newborn to 15-year-old age bracket, and the 2010 census showed a worsening of this skewed gender ratio. By 2014, there will be 40–60 million "missing" women. In India, 500,000 fewer girls than boys are born each year. This has resulted in a "deficit" of 10 million females over two decades. And, similar to China, the 2010 census shows that the gender ratio in India is worsening. What will the "extra" men do for wives?

China and India are trying to remedy the "missing girls" problem. They can't fix the unbalanced sex ratios of the

(Continued)

Life Application

Sex Selection (*Continued*)

Male children are preferred in many cultures for varying reasons. Both China and India engage in sex selection, and the shift in sex ratio has been dramatic in both countries.

generations that are already born, but they are taking steps to rebalance sex ratios for future generations. In India, for example, prenatal sex determination has been illegal since 1994. Also, the practice of paying dowries is slowly being shifted to "bride money," in which the groom pays the bride's family. In China, steps have been taken to make it more desirable to have a baby girl, including elimination of school fees for girls; the awarding of an annual pension to couples who limit their offspring to two girls; and the preferential treatment of parents of girls in health care, housing, and employment. But even if these efforts are successful, it will take decades to see the effects. ::

How Can Stem Cells and Cloning Be Used to Alter People?

The case of Carrie Buck and the modern development of gene therapy technologies illustrate our long-standing confidence, perhaps misplaced, that we really do know how to define desirable and undesirable traits. The eugenics movement in the United States is a terrible lesson about what can happen when actions go beyond actual knowledge and are fueled by prejudice (particularly targeted at the poor).

Gene therapy is not the only technology that is being developed to permanently change the biology of an individual at a very fundamental level. The use of stem cells and cloning also offer the hope of curing disease, as well as the risk that we might produce outcomes with unintended consequences. Do we have adequate knowledge to develop and use these technologies safely? Do we know enough to judge whether we should?

Stem Cells

The basic common feature shared by all **stem cells** is the ability for self-renewal. At each division, a stem cell can give rise to another stem cell as well as a more specialized cell (**FIGURE 8.13**). But not all stem cells are the same. They come from varied sources and differ in their developmental flexibilities. Some stem cells can be directed to make any type of cell in the body, whereas others are capable of producing only a much more limited range of cells.

How Stem Cell Research Works The hope engendered by stem cell research is that scientists will be able to develop methods to use stem cells to replace neurons lost by Parkinson's disease or spinal injury; replace diseased pancreatic cells and cure diabetes; or repair heart damage by adding new cells to the heart, to cite some examples. The concerns about stem cells center mostly on the sources of the stem cells themselves.

Stem cells can be collected from many sources: early embryos; fetal cells; umbilical cord blood; and babies, children, or adults—these are usually referred to as **adult stem cells**. Stem cells collected from early embryos have the greatest developmental flexibility, and those from babies, children, or adults have the least.

Stem Cells in Human Development Embryo development starts with a single cell, the zygote, which was produced when an egg and sperm fused together (Chapter 3). That cell is capable of producing any of the cells in the developing embryo; it is **totipotent**. This initial cell undergoes divisions and up until

stem cell :: a cell that can renew itself and give rise to other cells

adult stem cell :: a cell from an adult that can renew itself, but that is limited in its flexibility

totipotent :: a cell that is capable of producing any of the cells in a developing embryo

Human embryonic stem cells can be isolated from embryos that are five to seven days old; these cells are **pluripotent**, or capable of producing any of the cells of the body.

Adult stem cells can be isolated from infants, children, or adults; they generally have less development flexibility than cells collected from early embryos.

Single-cell embryo	Three-day embryo	Five- to seven-day embryo	Four-week embryo	Six-week embryo	Infant	Child	Adult
Embryo development starts with a single cell, which is **totipotent**.	Until the eight-cell stage, each cell of the embryo is theoretically capable of producing a baby.	After the eight-cell stage, the embryo forms an **inner cell mass (ICM)** that will develop into the embryo, fetus and eventual baby, and the **trophoblast** that will produce support structures such as the placenta.		Human embryonic germ cells can be isolated from embryos or fetuses that are six to nine weeks old; these cells produce eggs and sperm.			

FIGURE 8.13 A stem cell is self-renewing, and at each division it can give rise to another stem cell as well as a more specialized cell.

human embryonic stem cell :: a cell that comes from an inner cell mass and can be isolated from an embryo that is five to seven days old and grown in a lab

pluripotent :: a cell that is capable of producing any of the cells in the body but none of the support structures needed for the development of a baby

human embryonic germ cell :: a cell that comes from a six- to nine-week-old embryo and fetus; these cells ultimately form the cells that produce eggs and sperm

the eight-cell stage, each cell of the embryo is totipotent—theoretically capable of producing a baby. After this stage, the embryo forms two populations: the inner cell mass (ICM) that in normal development will form the embryo, fetus, and eventual baby, and the trophoblast that will eventually produce support structures such as the placenta.

The ICM cells are the source of **human embryonic stem cells**. They can be isolated from embryos that are five to seven days old and grown in a lab. These cells are **pluripotent**, or capable of producing any of the cells of the body but none of the support structures needed for the development of a baby. They can't help form the placenta and, therefore, are not remotely capable of producing a baby. Nevertheless, embryonic stem cells have tremendous flexibility. (See *How Do We Know? How Human Embryonic Stem Cells Can Be Directed to Form Specialized Cells.*)

Rivaling embryonic stem cells in their potential are the **human embryonic germ cells** collected from six- to nine-week-old embryos and fetuses. During normal development, embryonic germ cells ultimately form the cells that produce eggs and sperm. Embryonic germ cells, too, are pluripotent; they behave similarly to embryonic stem cells in experiments.

As development unfolds, the versatility of stem cells becomes more limited. In adults, for example, many types of stem cells are restricted, meaning that at each division they can make either another stem cell or *one* type of specialized cell. For instance, certain blood stem cells are limited in this way. Stem cells collected from umbilical blood have less developmental potential than embryonic stem cells but more than adult stem cells.

Concerns About the Sources of Stem Cells Embryonic stem cells have tremendous potential for the medical treatment of ailments like Alzheimer's disease, heart disease, brain injury, blood cancers, immune system disease, and many others. As a consequence, many individuals support human embryonic stem cell research wholeheartedly. However, a significant number of people oppose human embryonic stem cell research because it destroys human embryos. Individuals have varying beliefs about when human life begins and also about what is and is not appropriate to do with and to human embryos (**FIGURE 8.14**).

FIGURE 8.14 Individuals have varying beliefs about when human life begins and also about what is and is not appropriate to do with and to human embryos. In 2006, President George W. Bush vetoed legislation that would have expanded federal funding for stem cell research. In 2009, President Barack Obama signed an executive order restoring full federal funding of stem cell research.

How Human Embryonic Stem Cells Can Be Directed to Form Specialized Cells

Scientists have been studying and manipulating the cells of mammalian embryos, especially mice, since the 1950s. This research has led to the understanding that embryonic cells are totipotent at the very earliest stages, but cells of the inner cell mass are pluripotent. Many details of how to direct embryonic stem cells to form particular specialized cells in culture have also been determined. This progress is excellent if all we want to do is to heal the illnesses and disabilities of mice, but what about humans? Can human embryonic stem cells be directed to make specific types of cells?

The answer is *yes*. The first published report of human embryonic stem cells directed to make specialized human tissues appeared in 2001. Daniel Kaufman and his colleagues at the University of Wisconsin–Madison succeeded in getting human embryonic stem cells to make precursor cells for blood cell production.

Kaufman isolated human embryonic stem cells and cultured them in media containing fetal bovine serum and either mouse bone marrow cells or mouse yolk sac endothelial cells.

The idea was to mimic the cocktail of chemical signals and cell communications that happen during normal development. To everyone's delight, the human embryonic stem cells produced hematopoietic precursor cells, which in normal embryos will make different types of blood cells. Kaufman cultured these hematopoietic precursor cells in various hematopoietic growth factors—the same molecules that direct blood cell development in the bone marrow. Amazingly, myeloid, erythroid, and megakaryocyte colonies grew. These give rise to white blood cells, red blood cells, and platelets, respectively—all the solid components of blood.

Kaufman's findings are important for two reasons. First, they add to our basic knowledge of how to manipulate human embryonic stem cells. Second, they may allow for the development of a new way to generate blood cells for transfusion and for transplantation to treat anemias and blood cancers. Given the shortage of the blood supply and the difficulty in finding bone marrow donors, this would be a welcome breakthrough. ::

① Kaufman isolated human embryonic stem cells.

② Kaufman cultured the cells in media containing fetal bovine serum plus either mouse bone marrow cells or mouse yolk sac endothelial cells.

③ Hematopoietic stem cells developed.

The hematopoietic stem cells produced myeloid, erythroid, and megakaryocyte colonies.

Myeloid cells → White blood cells

Erythroid cells → Red blood cells

Megakaryocytes → Platelets

The colonies produced white blood cells, red blood cells, and platelets, respectively.

The first published report of human embryonic stem cells being directed to make specialized human tissues occurred in 2001. Daniel Kaufman and his colleagues at the University of Wisconsin–Madison succeeded in getting human embryonic stem cells to make precursor cells for blood cell production.

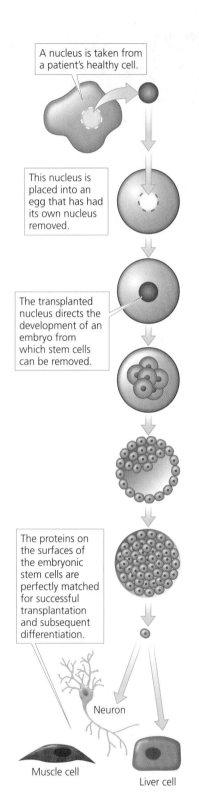

A nucleus is taken from a patient's healthy cell.

This nucleus is placed into an egg that has had its own nucleus removed.

The transplanted nucleus directs the development of an embryo from which stem cells can be removed.

The proteins on the surfaces of the embryonic stem cells are perfectly matched for successful transplantation and subsequent differentiation.

Neuron

Muscle cell

Liver cell

FIGURE 8.15 Therapeutic cloning is used to produce healthy embryonic stem cells for the medical treatment of individuals. In somatic cell nuclear transfer therapy (SCNT), a nucleus is taken from a somatic cell and placed in an oocyte to produce embryonic stem cells.

At issue are two questions: when does a human life begin, and what is appropriate to do *with* and *to* a human embryo? Neither question is easy to answer because many of the arguments are based in religious beliefs. And since we are a pluralistic society open and respectful to all faith traditions, it is difficult to envision how to resolve our differences. It may help to look at how human embryonic stem cells are collected.

Human embryonic stem cells can be collected from five- to seven-day-old embryos that are composed of 100–150 cells. Some argue it is acceptable to generate human embryonic stem cells by combining donated eggs and sperm in the lab expressly for this purpose. Others maintain that creating embryos as a source of stem cells is unacceptable, but using "extra" embryos left over in IVF clinics is fine. (There are more than 400,000 frozen embryos stored in the United States.) And some say that no source is acceptable.

People of all religions agree that human life is special and should be respected, valued, and protected. But they do not agree on whether embryonic stem cells should be used for research or for therapeutic reasons. Here are just *some* examples of the range of beliefs about this issue:

- According to Jewish teachings, a fertilized egg does not have "personhood." Therefore, it is acceptable to use it for stem cell collection for medical research or ultimately therapeutic treatment.
- Roman Catholics believe that an embryo is a human being from the moment of conception. Therefore, they maintain that human embryonic stem cell research is prohibited. The Catholic Church does, however, support adult stem cell research.
- Buddhist teachings do not directly address this question, and the community is divided.
- Although Hinduism teaches that life begins at conception, the religion does not have an official statement on the subject of stem cells.
- Protestantism covers a diverse range of religious views. The Episcopalian Church is in favor of stem cell research provided certain conditions are met: the embryos would have been destroyed anyway, the embryos were not created solely for research, and embryos are not to be bought or sold. The official positions of the United Church of Christ and the Evangelical Lutheran Church regarding using stem cells are *yes* and *no*, respectively. The Southern Baptist Church recently reaffirmed its opposition to the use of stem cells, while the Presbyterian and Methodist Churches affirmed their support for the use of stem cells for medical and therapeutic applications.
- Islamic teaching does not address this issue directly. Some Muslim leaders argue that termination of an embryo at any stage is unacceptable. Other Muslim leaders support using embryonic stem cells for research or eventually for medical therapies, provided cells are collected before "ensoulment"—the moment when an embryo or fetus attains full human rights. In either case, research on embryonic stem cells obtained from extra IVF embryos *or* from embryos created for research is acceptable.

Noting the lack of a clear consensus on the issue, the National Council of Churches officially neither endorses nor condemns the use of human embryonic stem cells. Perhaps considering several scientific facts may help to shed light on this question. First, human embryonic stem cells are isolated from embryos at such an early stage that no specialization or differentiation has occurred. In this view, the cells contain human genomes but are not persons. Second, human embryonic stem cells are not even "potential persons." The isolated cells could never give rise to an entire embryo because they can't produce any of the tissues needed to support the development of an embryo, fetus, or baby.

Is it possible to find a resolution for these dramatically varying perspectives? Should religion play a role in how we make research decisions? What do we do when religions come to different conclusions about issues that may affect everyone?

Cloning

The overarching goal of eugenics was to improve the biology of human beings. Carrie Buck's experience demonstrates "negative" eugenics—stopping the mating of so-called defectives. "Positive" eugenics is illustrated by germ-line gene therapy. Here the effort is to improve the genetic lineage for all descendants. One more potential manifestation of positive eugenics is human cloning, which is distinct from the gene cloning discussed in Chapter 7.

Just as gene therapy can be divided into somatic gene therapy (treatments meant to cure an individual) and germ-line gene therapy (efforts to fix "bad" genes for descendants), cloning can be divided into two categories. **Therapeutic cloning** refers to producing embryos as sources of healthy embryonic stem cells for the medical treatment of individuals. **Reproductive cloning** means creating an embryo with the intention of producing a baby. Any features present in the clone are intended to be passed on to future generations.

Therapeutic Cloning Using stem cells to treat or cure a disease can sometimes produce undesirable side effects. All cells have proteins on their surfaces that permit the body to differentiate between "self" and "nonself" to determine whether transplanted organs, tissues, or cells will be accepted or rejected. Human stem cells collected from an embryo will not have the same protein recognition markers as those of the patient's cells and therefore may be rejected when transplanted. The technical solution to this problem is therapeutic cloning done by **somatic cell nuclear transfer (SCNT)** (**FIGURE 8.15**).

In SCNT, a nucleus is taken from a healthy cell in the patient's body. This nucleus is placed into an egg that has had its own nucleus removed. The transplanted nucleus then directs the development of an embryo from which embryonic stem cells can be removed. Because the transplanted nucleus came from the patient, the proteins on the surfaces of the embryonic stem cells are perfectly matched for successful transplantation. Embryos created in this way are intended for the medical treatment of patients, not for reproduction.

Reproductive Cloning The express goal of reproductive cloning is to create embryos that will develop into offspring. As with SCNT, a nucleus taken from a "parent" is placed into an egg that had its nucleus removed. The embryo that results is implanted into a surrogate mother. An organism produced in this way will be genetically identical to the parent that donated the nucleus.

Several nonhuman animals—including a cat, a dog, cattle, a horse, a goat, a mouse, sheep, and a rabbit—have been cloned. But there seems to be little or no support for cloning humans. In fact, reproductive cloning of humans is illegal in more than 25 countries (but not in the United States; although some state laws ban the practice, no federal laws ban human reproductive cloning).

therapeutic cloning :: producing embryos as sources of healthy embryonic stem cells for the medical treatment of individuals

reproductive cloning :: creating an embryo with the intention of producing a baby

somatic cell nuclear transfer (SCNT) :: therapeutic cloning in which a nucleus is taken from a healthy cell and placed into an egg that has had its own nucleus removed; the transplanted nucleus then directs the development of an embryo from which embryonic stem cells can be removed

What Are the Benefits and Risks of Stem Cell Research?

When it comes to the medical role that stem cells can play in the future, there is generally more support for such research than support for cloning. What has the research revealed so far about the benefits and risks of using stem cells?

Benefits

Most of the experiments done to understand stem cells were performed on mice. Mouse embryonic stem cells have been directed to form pancreatic cells that make insulin, muscle cells that contract, and functional neural cells. Similarly, human embryonic stem cells have been directed to make specific cell types in culture.

Research in whole animals has also produced promising results. For example, when injected into the brains of rats that suffered from a disease akin to Parkinson's disease in humans, human embryonic stem cells differentiated into neurons. Parkinson's disease is characterized by the failure of certain neurons to secrete the chemical signal dopamine. The result is that brain cells deteriorate, with symptoms that include tremors, rigid movements, and slow speech. In rats, the neurons that formed from the injected human embryonic stem cells not only produced dopamine, but also reduced the Parkinson's-like symptoms. Similarly, human embryonic stem cells that differentiated into heart cells were able to repair a heart defect in pigs. There are many technical details to be worked out, but it does seem plausible that stem cells may be able to cure and treat disease.

Human embryonic stem cells have several other uses besides the treatment of disease. First, they are a basic research tool to learn more about normal human development and how birth defects occur. Second, by altering specific genes of embryonic stem cells, it will be possible to create human disease models at the cellular level instead of relying on mouse or other mammalian versions. Third, currently, pharmaceutical drugs are initially tested on nonhuman organisms. Embryonic stem cell cultures would allow scientists to test drug effects on specific types of human cells.

Risks

As with any new technology, stem cells are not without risk. First, the procedures are not completely safe as yet. Transplanted stem cells are not always perfectly matched to patients, and the cells may be rejected. For example, adult stem cells are routinely used to treat certain blood cancers. If these stem cells are rejected, the patient may develop host-versus-graft disease wherein the immune system attacks other parts of the body. Also, the dosage of embryonic stem cells needs to be carefully controlled or tumors may form. This occurred when neurons developed from human embryonic stem cells that were transplanted into rat brains.

Besides the safety issues surrounding the actual use of stem cells, the donation of eggs needed for SCNT is also potentially dangerous. A large number of eggs will be needed for research and implementation of medical treatments. To donate eggs, women must undergo hormone treatments so they will "superovulate," or release multiple eggs. About 5% of women who go through this process develop ovarian hyperstimulation syndrome; 1–2%

experience a severe form of this syndrome, leading to possible kidney failure, lung failure, shock, or ruptured ovaries. Although it might seem that a woman fully informed about the risks can make her own decision about whether to donate eggs, some people worry that poor women might be exploited or pressured into selling eggs.

Another concern about stem cell and cloning technologies is that they could be used for enhancement rather than for medical or therapeutic purposes. Even individuals supportive of therapeutic cloning worry that the knowledge gained in this research could be used to make human reproductive cloning a reality.

Finally, some people are concerned that these new technologies could extend the human life span beyond what is beneficial to society. They argue that having an overly long life, relatively free of worry about "unfixable" ailments, might lead to a less meaningful life since there is less concern about dying any time soon. Also, an able-bodied healthy but older generation would not have to get out of the way for the next, so perhaps innovation would suffer. Or maybe we need to rethink how we shape our lives so that fulfillment is possible at *any* age.

8.6 What Other Challenges Result from Advances in Medical Technology?

This leads us back to where we began—with Carrie Buck. A group of powerful people used the claim of scientific expertise, political power, and ideas about what was desirable or normal to put an end to her germ line forever. They made sure that no more "imbeciles" would be born in the Buck family.

The eugenics law, with mandatory sterilization, was not unique to Virginia. More than 30 states enacted them by the 1930s. And the laws remained on the books for several decades. For example, North Carolina's eugenic sterilization law, although not enforced, was updated in 1973 and 1981. Also, forced sterilizations continued at least into the 1970s: between 1972 and 1976, 3406 women and 142 men were sterilized in just four hospitals in a program of Native American sterilization implemented by Indian Health Service physicians.

Today we don't define "desirable" in as coarse a manner. Yet we do make judgments about which traits and physical characteristics are good and which are to be avoided. And we still face challenges over how to avoid the undesirable characteristics and how to pay for the privilege.

Privacy

The ability to identify specific disease genes and other DNA sequences in people has the potential to create a new body of information about each of us (**FIGURE 8.16**). What disease genes do I carry? What risk does my genome pose for me or my children? (See *Technology Connection: Who's the Daddy?*) Do my genes suggest that I have a predisposition for cancer, heart disease, schizophrenia, or alcoholism? Who should be allowed access to this private information about me? My doctor? Employer? Insurance company? The police or government?

Blood from the heel of a newborn baby is taken via a Guthrie test that screens for a rare genetic disease, phenylketonuria (PKU), which affects around 1 in 16,000 people and can cause severe brain damage.

According to the Health Insurance Portability and Accountability Act (HIPAA), patients have significant legal rights and control over their personal medical information and records.

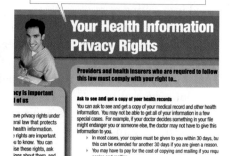

FIGURE 8.16 Genetic testing is available for many diseases and disorders. However, such testing and results raise concerns about privacy and who should be allowed access to this private information.

Technology Connection

Who's the Daddy?

Although it's generally easy to identify a baby's mother at birth, it can sometimes be a challenge to identify the father. In 1944, for example, the actress Joan Berry sued Charlie Chaplin, a famous silent film star, for paternity. Chaplin denied that he was the father of her baby, and the case went to court. Chaplin, Berry, and the baby all had their blood types tested. The results were unequivocal—Chaplin was not the father. Three physicians testified at the trial on Chaplin's behalf, but the judge refused to allow blood type as evidence. Chaplin was judged to be the father and ordered to pay child support. This case led to a change in the law regarding paternity and blood tests. But blood tests can only *rule out* paternity, not confirm it. More informative tests were needed to identify the match with more certainty.

Fast forward 60 years. Paternity testing is now almost foolproof, thanks to the development of DNA testing in the late 1980s. Here is how it works.

Each of us has 46 chromosomes—23 inherited from our mother, 23 from our father. Consequently, each of us has a unique combination of chromosomes. A comparison of the DNA base sequence (A, G, T, and C) of any two human beings is more than 99.9% identical. The 0.1% translates into approximately 3 million bases that are different from person to person. The sequence of this 0.1% of DNA is unique to an individual. DNA testing examines these sequences to determine identity. In the case of paternity testing, DNA samples are taken from the mother, baby, and potential father. The DNA is chopped up into pieces using enzymes that break the DNA

strand at specific sequences of bases. The fragments that result are particular to an individual. Next, these DNA fingerprints from the baby, mother, and potential father are compared to each other to see whether they match. Paternity can be confirmed or ruled out with greater than 99.9% accuracy.

Why go through all this trouble to learn the identity of a father? There are several good reasons: to establish legal rights such as eligibility for the father's Social Security or veterans' benefits; to obtain family medical history for better medical care; and to establish a caring and supportive relationship between father and child. ::

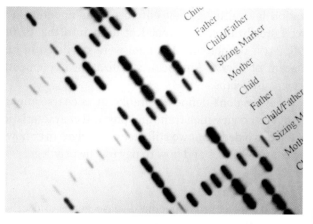

Paternity can be confirmed or ruled out with greater than 99.9% accuracy. In both visible paternity investigations on this autoradiograph, the supposed father is "included," meaning that his DNA bands match those of the child.

Accessibility

If the much-hoped-for medical cures from stem cells and therapeutic cloning become a reality, will they be available to everyone? And if so, how? These cures are likely to be expensive, so it is hard to imagine how poor people will be able to gain access. Even today, in countries with privatized health care like the United States, poor people are unable to obtain the same medical treatment enjoyed by the wealthy. In countries where there is not a well-developed medical system with hospitals, clinics, and an adequate number of physicians and nurses, there is little or no access to the medical technologies that already exist.

In fact, it is reasonable to ask whether spending money on the development of advanced technologies like stem cells and therapeutic cloning is a good use of resources. Would it be wiser to provide basic health care to all? Are there adequate resources to do both?

Danger of a New Eugenics Movement

We have established an entire system of genetic counseling and prenatal and postnatal screening. Is our scientific knowledge more accurate today compared to Carrie's time? Undoubtedly, yes, but there is still a great deal to learn. Has our judgment improved since the 1920s? Or are we creating a new eugenics movement? These questions are harder to answer.

Genetic screening makes sense to most people. The results provide parents with information about their child's potential future. If an embryo or fetus has a "problem," the parents can decide whether they wish to continue the pregnancy. In effect, they are deciding which traits are acceptable and which are not. Not everyone agrees about these judgments.

In a survey conducted by the Shriver Center across 37 countries (the results of which were published in 2002), 82% of U.S. geneticists supported the abortion of fetuses with Down syndrome (**FIGURE 8.17**) and 92% thought it advisable to abort fetuses with severe spina bifida. Abortion of fetuses with

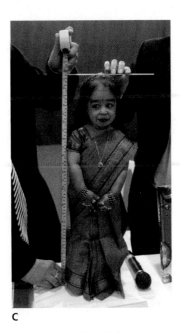

A   B   C

FIGURE 8.17 In a survey conducted by the Shriver Center across 37 countries, the majority of U.S. geneticists advised the termination of pregnancy for fetuses with Down syndrome (A), spina bifida (B), or dwarfism (C). The public generally did not agree.

achondroplastic dwarfism was advised by 56% of the geneticists. However, the general public was not so eager to support the termination of fetuses unless there was the certainty of extremely severe mental retardation.

Most people would probably agree on the merits of prenatal screening for the detection of invariably fatal genetic diseases, but even here the choice is not easy. What if the fetus has a fatal genetic disease like Huntington's, which generally does not reveal symptoms of neurological degeneration until the third or fourth decade of life? If a person dies at 50 years, was that life not worth living? Should such a person not have children?

At its extreme, genetic counseling and selective abortion could create a world in which wealthy families have less risk of having disabled children—simply because they can afford the tests. This world could be divided into genetic "haves" and "have-nots." Or we could try to use these tools more equitably. We could recognize that because of mutation, there will always be genetic disease and that meaningful, satisfying lives come in many forms.

 # Biology in Perspective

Most people are confused about what genes can and can't do. At its most basic, the DNA of genes has the information to direct the production of a protein. The protein has some particular function in a cell. Somehow this simple scenario plays a critical role in how we look, function, and act. It is also important for influencing whether we are healthy or sick.

Genes don't act alone, however. They form interacting networks. The proteins that are produced may in turn regulate gene function, and they may work together with other proteins too. The whole process is sensitive to the environment, including many of the behavioral and lifestyle choices we make. We have learned so much about gene structure and function that it is now possible to change the genetic makeup of cells, organisms, and even entire families. It is important to recognize that for every question answered by discovery, new ones, often unanticipated, are uncovered. As much as we know, there is even more to learn.

The ability to manipulate genes in people, collect stem cells from human embryos, and clone human embryos has crowded us into a space where science is forcing us to define ethical and moral boundaries to guide this work. The question remains: how do we define universal ethical standards so we can decide whether or how to move forward?

Chapter Summary

8.1 Do Complex Human Characteristics Have a Genetic Basis?

Assess the legitimacy and risks of genetic determinism to society.

- Genetic determinism is the idea that genes determine, direct, or cause all traits.
- In the early 20th century, eugenics laws designed to improve the "human stock" forcibly sterilized people perceived as defective.
- In contrast to genetic determinism, we now know that traits are the result of the interaction of the environment and genetics.
- We are beginning to be able to identify maladaptive genes; gene therapy and cloning have the potential to cure many diseases but also hold the inherent risk of defining a new "normal."

8.2 What Is Gene Therapy?

Distinguish between somatic and germ-line gene therapies.

- Somatic gene therapy cures genetic disorders by inserting normally functioning genes into patients.
- Germ-line gene therapy alters the genes in germ cells (eggs or sperm) prior to fertilization.

8.3 What Are the Benefits and Risks of Genetically Altering Humans?

Contrast the benefits and risks of somatic and germ-line gene therapies.

- Somatic cell therapy has the potential to cure many genetic disorders by replacing specific cells in an individual, but those changes are unlikely to be passed on to offspring.
- Germ-line gene therapy can eliminate maladaptive genes from embryos so that all of the cells of the individuals have the trait.
- For both therapies, it can be difficult to predict the impact of new or changed genes on other genes.
- Some potential risks specific to germ-line gene therapy include the loss of potentially valuable variation through designer babies and the development of an enhanced genetic class.

8.4 How Can Stem Cells and Cloning Be Used to Alter People?

Compare the different methods and uses of stem cells and cloning.

- Stem cells offer the promise of curing many diseases because they can replace damaged, dysfunctional, or missing cells.

- Totipotent cells are taken from embryos up to the eight-cell stage and can develop into any cell type. Pluripotent cells are taken from the inner cell mass of a five- to seven-day-old embryo and can differentiate into many but not all cell types and consequently cannot grow into a baby.
- Human embryonic germ cells are cells taken from six- to nine-week-old embryos; germ cells form eggs or sperm.
- Cloning is the creation of a genetically identical copy of an organism. In therapeutic cloning, healthy embryos are created for medical treatment. Reproductive cloning creates embryos to form a baby.

8.5 What Are the Benefits and Risks of Stem Cell Research?

Use information on risks and benefits to evaluate the ethics of stem cell research.

- Human embryonic stem cells may be the key to curing many diseases.
- We currently use stem cells to learn about human development, and in the future we may have cells that model disease and enable us to test drugs, reducing the use of laboratory animals.
- Technical risks for stem cells include possible rejection by patients, cancer, and tumors.
- Egg donation to generate embryos requires hormone treatments that can be dangerous to the donor.
- Stem cells and cloning could be used for enhancement, rather than medical or therapeutic purposes, and may extend life beyond the normal life span, putting a possible burden on society.

8.6 What Other Challenges Result from Advances in Medical Technology?

Assess the ethical challenges we face as a result of advances in genetic technology.

- Access to genetic screening data could be used by governments, insurance companies, or corporations to discriminate against individuals.
- At the other extreme, individuals with resources to genetically enhance themselves and their children may create a genetically enhanced class.
- Our ability to identify genetic traits is more advanced than our ability to fix genetic defects.

Key Terms

adult stem cell 231
eugenics 219
genetic counseling 228
genetic determinism 220
genetic essentialism 220
genome 220
germ-line gene therapy 224

hemophilia 227
human embryonic germ cell 232
human embryonic stem cell 232
immune response 224
lymphocyte 224
pluripotent 232
reproductive cloning 235

somatic cell nuclear transfer (SCNT) 235
somatic gene therapy 223
stem cell 231
therapeutic cloning 235
totipotent 231

Review Questions

1. What is genetic essentialism?
   a. The notion that our genes determine, direct, or cause everything about us
   b. The notion that being human means having a human genome
   c. The notion that there are just a few essential things about our genes that make us human
   d. The notion that there are several things about our genes that make us human
   e. The notion that all living organisms are unique

2. What is genetic determinism?
   a. The idea that our genes determine, direct, or cause everything about us
   b. The idea that being human means having a human genome
   c. The idea that there are just a few essential things about our genes that make us human
   d. The idea that there are several things about our genes that make us human
   e. The idea that all living organisms are unique

3. Describe the process of somatic gene therapy.

4. Describe the process of germ-line gene therapy.

5. What are four ethical concerns associated with germ-line gene therapy?
   a. Safety, distinguishing between medical treatment and enhancement, privacy, and changing how we think about the sick and disabled
   b. Safety, accessibility, privacy, and changing how we think about the sick and disabled
   c. Privacy, accessibility, safety, and difficulty distinguishing between good and bad genes
   d. Privacy, safety, a misunderstanding about what genes can and can't do, and accessibility

   e. Safety, difficulty distinguishing between good and bad genes, distinguishing between medical treatment and enhancement, and changing how we think about the sick and disabled

6. How do the benefits and risks of somatic gene therapy and germ-line gene therapy compare to one another?

7. What are stem cells?

8. What is the difference between totipotent and pluripotent stem cells?
   a. Totipotent cells are capable of producing any of the cells in the body but none of the support structures needed for a baby's development; pluripotent cells are capable of producing any of the cells in a developing embryo.
   b. Totipotent cells will form the embryo, fetus, and eventual baby; pluripotent cells generally produce support structures such as the placenta.
   c. Totipotent cells are capable of producing any of the cells in a developing embryo; pluripotent cells are capable of producing any of the cells in the body but none of the support structures needed for a baby's development.
   d. Totipotent cells produce the support structures of the embryo such as the placenta; pluripotent cells form the embryo, fetus, and eventual baby.
   e. There is no real difference between totipotent and pluripotent stem cells; they are synonyms.

9. Describe the process of therapeutic cloning using somatic cell nuclear transfer (SCNT).

10. How do the objectives of therapeutic and reproductive cloning differ?

11. What are the potential benefits that may derive from (a) gene therapy, (b) the use of stem cells, and (c) cloning?

12. What are the potential risks that may derive from (a) gene therapy, (b) the use of stem cells, and (c) cloning?

13. What are some new challenges that result from our new medical technology?

The Thinking Citizen

1. Considering the question from a strictly scientific standpoint, could eugenics be used effectively to improve humans? Why or why not?

2. Explain, with examples, why it is more difficult to correct gain-of-function mutations than loss-of-function mutations in patients.

3. Is there a genetic basis for poverty? Why does it often seem to "run in families"?

4. How do you define "normal"? Explain your reasoning.

5. Should we try to eliminate all genetic defects? Why or why not? Is this possible to do? Why or why not?

The Informed Citizen

1. Should the reproductive cloning or germ-line gene therapy of humans be permitted? Why or why not?

2. In what ways are modern methods to manipulate genes in people different from eugenics? In what ways are they similar?

3. Is it acceptable to use gene therapy for enhancement? Would it be appropriate to insert genes so that a person could be taller, stronger, faster, or prettier? Where do we draw the line between treatment and enhancement? Is being short a medical condition? What about being bald?

4. What solutions can you suggest to solve the problem of providing equal access to medical treatments and technologies for all? Is it possible to do so?

5. What do you think are the potential consequences of a world in which there are genetic "haves" and "have-nots"?

6. What do you think are the potential positive and negative consequences of longer life spans?

7. Given that we have finite resources, should we spend our money on basic health care for all right now or on research to develop the cures of the future? Explain your reasoning.

8. Who should have access to your genetic medical record? Why?

9. What role, if any, should religion play in medical research decisions? How do we resolve differences in judgments about issues that affect all of us when these opinions are based in religion?

CHAPTER LEARNING OBJECTIVES

After reading this chapter, you should be able to answer the following questions:

9.1 How Does Your Body Reflect an Evolutionary History?
Provide three examples demonstrating the effect of evolutionary history on human structure or function.

9.2 What Convinced Darwin of the Fact of Evolution?
Detail the discoveries that led Darwin to change his mind about evolution and describe the steps of natural selection.

9.3 How Do Humans Adapt to Their Environment?
Use specific examples to describe how humans adapt to their environments.

9.4 How Does Natural Selection Produce Adaptations?
Use natural selection to describe the process whereby a population adapts to a specific environment and outline ways that adaptation might be limited.

9.5 What Are Random Events in Evolution?
Describe the potential role of random processes in evolution and differentiate natural selection from genetic drift.

9.6 What Is the Evidence for Speciation?
Summarize the evidence for speciation.

9.7 How Do New Species Arise?
Outline the stages by which new species arise from a single species.

9.8 Why Is It So Difficult for the Public to Accept Evolution?
Assess reasons why some public resistance to evolution still exists despite overwhelming scientific evidence.

Evolution helps us understand many of the most prevalent facts of nature: why organisms fit in so well with their environments, why there are so many species, and why some species are so similar while others are not.

CHAPTER 9

Evolution

How Do Species Arise and Adapt?

Becoming an Informed Citizen ...

Evolution is the central unifying theme of biology; it explains how all living organisms are related and how existing species adapt to their environments and new species arise. For the majority of scientists, evolution is an accepted fact of nature; yet, in the United States, controversy and confusion over the theory remain.

evolution :: theory that explains how all living organisms are related and how existing species adapt to their environments and new species arise

n Chapter 1, we pointed out that **evolution**—*the theory that explains how all living organisms are related and how existing species adapt to their environments and new species arise—is the central unifying theme of biology. All of the biological concepts we discuss in this book and more are shaped by evolution. Our discussion so far has focused on* how *these biological processes work. Evolution helps us understand* why *they work the way they do. Evolution also helps us understand many of the most prevalent facts of nature: why organisms fit in so well with their environments, why so many species exist, and why some . species are so similar while others are not.*

In this chapter, we complete the story of what evolution is and how it works. We point out interesting and puzzling aspects of the natural world and show how they can be accounted for by evolution operating over long periods of time. Understanding evolution is vital to understanding nature, biology, and medicine. For the majority of scientists and many other people, the idea of evolution is no more controversial than any other fact of nature. Still, controversy (and confusion) over evolution is common in the United States, and we will examine why.

case study

Lactose Intolerance and the Geographic Variation of Human Traits

Do you like ice cream? How about a nice grilled cheese sandwich or yogurt for lunch, or perhaps some milk and cookies before bed? If you enjoy this sort of thing, it might surprise you to learn that for *most* adults, eating food like this has disastrous consequences: intestinal cramping, bloating, diarrhea, and embarrassing flatulence.

Dairy products contain a sugar called lactose. Humans—and all mammals—drink their mother's milk as infants, and most infants are able to break down lactose in breast milk. But the majority of humans lose this ability around two years of age, around the time they are weaned. Worldwide, about 75% of us can't digest dairy products, and only 25% can eat that ice cream sundae without suffering uncomfortable side effects.

Lactose is broken down by an enzyme called lactase (**FIGURE 9.1**). Almost all of us can make lactase, but its production is controlled by another gene. In humans, this gene shuts down lactase production in adults. But there is a recessive **mutation**, or change in DNA, that allows lactase production to continue on into adulthood. This mutation is not common worldwide, which is why most adults can't digest dairy products. Interestingly, this mutation is common in two groups of people: Northern Europeans and their descendants, and several ethnic groups from equatorial Africa (**FIGURE 9.2**). Using genetic techniques, scientists can estimate when this recessive mutation first appeared. Among Europeans, it became common about 10,000 years ago. How can we explain the geographic distribution and age of this mutation?

For some genetic traits that vary geographically, the consequences are more severe than indigestion. For example, one way to treat malaria is to give

mutation :: change in DNA

Lactose metabolism

The enzyme lactase acts on the lactose, converting it into the sugars galactose and glucose.

CH_2OH ... CH_2OH ... Lactase ... CH_2OH ... CH_2OH

Lactose → Galactose + Glucose

FIGURE 9.1 Lactose, the sugar found in dairy products, is broken down by the enzyme lactase.

the patient "oxidizing drugs." These drugs put metabolic stress on red blood cells—the cells that the malaria parasite infects. The stress is bad for the blood cells, but it's even worse for the parasite. The treatment is calibrated to kill a maximum number of parasites and a minimum of red blood cells. But when this treatment is given to people of African and Mediterranean descent, it kills about 10% of these individuals. People from other ancestries are not subject to the potential lethal effects of these drugs. Why are people from areas with a high rate of malaria so susceptible to this drug treatment, while people from other areas are not?

There are many traits that show geographic **variation**, which means the traits occur in people in some regions of the world but not in others: skin color, susceptibility to various types of cancer, cholesterol levels, and the ability to metabolize alcohol, to name just a few. Why do these differences occur, and why are they tied to geography? All of these traits are examples of relatively recent human evolution, as our ancestors adapted to their local environment in different parts of the world.

variation :: traits that differ, often by geographic location

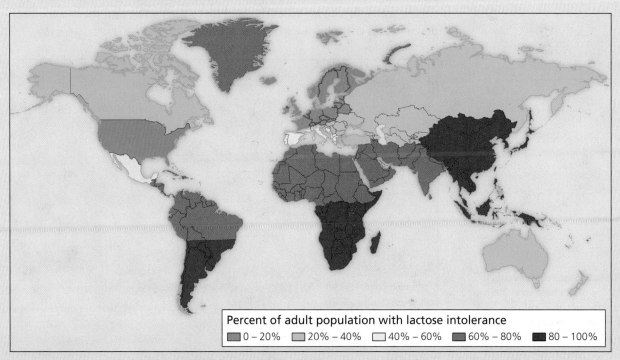

Percent of adult population with lactose intolerance
0 – 20%   20% – 40%   40% – 60%   60% – 80%   80 – 100%

FIGURE 9.2 Worldwide, 75% of adults have a gene that shuts down the production of lactase, making digestion of dairy products difficult. Even though this map depicts lactose intolerance as being low overall in the United States, there are large populations of people who are lactose intolerant.

9.1 How Does Your Body Reflect an Evolutionary History?

Suppose you wanted to learn about the borders of the nations of the Middle East (**FIGURE 9.3**). Some of these borders make perfect sense, as they follow natural features like mountains or perhaps rivers. But many other borders appear arbitrary and don't seem to have an obvious basis, geographical or otherwise. How can you explain why the borders are where they are?

To fully understand, you would need to study history. You would have to learn about the founding of Israel after World War II, colonial rule in Iraq by the British following World War I, and a series of other events reaching far back into history. If you want to understand why things are the way they are, including political borders, you need to consider not just the current situation but also the historical events that caused it.

Life is like this, too. Significant events in the history of life help to explain current life. Let's look at some examples from a familiar source: your own body.

Human (and Mammalian) Testes Hang Loose

Given male testes' importance in reproduction and sensitive nature, it is no surprise that most male vertebrates keep their testes protected inside their bodies. Mammals, and therefore humans, are exceptions to this rule. As you know, our testes are contained within an external scrotum adjacent to the penis (Chapter 6). This seems to be a convenient arrangement: the testes produce sperm, and the penis delivers them, all in one nicely localized package.

Things are not exactly as they seem, however. The vas deferens, or sperm duct, does not go directly from the testes to the penis. Instead, it leaves the scrotum, travels anteriorly, curves around the pelvis, loops around the ureter (which transports urine from the bladder to the penis), and then heads back down to join with the penis (**FIGURE 9.4**).

FIGURE 9.3 Significant events in the history of life help to explain current life, just as significant past world events help to explain current world affairs. After World War I, the victorious countries dissolved the Ottoman Empire and redrew the boundaries of countries in the Middle East, setting the stage for many of the current hostilities and conflicts.

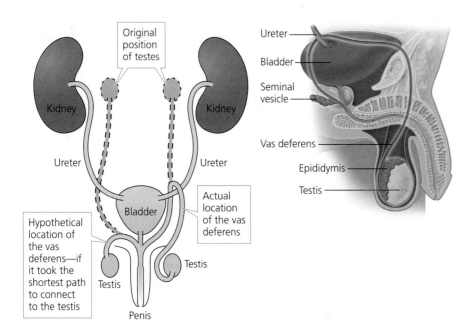

FIGURE 9.4 **The vas deferens, or sperm duct, takes a circuitous route from the testes to the penis rather than a more direct one.** There is no particular functional reason for where the vas deferens is; rather, it is a legacy from earlier ancestors.

Such a circuitous route! The vas deferens may be over 10 inches long, even though only a few inches would be needed. Our bodies do make use of this detour, however. Sperm cells mature during their passage through the vas deferens, and this route allows more room for the seminal vesicle and prostate gland that add seminal fluid to the sperm. Even so, it is hard to imagine that a more efficient route for sperm could not be found. And there is still the question of why the testes in mammals are external in the first place.

If we recognize that life has a history, and that all organisms are related because of this history, some pieces of the puzzle fall into place. In vertebrates (which include humans and all other mammals), the testes form inside the abdominal cavity near the liver during embryological development. In most vertebrates, that is where they stay, protected from harm.

However, mammals maintain a constant high body temperature, and sperm cells cannot develop properly at these temperatures. Male mammals have developed an external scrotum to house the testes and keep them cooler than the rest of the body. But in mammals, the testes form initially far from the scrotum, just as they did in our nonmammalian ancestors. This problem is solved by having the testes "descend" from the abdomen into the scrotum, and as they travel they drag the vas deferens along with them. In the process, the vas deferens loops over the ureter, the outlet tube from your bladder, and settles into the circuitous route just described. There is no particular functional reason for where the vas deferens is. It is a legacy from earlier ancestors, a product of life's historical events.

You Can Get Scurvy, but Your Pet Can't (Unless You Have Guinea Pigs)

Scurvy is a condition caused by a lack of vitamin C in your diet. It is not a pretty sight. Scurvy results in spongy, bleeding gums and eventual tooth loss, and in advanced cases patients develop open, pus-filled wounds (**FIGURE 9.5**). Scurvy has been recognized since ancient times, and many herbal cures are known, all of which are rich in vitamin C. British royal surgeon James Lind is credited with discovering in the 1750s that scurvy could be prevented by

FIGURE 9.5 **Scurvy is a condition caused by lack of vitamin C that results in spongy, bleeding gums and eventual tooth loss.** The vast majority of mammals do not need vitamin C in their diets because their cells can make their own vitamin C, but humans inherited a pseudogene that destroyed this function.

FIGURE 9.6 Vitamin C is made from a modified form of glucose. A series of enzymes alter this molecule in an assembly-line process, which results ultimately in the production of vitamin C.

eating citrus fruit, which resulted in the British Navy's policy of stocking its ships with limes and other fruits.

You know you need to add vitamin C to your diet. But you may not know that this is unnecessary for your cat, dog, or horse. In fact, the vast majority of mammals do not need vitamin C in their diets because their cells can make their own vitamin C. Vitamin C is made from a modified form of glucose, a simple sugar. A series of enzymes alter this molecule in an assembly-line process, which results ultimately in the production of vitamin C (**FIGURE 9.6**).

Most mammals have working copies of all enzymes, but humans (and all other primates) are missing the enzyme that performs the last step. However, humans are not missing the gene that makes this enzyme. The gene is in our chromosomes, but it carries a mutation that renders the enzyme it makes nonfunctional. All other primates carry exactly the same mutation in this gene and therefore lack the ability to make vitamin C.

Genes carrying mutations that make their products nonfunctional are known as **pseudogenes**. Modern molecular techniques allow us to examine the DNA sequence for many or all of an organism's genes. It turns out pseudogenes are surprisingly common. Some estimates for mammals run as high as 20,000 pseudogenes in each species—almost as many pseudogenes as regular, functional genes. What does the presence of a specific pseudogene shared only by primates, including humans, say about the genetic relatedness among the members of this group?

A mutation that prevented a species from making its own vitamin C would be lethal in most species, unless that species happened to have a diet already rich in vitamin C. Primates eat a lot of fruit and therefore get plenty of vitamin C. The fact that all primates have the same pseudogene indicates that the mutation that destroyed its function occurred in the original ancestor species of all the primates. We have all inherited this pseudogene from this ancestor.

pseudogene :: a gene carrying a mutation that makes its products nonfunctional

Your Eye Is Organized Backward

The human eye is essentially identical in structure to the eyes of all vertebrates. It is a remarkable organ, one that works exceedingly well and plays an important role in the life of all vertebrates that are able to see. But the vertebrate eye does have one peculiar feature: it apparently is organized backward.

The light you see enters your eye and hits the photoreceptors—rods and cones—in the retina. These receptors send impulses to nerves that carry the

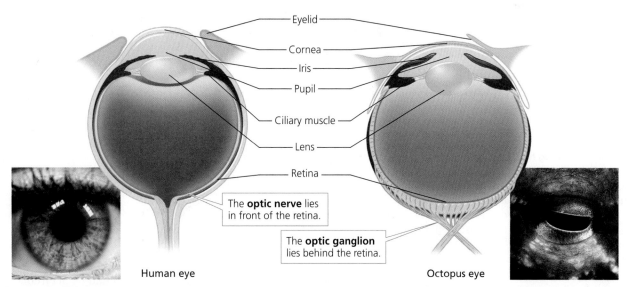

Eyelid

Cornea

Iris

Pupil

Ciliary muscle

Lens

Retina

The **optic nerve** lies in front of the retina.

The **optic ganglion** lies behind the retina.

Human eye

Octopus eye

FIGURE 9.7 The human eye is nearly identical in structure to the eye of certain vertebrates, except that it is organized backward. In humans, the optic nerve lies in front of the retina, resulting in a blind spot, whereas in the eyes of a few types of invertebrates, such as octopi, the optic nerves are behind the retina.

signal to the brain for interpretation. However, these transmission nerves lie *in front of* the retina itself. Light must filter through these nerves in order to hit the retina. Furthermore, these nerves must at some point exit the eye to reach the brain, resulting in a hole, or "blind spot" in the retina.

We cannot fault the eye for its functionality, but the design seems odd. Squid, octopi, and cuttlefish have eyes that work as well as ours, but in these species the transmitting nerves lie behind the retina (**FIGURE 9.7**). No light filtering occurs, and there are no shadows or need for a blind spot in the retina, which means that in some ways they actually see better than we do.

The human eye is an example of excellent functionality supported by a less-than-excellent design. It is likely that the eye developed over a long period of time through a number of small steps. Each step would improve the function of the eye a bit. Each new step would have to be based on earlier ones—which means that each step *constrains* what the next step can be. Over time, the eye could become highly functional, but its structure would reflect the historical constraints that led to it. It wouldn't be a "perfect" organ. The end result of such a process is exactly what we see in the human eye: excellent functionality stemming from a historically constrained structure.

The point of this section is that life does indeed have a history. If we recognize this fact, we can understand many features of life (for example, the structures of our own bodies) that otherwise make little sense. Recognizing that life has a history, and that historical changes and relationships account for the features of life we observe today, is the same thing as recognizing that life has evolved. This is one line of evidence (and there are many others) that has convinced biologists that evolution is simply a fact of nature.

9.2 What Convinced Darwin of the Fact of Evolution?

In the early 1800s, the prevailing scientific view of nature was that God created all forms of life in much the same form as they appear today. This view

FIGURE 9.8 During his five-year voyage aboard the H.M.S. *Beagle* (1831–1836), Darwin made extensive studies of the plants and animals living in South America. He began to see evidence that eventually would lead him to the idea of evolution.

was not universally held, however. Ideas about evolution had been circulating for quite some time, and advances in geology pointed to a much more dynamic, and older, Earth. But as a young man, Charles Darwin (1809–1882) held to the prevailing view.

The Voyage of the *Beagle*

Darwin's perspective began to change during his five-year voyage aboard the H.M.S. *Beagle* (1831–1836) (**FIGURE 9.8**). In his role as ship's naturalist, Darwin made extensive studies of the plants and animals living in South America. As Darwin learned more about this new part of the world, he began to see evidence that eventually would lead him to the idea of evolution: that species could change and new species could arise from existing ones.

One of these lines of evidence involved **biogeography**, the study of where different species live throughout the world. South America has rain forests, as do Africa and Southeast Asia. South America also has extensive prairies, foothills, and mountain habitats like those found on other continents. If species were created specifically to live in a particular habitat like a rain forest, you might expect that species in different rain forests would be related, or perhaps that the same species would be found in all of them.

Instead, what Darwin found was that the rain forest species in South America were more similar to the non–rain forest species of South America than they were to species from other rain forests (**FIGURE 9.9**). The same held true for other habitat types. Apparently, living on the same continent was a better predictor of relatedness than the habitat in which a species lived. Darwin also discovered that the fossil species found in South America bore a close relationship to the current living species, though some modifications had occurred (**FIGURE 9.10**).

Darwin reasoned that one way to explain these facts is to infer that species have a history. They have changed over time, which accounts for the fossil evidence of both relatedness and differences of past and current species. As successful species spread out across a continent, they change in ways that help them to survive in the new habitats they encounter. Species on the same continent may, therefore, be related even though they live in different habitats. Darwin's key inferences in both cases were that species change over time, and that these changes take a long time to occur.

Another line of evidence to support Darwin's growing consideration of evolution came from his celebrated visit to the Galapagos Islands, about

biogeography :: the study of where different species live throughout the world

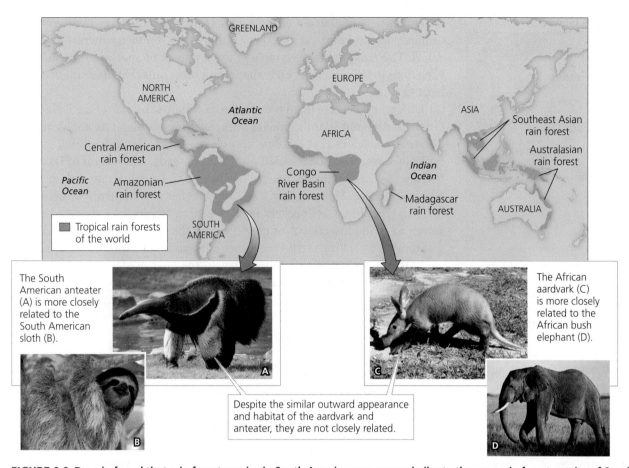

FIGURE 9.9 Darwin found that rain forest species in South America were more similar to the non–rain forest species of South America than they were to species from other rain forests. Even though they might look similar, the aardvark found in an African rain forest and the anteater found in a South American rain forest are not closely related.

The South American anteater (A) is more closely related to the South American sloth (B).

The African aardvark (C) is more closely related to the African bush elephant (D).

Despite the similar outward appearance and habitat of the aardvark and anteater, they are not closely related.

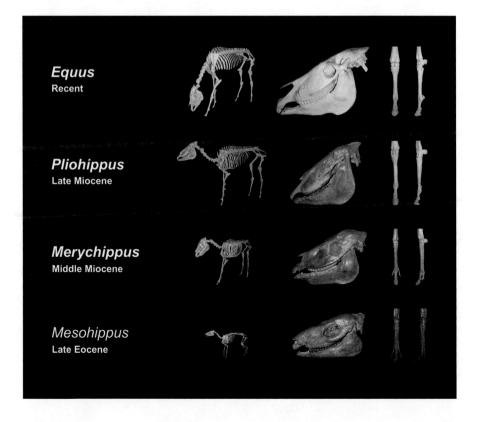

Equus
Recent

Pliohippus
Late Miocene

Merychippus
Middle Miocene

Mesohippus
Late Eocene

FIGURE 9.10 Darwin discovered that the fossil species found in South America bore a close relationship to the current living species, though some modifications had occurred. The fossil record of horse evolution (although not studied by Darwin) also shows modification of ancestral species.

GALAPAGOS ISLANDS

FIGURE 9.11 Another line of evidence to support Darwin's consideration of evolution came from his celebrated visit to the Galapagos Islands, about 600 miles west of Ecuador.

endemic :: a species that is specific to a location and found nowhere else in the world

Darwin's finches :: the finches on the Galapagos Islands that occupied different habitats and developed different behaviors

600 miles west of Ecuador (**FIGURE 9.11**). On the islands, Darwin discovered a number of **endemic** species, species that were found nowhere else in the world. The most famous example is the group of 14 species of finches with variations in bill structure, body shape, and short tails that made them unique (**FIGURE 9.12**).

Different species of finch were found on different islands of the Galapagos. On these different islands, finches occupied different habitats and developed different behaviors that in some cases were decidedly unfinchlike. Finches typically have strong beaks used for crushing seeds, and several of **Darwin's finches** (as the Galapagos finches have come to be called) used their beaks for this purpose. However, some of these finches also ate fruit, some drilled holes in trees like woodpeckers in their search for grubs, some used cactus spines to probe insects out of crevices, and the sharp-beaked ground finch was observed pecking the flesh of other animals and feeding on their blood.

In the published record of his journey, *The Voyage of the* Beagle, Darwin writes about these finches: "Seeing this gradation and diversity of structure in one small, intimately related group of birds, one might really fancy that from an original paucity of birds on this archipelago, one species had been taken and modified for different ends."

Darwin's Theory of Natural Selection

As Darwin considered what he learned on the *Beagle*, he came to an inescapable conclusion: species change over time, even to the point of becoming new species. There is an important lesson about science here: Darwin changed his mind. He began his career convinced that nature had been created as we see it today. The evidence he found did not support this view, however, so he

FIGURE 9.12 The 14 species of Galapagos finches (or Darwin's finches, as they are called) are endemic species, which means they are found nowhere else in the world. The finches evolved in terms of body shape, tail size, and bill structure (shown categorized by diet) to adapt to their environment.

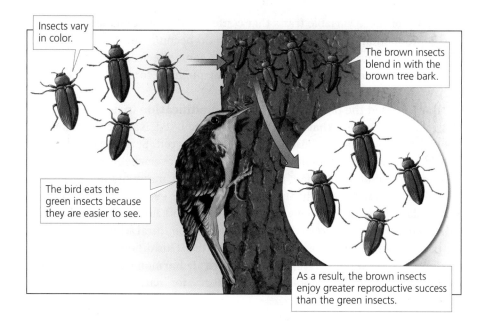

FIGURE 9.13 Natural selection is based on three observable facts of nature: individuals in a population vary, at least some of these variations are inherited, and many more offspring are born than survive to adulthood.

changed his views to accommodate the evidence. This is the way science works: evidence trumps belief and prior views.

Darwin realized his new ideas were radical. He understood the need to compile evidence that species change and to discover the mechanism by which the change occurred. He spent the next 20 years on this task, and in 1859 he published his most famous and influential book: *The Origin of Species by Means of Natural Selection or the Preservation of Favoured Races in the Struggle for Life*.

Natural selection, the mechanism whereby individuals with certain heritable traits have an increased chance of surviving and producing offspring, is one of those rare ideas that is both simple and earth-shattering. The pieces of this idea were available for all to see. It is a tribute to Darwin's intellect that he was able to put all the pieces together and understand their significance.

Natural selection is based on three observable facts of nature (**FIGURE 9.13**):

1. Individuals in a population vary; they differ from one another (variation).

2. Variations are inherited; parents pass some of their variation on to their offspring (inheritance).

3. In any population, many more offspring are born than survive to adulthood (differential survival and reproduction).

Point 3 raises an important question: if many (in some species, most) of the offspring die, which ones survive? Perhaps mortality is simply random, and every individual has the same chances. But if individuals vary (point 1), maybe some are better at survival than others, at least in their particular environment. Certain individuals with particular variations would have a greater chance of survival and therefore a greater chance of reproducing. And when they reproduced, these individuals would pass on to their offspring the variations that helped them survive (point 2).

The logic was simple, but the conclusion was startling: species would change over time. Each generation would differ from the one before it, and it would differ in a specific way. The next generation would have a higher

natural selection :: the mechanism whereby individuals with certain heritable traits have an increased chance of surviving and producing offspring frequency of those variable traits that conferred a greater chance of survival and reproduction. Over time, a species would become better adapted to its environment. And if two populations of the same species, perhaps living in different areas, adapted independently, they might become so different that they would split into two different species.

Darwin used the phrase **descent with modification** to describe the fact that species changed—that is, that evolution occurred. Natural selection was his theory about *how* evolutionary changes occurred. Scientists at the time considered Darwin's evidence for evolution to be overwhelming, and many recognized evolution as a fact of nature. However, many did not agree that Darwin had found the right theory to account for evolution. Darwin could not explain convincingly why organisms vary and how the variations are inherited—he had no knowledge of genetics. Also, natural selection portrayed nature as limited, cold-hearted, and dog eat dog. This differed radically from the prevailing view of nature as a bountiful, harmonious world created by God. It would take another half century for natural selection to be accepted.

The Modern Synthesis

The early part of the 20th century witnessed many advances in biology. Mendel's work with pea plants (Chapter 4) was rediscovered, and a science of genetics emerged. It soon became clear that genetics solved several of the problems that plagued Darwin. Traits were controlled in part by genes. Genes existed in several different forms, or alleles, which helped to explain why organisms vary. And genes passed from parent to offspring according to genetic rules of inheritance. Natural selection explained how **allele frequency**, the number of alleles present in a population, changes over time. The integration of genetics and evolutionary biology was known as the **modern synthesis**, and one of its main architects was Sir Ronald Aylmer Fisher (see *Scientist Spotlight: Sir Ronald Aylmer Fisher*).

One of the discoveries from the modern synthesis was that mathematical equations could be developed that predicted how a population would evolve under a particular set of circumstances. These equations could predict, for example, how rapidly a beneficial allele would increase in frequency throughout a population, or how long it would take for a disadvantageous allele to be eliminated (**FIGURE 9.14**). Scientists tested these mathematical predictions experimentally in the lab and in the field. They found that natural selection

descent with modification :: a phrase used by Darwin to describe the fact that species changed and that evolution occurred

allele frequency :: the number of alleles present in a population

modern synthesis :: the integration of genetics and evolutionary biology; one of its main architects was Sir Ronald Aylmer Fisher

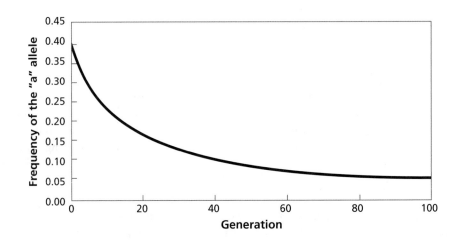

FIGURE 9.14 One of the discoveries from the modern synthesis was that mathematical equations could be used to predict how a population would evolve under a particular set of circumstances. For instance, equations could be used to predict how long it would take for a disadvantageous allele to be eliminated from a population.

Scientist Spotlight

Sir Ronald Aylmer Fisher (1890–1962)

R. A. Fisher was one of the main architects of the modern synthesis. Fisher was born February 17, 1890, in London. He showed talent in mathematics as a boy, and in 1909 he entered Cambridge University to study mathematics and astronomy. He graduated with distinction in 1912. He tried to enlist in the army during World War I but was rejected due to poor eyesight. Instead, he taught mathematics and physics at local schools. Fisher married Ruth Eileen in April 1917. Eileen was barely 17 at the time, and because her mother objected to her early marriage, they married in secret. Fisher and Eileen had two sons and seven daughters, though one died in infancy.

In 1919, Fisher accepted a job as statistician at Rothamsted Agricultural Experiment Station, where he developed statistical techniques to help agriculturists design and interpret experiments involving crop breeding. In 1930, Fisher was appointed Galton Professor of Eugenics at University College. He published several influential books, including *The Genetical Theory of Natural Selection* in 1930, and has been described as "Darwin's greatest twentieth-century successor." He also developed a number of important statistical tests that scientists use to analyze data.

It is a stunning achievement for any one person to reach the pinnacle of two different fields, but that is exactly what Fisher did in biology and mathematics. He was elected to the Royal Society in 1929, and he received the society's Royal Medal in 1938 and its Darwin Medal in 1948. Queen Elizabeth dubbed him a Knight Bachelor in 1952. Fisher retired in 1957 and moved to Adelaide, Australia, where he died July 29, 1962.

Fisher was an intense and brilliant man whose temper frequently got the better of him. He was intolerant of

R.A. Fisher was one of the main architects of the modern synthesis—the integration of genetics and evolutionary biology.

mistakes in other people, and his teaching and writing were difficult for most people to understand. His contributions to statistics and biology were significant, far reaching, and long lasting.

Fisher's interest in eugenics (see Chapter 8) began at an early age. At Cambridge, he formed the Cambridge Eugenics Society and became its first president. His wife shared his views on eugenics, and because both felt they had desirable traits worth promoting, they decided to have a large family. Fisher was a fierce opponent of contraception, especially among the educated and wealthy. He was concerned that members of the educated and upper classes weren't having enough children and would soon be overrun by more rapidly breeding members of lower classes. Fisher is remembered as a brilliant scientist, but on the personal level most people today would regard his views as extreme. ::

worked just as Darwin thought it did, and just like the equations said it should. It took almost 70 years, but by the 1940s scientists accepted natural selection as an important and powerful scientific theory, supported by mountains of theoretical and empirical evidence.

9.3 How Do Humans Adapt to Their Environment?

Let's return to the examples in our case study. How can we explain the geographic distribution and the age of the recessive mutation that allows lactose production to persist into adulthood? And how can we explain why some people of African and Mediterranean descent are killed by certain treatments for malaria when they come from regions where malaria is prevalent?

Lactase Persistence

We know that mutations occur randomly. The particular mutation that enables lactase production to continue into adulthood has occurred many times throughout the 200,000-year history of our species, and it has occurred across the globe, everywhere humans live. But the mutation didn't "take," that is, it didn't increase in frequency, until it happened to occur in Northern Europe about 10,000 years ago. Why?

Consider what might happen to the mutation if it occurred in a typical settlement of our earliest ancestors. For an infant feeding on breast milk, lactase production is essential. But after the child is weaned, there is no more milk to drink. There are no domesticated dairy cows, no cheese, no butter, and no yogurt—nothing but game, nuts, maybe berries, and other available plant material. Under these conditions, a mutation that allowed an adult to produce lactase would not provide any advantage. It could even be the case that lactase interfered with the digestion of nondairy food. Regardless, there would be no particular reason for the mutation to increase its frequency in the population.

The picture changes dramatically when domesticated cows and goats are around. Their milk provides an untapped reservoir of food for humans who can digest it. In times of famine (and there were plenty of those), an adult who could produce lactase might survive and pass the mutation on to his or her offspring. An adult lacking the mutation might starve. Under these conditions, the mutation for lactase persistence would confer an advantage and likely would increase in frequency throughout a population that kept dairy animals.

When and where did the domestication of dairy animals first occur? It's no surprise that one center of dairy farming originated in Northern Europe about 10,000 years ago (**FIGURE 9.15**). Eventually the mutation for lactase persistence occurred in this population and increased in frequency. Among Northern Europeans and their descendants in North America and Australia,

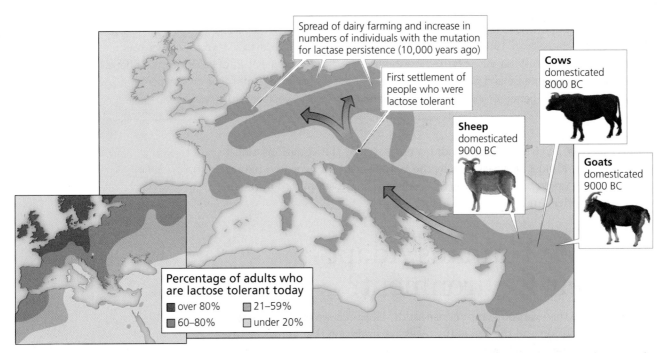

FIGURE 9.15 The mutation for lactase persistence originated in many places in the world but became established and increased its frequency in places where dairy farming thrived. One center of dairy farming originated in Northern Europe about 10,000 years ago, and among Northern Europeans and their descendants in North America and Australia, almost all individuals can digest lactose as adults.

Prevalence of G6PD deficiency

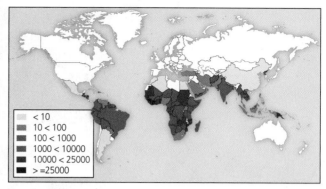

Prevalence of malaria infection

FIGURE 9.16 **In areas where malaria is prevalent, an adaptation—a deficiency in the enzyme glucose-6 phosphate dehydrogenase (G6PD)—reduces a red blood cell's ability to use sugar as an energy source and hinders reproduction of any infecting malarial parasites.** This is the same effect and result achieved using the oxidizing drugs we have developed to treat malaria.

almost all individuals can digest lactose as adults. The Tutsis and Fulanis of equatorial Africa have a high frequency of persistent lactase, and they are also among the few African people who have a long history of herding dairy cattle.

Adult-persistent lactase is a human **adaptation**—a change by a population to better align with specific environmental conditions. It is an example of how natural selection acts on human populations and causes evolutionary change.

adaptation :: a change by a population to better align with specific environmental conditions

Malaria and Oxidizing Drugs

In the opening case study, we also described a situation in which some people of African and Mediterranean descent are killed by certain treatments for malaria. Given that these people come from regions where malaria is a problem, why would they respond this way to treatment?

The answer is that malaria, like all major diseases, sets up conditions for natural selection in our species. In Chapter 4, we discussed one human adaptation to help fight malaria: the sickle cell trait. But in high-malaria areas, a whole host of adaptations to fight malaria have evolved.

One of these adaptations is a mutation—a deficiency in the enzyme glucose-6-phosphate dehydrogenase (G6PD)—that reduces a red blood cell's ability to use sugar as a metabolic energy source (**FIGURE 9.16**). This mutation puts the red blood cell under metabolic stress and hinders reproduction of any infecting malarial parasites—the same effect and result as the oxidizing drugs we have developed to treat malaria.

But when a person with this genetic adaptation takes an oxidizing drug to treat malaria, the red blood cells are put under a double dose of metabolic stress. Many red blood cells may burst, and the patient may die. It's interesting to see that humans have adapted just like other animals. But it's even more important to understand that these adaptations, and our evolutionary history in general, have medical consequences that we need to consider to provide good care.

9.4 How Does Natural Selection Produce Adaptations?

We have presented several examples of human adaptations. Lactase persistence is advantageous in populations where dairy animals are kept, and the

FIGURE 9.17 Peter and Rosemary Grant have been studying the Galapagos finches continuously since 1973.

metabolic stress adaptation is advantageous in regions with malaria, although it can be lethal with the wrong medication. Examples like these provide evidence that humans evolve and adapt just like other organisms. However, they do not really show the details of how natural selection acts to produce adaptation. Detailed studies of natural selection take a great deal of time and effort—over 40 years, in the case of Peter and Rosemary Grant's work.

The Grants' 40-Year Study of Natural Selection

Peter and Rosemary Grant have been studying the Galapagos finches continuously since 1973 (**FIGURE 9.17**). The work we describe here took place between 1975 and 1978 on Daphne Major, which, despite its name, is one of the smallest islands in the archipelago. The medium ground finch (**FIGURE 9.18**) lives on Daphne Major, with an average population size of 1200 birds. With great patience, the Grants captured every finch and gave them each distinctive leg bands. Scientists on the island know each bird individually. They can tell which birds mate and with whom. They know who the parents are and who the young are, so they are able to see how genetic traits are passed on. They can tell when each bird is born and when it dies. The Grants' 40-year effort to study the medium ground finch and other inhabitants of the Galapagos is a remarkable achievement, and one that has provided numerous tests of natural selection.

Medium ground finches are seed eaters, and they use their beaks to pick up and crush the seeds they eat. Beak size is variable in this finch. Birds with larger beaks are able to crush and eat larger, tougher seeds. Birds with smaller beaks are better able to handle smaller seeds. Daphne Major is a dry volcanic island with a typical rainfall of only 5 inches per year. The vegetation is mainly dry forest, scrub, and cactus. Still, Daphne Major supports a diversity of plants, and a variety of seeds are typically available for the finches to eat.

Between 1976 and 1977 a drought hit Daphne Major, and less than an inch of rain fell during the "rainy" season. This drought severely reduced the number of seeds available, and over the course of the drought more than 80% of the medium ground finches died (**FIGURE 9.19**). The few soft seeds produced were quickly eaten, and for much of the drought, mostly large, hard seeds were available (**FIGURE 9.20**).

FIGURE 9.18 The medium ground finch lives on Daphne Major. Medium ground finches are seed eaters, and they use their beaks (which are variable in size) to pick up and crush the seeds they eat.

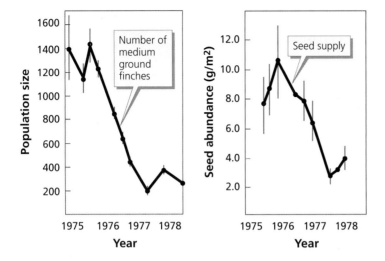

FIGURE 9.19 Between 1976 and 1977, a drought hit Daphne Major. This drought severely reduced the number of seeds available, and over the course of the drought more than 80% of the medium ground finches died.

The drought on Daphne Major set the stage for a decisive test of Darwin's theory of natural selection. We can see that there is variation in beak depth in the finch population. The Grants analyzed data (not reported here) to verify that the variation is inherited: parents with larger beaks tend to have offspring with larger beaks. The drought reduced the food supply and set up a struggle for survival, and many finches died.

The Grants routinely capture their finches and take measurements of their weight, size, and beak depth. **FIGURE 9.21** shows the distribution of beak depths before and after the drought. As you can see, the average beak depth grew from 9.4 mm to 10.1 mm, an increase of 7% in just a few seasons. In an environment with mostly large, hard seeds, birds with bigger beaks have the advantage. This change in average beak depth was due to a higher rate of mortality among smaller-beaked birds.

To demonstrate evolutionary change and adaptation, however, we need to look at changes that occur in the next generation. **FIGURE 9.22** shows the distribution of beak depth for birds that hatched in 1976, before the drought, and for birds that hatched after the drought in 1978. Again, beak depth grew, from about 8.9 mm to 9.7 mm, a 9% increase. This demonstrates that not only did larger-beaked birds survive the drought better, but they also passed on their alleles for larger beaks to their offspring. Evolution occurred in the population of medium ground finches. Over several seasons, the average beak depth increased as the population adapted to drought conditions.

FIGURE 9.20 For much of the drought, mostly large, hard seeds were available, and the average size of the birds increased during that time as well.

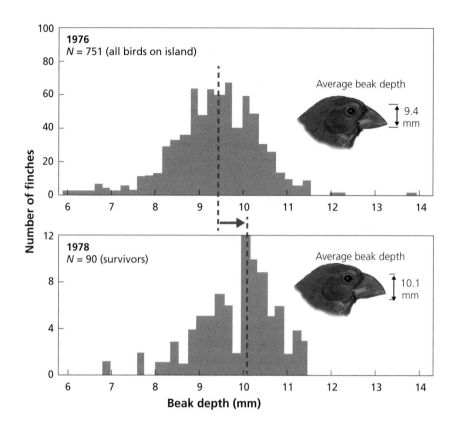

FIGURE 9.21 Between 1976 and 1978, the average beak depth of the finches grew from 9.4 mm to 10.1 mm, an increase of 7%. This change in average beak depth was due to a higher rate of mortality among smaller-beaked birds.

It is important to note that no individual finch changed during the drought: once a finch reaches maturity, its beak depth stays the same. Adaptation is a property of populations, not individuals. It is also important to note that adaptation is not necessarily a unidirectional process. Droughts are common on Daphne Major, and they are interspersed with wetter seasons. During drought, food is scarce, larger seeds predominate, and beak depth increases. During wetter years, the reverse is true: small seeds predominate and beak depth decreases. Over time, beak depth fluctuates both up and

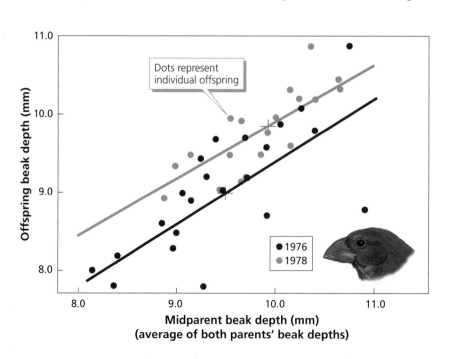

FIGURE 9.22 Larger-beaked ground finches survived the drought better, and they also passed on their alleles for larger beaks to their offspring. The beak depth for ground finches that hatched in 1976 before the drought was about 8.9 mm, and the beak depth for ground finches that hatched after the drought in 1978 was about 9.7 mm, a 9% increase.

down. As is true of evolution in general, there is no overall direction, simply a response to local conditions.

Fitness and Natural Selection

Fitness is a simple-sounding term that can have several different meanings. We say that an organism "fits" its environment well if it has adaptations that promote its survival and reproduction. Evolutionary biologists often use the term to refer to an organism's reproductive potential or chances of survival. It is common to say that natural selection selects the fittest individuals, although this is sloppy logic because it doesn't really say what fitness is—other than that it is what natural selection selects.

The modern synthesis brought together evolutionary biology and genetics, and this suggested a specific definition of fitness. Fitness is the number of alleles, or forms of a gene, an organism passes on to the next generation. This is a clear definition, and it has a nice scientific feature: under many conditions, fitness can be measured in the laboratory or in the field. The Grants, for example, were able to measure finch fitness precisely.

Natural selection acts to increase fitness in a population. Adaptations increase survivorship and/or reproduction, which in turn increases the number of genes an organism leaves in the next generation. This helps to explain the tendency for parents to care for their offspring. By so doing, they are helping to pass on their genes.

According to the rules of sexual reproduction, your children share half their genes with you. But your children are not the only individuals with whom you share genes. You share on average half your genes with your siblings, about a quarter with your grandchildren, and an eighth with your cousins (**FIGURE 9.23**). Our definition of fitness rewards passing on genes. But it doesn't matter who happens to have those genes.

fitness :: the number of alleles, or forms of a gene, an organism passes on to the next generation

FIGURE 9.23 An organism maximizes its fitness by passing genes on to the next generation, but it does not matter who happens to have those genes. You share on average half your genes with your children and your siblings, about a quarter with your grandchildren, and about an eighth with your cousins.

kin selection :: selection for behaviors that help your relatives survive and reproduce

Although we pass on our genes directly through our children, we also share genes with all of our relatives. Therefore, there is an evolutionary advantage to helping your siblings, grandchildren, and cousins pass on their genes as well. Such activity is common in humans, and also in many other animal species. If fitness were limited to genes passed on directly through your own descendants, it would be hard to explain why you would help anyone else. But if fitness is more inclusive, and can refer to genes passed on by your kin, this type of social behavior becomes easier to explain. Selection for behaviors that help your relatives survive and reproduce is known as **kin selection**.

Limits of Natural Selection

The more you learn about adaptation, the stronger the temptation becomes to infer that given enough time, natural selection can do almost anything. To see why this is not the case, it is important to understand the factors that limit and constrain what natural selection can do.

Natural selection can't anticipate or plan for future events. It therefore can't have goals or make progress toward goals. Natural selection responds only to the current situation, and whenever that changes, the traits that are selected change as well.

This is a photo-edited image; pigs do not have wings!

FIGURE 9.24 Natural selection can only act on the variation present in a population. While it might be useful for pigs to have wings, if the available range of genetic variation doesn't give natural selection something to work with, they never will.

Natural selection can only select from the variation that is present in a population. This variation is produced by random processes like mutation; natural selection can't produce specific variation, even if it would be very useful. It might be useful for pigs to have wings, but if the available range of genetic variation doesn't give natural selection something along these lines to work with, they never will (**FIGURE 9.24**). Natural selection works under historical constraints as well. We inherit a lot from our ancestors and from our ancestral species. This inheritance constrains the types of organisms we can be.

The fact that natural selection has limitations and constraints is not surprising. Any natural process must work this way, since there is no agency to guide it. The consequence of these limits and constraints is that organisms will have features that are not perfect or optimal. We discussed three such features of the human body in Section 9.1: the male testes that exist outside of the body, the missing enzyme to produce vitamin C, and the backward organization of the eye. Design imperfection in features like these, even though they function well, provides some of the strongest evidence that the features were shaped by a natural process like selection rather than being created by a supernatural designer.

9.5 What Are Random Events in Evolution?

Natural selection is not the only way that evolution occurs. Natural selection accounts for one particular type of evolution: adaptation. But random events can also cause evolution. Random events don't change a population in any particular way; they just cause change.

Mutations are random events. Recall that mutations are random mistakes made in copying DNA when a cell divides. Although mutations are rare, organisms have so much DNA that virtually every newborn individual will have a couple of mutations. Random mutation is the first major cause of genetic change in populations and is one of the reasons populations have the variation required for natural selection to work.

Natural selection will increase the frequency of beneficial mutations and reduce the frequency of harmful mutations in a population. But some mutations are neutral: neither beneficial nor harmful. Natural selection does not act on these mutations. Still, the frequency of these mutations does not simply stay the same year after year. Its frequency changes randomly over generations, like a series of coin flips: heads it increases, tails it decreases. This process of random changes in the frequency of an allele is called **genetic drift**, and it is a second major cause of genetic change in populations (**FIGURE 9.25**).

One special case of genetic drift is **founder's effect**. Often a few individuals of a population leave (or are forced to leave) and begin a new population. For example, a flood or forest fire might make a small population of squirrels seek a new place to live. By chance, it may happen that these founders differ genetically from the main population. If so, the new population starts off being genetically different from the old. With a different genetic starting point and different conditions in its new habitat, the new population may adapt in new and different ways (**FIGURE 9.26**).

genetic drift :: random changes in the frequency of an allele

founder's effect :: a special case of genetic drift in which a few individuals of a population leave or are forced to leave and begin a new population

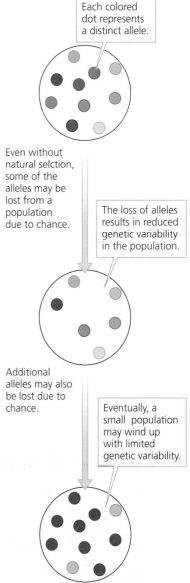

Each colored dot represents a distinct allele.

Even without natural selction, some of the alleles may be lost from a population due to chance.

The loss of alleles results in reduced genetic variability in the population.

Additional alleles may also be lost due to chance.

Eventually, a small population may wind up with limited genetic variability.

FIGURE 9.25 Genetic drift leads to random changes in the frequency of an allele as a consequence of random or chance events. It can be a major cause of genetic change in populations.

Ancestral population

Some salamanders migrated east.

Some salamanders migrated west.

The salamanders found themselves separated by mountains and could not easily mingle.

Genetic drift

Eventually the salamanders moved far enough south and past the mountains that reconnection was possible.

The salamanders had changed so significantly that they could no longer mate.

FIGURE 9.26 In a special case of genetic drift known as founder's effect, a few individuals of a population may leave (or be forced to leave) and begin a new population. If by chance these founders differ genetically from the main population, the new population may adapt in new and different ways.

Founder's effect has been important in human evolution. As our earliest ancestors dispersed from Africa, they traveled as small groups and founded small settlements. When a settlement got too large, a small number of people would leave to start a new settlement. The native residents of Asia, Australia, Europe, and the Americas all descended from small groups of people that started off genetically different from one another. Combined with adaptation to the different environmental conditions in these regions, founder's effect helps to explain why people from these regions differ in skin color, height, and even in the prevalence of certain diseases such as sickle cell disease (see Chapter 4).

There are a number of more recent examples of founder's effect, too. One involves the Amish of eastern Pennsylvania (**FIGURE 9.27**). The Amish descend from a small group of perhaps 200 German immigrants, dating back to the early 1700s. The Amish practice a traditional lifestyle and tend to marry among themselves. As a result, the genetic makeup of the Amish comes largely from their small group of ancestors.

Two of the early Amish immigrants were Samuel King and his wife, who arrived in Pennsylvania in 1744. One or both of the Kings carried a mutation for Ellis-van Creveld syndrome (**FIGURE 9.28**). Ellis-van Creveld syndrome has several symptoms, usually seen in newborns: polydactyly (more than five fingers or toes), a narrow chest, signs of dwarfism, and several types of heart defects. The Kings were only one couple, but in a small population a single couple can have a large genetic impact. They passed the mutation on to their offspring, and as other Amish married into the family, the mutation increased in frequency. Today 5 of every 1000 infants of the Old Order Amish in Pennsylvania have the syndrome, compared to 1 in 60,000 in the general population. Ellis-van Creveld syndrome is not an adaptation, and there is no particular reason why the Amish have it—it was just the luck of the draw when the Kings moved in.

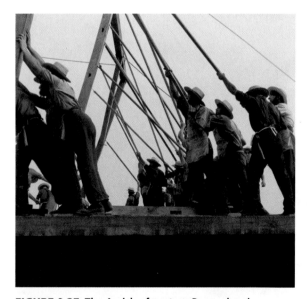

FIGURE 9.27 The Amish of eastern Pennsylvania descend from a small group of perhaps 200 German immigrants, dating back to the early 1700s. They practice a traditional lifestyle and tend to intermarry, so their genetic makeup comes largely from their small group of ancestors.

FIGURE 9.28 One of the early Amish immigrants carried a mutation for Ellis-van Creveld syndrome, which may result in infants who are polydactyly (having more than five fingers or toes). Today, 5 of every 1000 infants of the Amish in Pennsylvania have the syndrome, compared to 1 in 60,000 in the general population; this is an example of founder's effect.

9.6 What Is the Evidence for Speciation?

There is no doubt that species change genetically—that is, that they evolve. Scientists like the Grants and many others have painstakingly documented the actions of natural selection. They have shown that natural selection causes genetic changes in a population that allow it to better adapt to the environment. Similarly, lactase persistence in adults was an advantage in populations where dairy foods were staples. Random genetic changes occur as well. But a changed species can still be the same species. Do we have evidence that brand new species can appear through evolution?

The answer is yes. Scientists have a good deal of evidence to show that new species branch off from existing species, a process called **speciation**. We also have evidence that speciation occurs regularly and continuously under the right conditions. It's true that speciation can take thousands of years, which makes direct observation of the process from start to finish a difficult or impossible task. But when speciation occurs, it leaves behind many clues. Scientists put together and analyze these clues, much the same way a detective uses clues to solve a crime, even if no one witnessed it.

speciation :: a process by which species branch off from existing species to form new species

Biological Species Concept

Two individuals are considered to be in the same species if they can interbreed. This has several important biological consequences. Individuals of the same species are genetically linked: they share the same set of genes. If a mutation (or an adaptation) occurs in one individual, it could increase its frequency in different families of the same species by interbreeding. On the flip side, individuals of different species cannot interbreed. The species are genetically isolated. Because they can't interbreed, mutations in one species cannot spread to the other. Each species follows its own independent evolutionary path. This is called the **biological species concept**, and it defines species as genetically distinct, independently evolving units.

biological species concept :: it defines species as genetically distinct, independently evolving units

The biological species concept works well in many cases. Dogs and cats look very different. They are different species, and they can't interbreed. **FIGURE 9.29** shows four corn snakes. They look very different as well, but they can interbreed. Because they can, biologists consider them to be variant forms of the same species.

But the biological species concept doesn't work in all cases. Dogs and wolves can interbreed, as can lions and tigers, whales and dolphins, and

Cats and dogs are distinct species. Although they may look cute together in this photo, they cannot interbreed.

Although each of these corn snakes looks very different, they are all the same species and can interbreed.

FIGURE 9.29 According to the biological species concept, each species follows its own independent evolutionary path, and individuals of different species cannot interbreed.

A Zorse—male zebra mates with female horse; offspring are generally sterile.

B Wholphin—male whale mates with female dolphin; offspring are generally sterile.

C Liger—male lion mates with female tiger; offspring are generally sterile.

FIGURE 9.30 The biological species concept does not apply in all cases, however. Zebras and horses can breed to form a zorse (A), whales and dolphins can breed to form a wholphin (B), and tigers and lions can breed to form a liger (C), among many other examples.

horses and zebras, if they get the opportunity (**FIGURE 9.30**). Different plant species interbreed regularly, producing hybrids. Some one-celled organisms reproduce asexually and generally don't "breed" at all in the regular sense. Breeding is also not an option for a fossil. Scientists can't use the biological species concept for a species identified only from fossils.

We'll use the biological species concept for our discussion, recognizing that it doesn't apply universally to all organisms. In a way, the fact that species are difficult to define is good evidence that new species arise when an existing species splits in two. If speciation is ongoing, you'd expect to see some cases of species just starting to split, others that are far along in the process, and everything in between. This explains why similar species like dogs and wolves can still interbreed (though at some point in the future they may lose this ability), while others like horses and donkeys technically can interbreed, but their offspring are sterile. If every species was created separately, you'd expect that it would be much easier to define a species and tell one apart from another.

Evidence for Speciation

One intriguing "historical" record lies in the sequence of nucleotides in an organism's DNA. DNA is passed down through the generations relatively intact, but over time mutations accumulate that change, add, or remove nucleotides. If we compare the DNA sequence for a specific gene in two different species and find that there are few differences, we conclude that the species are closely related. More differences indicate more distant relatedness. By taking into account the various degrees of similarity among a group of species, it is possible to construct an evolutionary tree that shows one way in which the group may have evolved. (See *How Do We Know? Constructing Evolutionary Trees.*)

A group of scientists performed a study of this type on 17 species of snapping shrimp, marine crustaceans that resemble crayfish or small lobsters. All of these species of shrimp came from the waters off the coast of Panama. The scientists extracted DNA from each species, determined the sequences of specific regions, and constructed an **evolutionary tree**, or a diagram that took into account the degrees of similarity among the species based on their

evolutionary tree :: diagram that takes into account the degrees of similarity among a species based on their physical characteristics and genes

How Do We Know?

Constructing Evolutionary Trees

One of the most important consequences of speciation is that all species share a common ancestor. For closely related species, the common ancestor was relatively recent, while more distantly related species have a common ancestor further back in time. An evolutionary tree is therefore very similar to a family tree, or pedigree. You and your siblings are closely related because your common ancestor is your parent. You and your cousin share a grandparent as a common ancestor, but you are less closely related because the grandparent lived before your respective parents.

Evolutionary biologists construct evolutionary trees to study the lineages of species just as genealogists construct trees to study families. There are many ways to construct an evolutionary tree. One of the best ways is to group together species whose DNA is most similar. This makes sense because relatedness, whether between species or individuals in a family, is determined by how similar the DNA is.

DNA sequences can be represented by strings of four characters: A, T, C, and G. Each character represents one of the four nucleotides present in DNA. We start with the assumption that the species under study all had a common ancestor, with a single ancestral DNA sequence. The sequences we observe in the species differ from that of the ancestor because they have undergone mutations. Mutations include exchanging one nucleotide for another. They also include the loss or insertion of a nucleotide—the DNA sequences in the descendants can be different lengths. Finally, we assume that species that share mutations are more closely related than those with different mutations.

We can construct an evolutionary tree by grouping together the species with the most similar DNA, adding in species with slightly less similar DNA, and continuing until all species are part of a group. The process is tedious and mathematical and requires considerable trial and error to get it right. That's why we have developed computer programs to do the work for us. The simple example in the figure illustrates some of how the process works.

Step 1 of the figure shows a DNA sequence from five species. You can see that there are similarities and differences between the DNA sequences. Step 2 shows an alignment: "blanks" that represent the loss of a nucleotide have been added to make the sequences match up better. Once the program computes the alignment, it forms a tree by going through the grouping process just described, placing the most similar DNA sequences on the closest branches in the tree. Step 3 shows the tree that results from this particular set of DNA sequences. ::

① DNA sequence from five species.

Species S1: **CATCGGAATC**
Species S2: **AATCCGATG**
Species S3: **ACTCTGATTC**
Species S4: **AACGGATC**
Species S5: **ACTCGGATTC**

② An alignment: gaps representing the loss or addition of a nucleotide have been added to make the sequences match up better.

Species S1: **CATCGGAATC**
Species S2: **AATCCGA TG**
Species S3: **ACTCTGATTC**
Species S4: **AA CGGA TC**
Species S5: **ACTCGGATTC**

③ Once the program computes the alignment, it forms a tree, placing the most similar DNA sequences on the closest branches.

Species S1: **CATCGGAATC**
Species S2: **AATCCGA TG**

S1 and S2 are more closely related to each other than either is to S4.

Species S4: **AA CGGA TC**
Species S3: **ACTCTGATTC**
Species S5: **ACTCGGATTC**

Evolutionary biologists construct evolutionary trees to study the lineages of species just as genealogists construct trees to study families. One of the best ways to construct a tree is to group together species whose DNA is most similar.

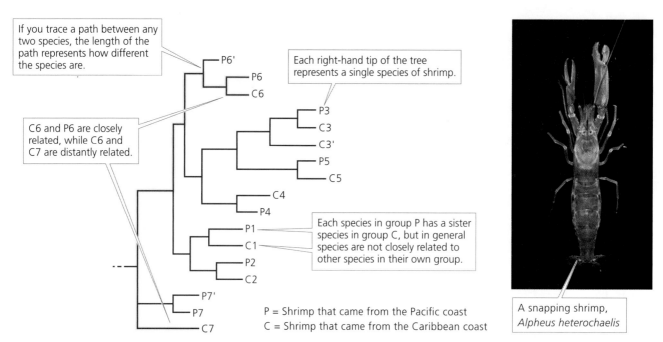

If you trace a path between any two species, the length of the path represents how different the species are.

Each right-hand tip of the tree represents a single species of shrimp.

C6 and P6 are closely related, while C6 and C7 are distantly related.

Each species in group P has a sister species in group C, but in general species are not closely related to other species in their own group.

P = Shrimp that came from the Pacific coast
C = Shrimp that came from the Caribbean coast

A snapping shrimp, *Alpheus heterochaelis*

FIGURE 9.31 An evolutionary tree is a diagram showing the degrees of similarity among species based on DNA sequences in genes. A group of scientists studying 17 species of snapping shrimp from waters off the coast of Panama created an evolutionary tree and discovered that not all shrimp were closely related, despite living in close proximity.

physical characteristics and genes (**FIGURE 9.31**). Each right-hand tip of the tree represents a single species of shrimp. If you trace a path between any two species in the tree, the length of the path represents how different the species are with respect to DNA sequence. The 17 species of shrimp are divided into two subgroups, labeled P for the shrimp that came from the Pacific coast and C for the shrimp found off the Caribbean coast. The species sort out mainly into pairs, with one member from P and one from C in each pair. This indicates that each member of the pair is more closely related (according to DNA sequence similarity) to the other than either is to any other species. Each species in group P has a sister species in group C, but in general species are not closely related to other species in their own group. (For a more detailed discussion of the process the scientists used to extract the DNA from each species of shrimp, see *Technology Connection: Genbank*.)

The scientists uncovered a puzzle. Pacific coast shrimp are not closely related, even though they live in close proximity. The same holds true for the Caribbean shrimp. However, each species is closely related to a species found on the other side of the Isthmus of Panama.

Geology helps us to solve this puzzle. The Isthmus of Panama is about 3 million years old. Prior to its rising, North and South America were not connected, and the Caribbean and the Pacific Oceans were confluent in this region, meaning they flowed together and were not separated by a landmass (**FIGURE 9.32**, p. 272). We can assume at this time that there were only six or seven species of shrimp, which we could label 1–7 without any Ps or Cs. After Panama rose from the sea, each of these species ended up separated, with about half their population in the Caribbean and the other half in the Pacific and with no chance of contact between them.

At this point, the process Darwin described as descent with modification began to occur. Each half of the original species began to diverge genetically from the other, adapting to different conditions on either side of the isthmus.

Technology Connection

Genbank

Using current technology, scientists can extract DNA from the cells of the organisms they study and rapidly determine the sequence of nucleotides in each segment of DNA. This leads to an interesting question: what should be done with all this sequence information?

The National Center for Biotechnology Information (NCBI) has the answer. They have assembled a cluster of supercomputers to host a huge database called GenBank, which stores the DNA sequences and other information as well. Any scientist can upload his or her sequence data to GenBank. There is an online form to record the necessary information, and the process is not much more involved than making an online purchase. In addition to DNA sequence data, GenBank also stores protein sequence data. It has over 15 million sequence records, and the size increases virtually every day.

Impressive as it is, GenBank is just the tip of the biotechnology iceberg. NCBI provides a number of online tools to help scientists access and analyze this storehouse of genetic information.

Entrez is a cross-database search program that allows you to look up sequence information by keyword search. If you want to know the DNA sequence for the human hemoglobin gene, simple go to Entrez and type in "human hemoglobin."

Another program, BLAST (Basic Local Alignment Search Tool), is also a search engine, but of a very different kind. If you give BLAST a DNA (or protein) sequence, it will search the entire GenBank database and return all the sequences that "match" yours. How does this work? A DNA sequence is represented by a long string of the characters ATG and C. The records in GenBank might be hundreds or thousands of characters long, and your search sequence can be long as well. Identical matches between the search sequence and a GenBank record are easy to detect—but hardly ever happen. Usually BLAST finds sequences in GenBank that are "close" to the one you submitted.

The concept of what it means for two strings of characters to be "close" is sophisticated and complex, an advanced version of what your spell checker does when it looks for alternatives to the words you misspell. The end result is that if you find the DNA sequence for human hemoglobin in Entrez, you can use this sequence to search GenBank using BLAST, and in a few seconds you will receive a record of all the hemoglobin genes from all the organisms whose hemoglobin gene is like ours. This can be the starting point for an in-depth study of the evolution of the hemoglobin gene.

NCBI also provides a number of programs to view DNA and protein sequences, a viewer to visualize protein molecules in 3-D with rotations, and tutorials on how to use and interpret all this information. And all of this is free to any scientist—or any person—anywhere in the world if you have an Internet connection. If you would like to learn more, visit the NCBI website: http://www.ncbi.nlm.nih.gov. ::

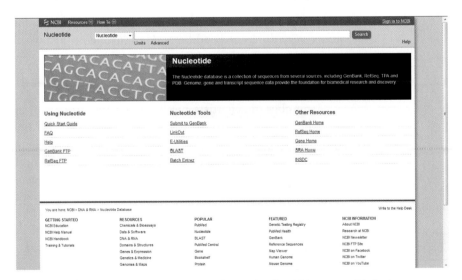

The National Center for Biotechnology Information (NCBI) has assembled a huge database called GenBank, which stores DNA sequences. Any scientist can upload his or her sequence data to GenBank.

10 million years ago

Ten million years ago, North and South America were not connected, and the Caribbean and the Pacific Oceans flowed together; there were likely only six or seven species of shrimp at that time.

3 to 5 million years ago

Three to five million years ago, the Isthmus of Panama rose from the sea connecting North and South America and likely separating the shrimp. Descent with modification occurred, and the species diverged genetically as they adapted to different conditions on either side of the isthmus.

Present day

Three million years later, these populations have diverged into separate species but still share the greatest degree of similarity with their sister species on the other side.

FIGURE 9.32 The geological history of the Isthmus of Panama explains why some species of Pacific coast snapping shrimp were more closely related to species of Caribbean coast snapping shrimp than they were to other species of Pacific coast snapping shrimp.

Three million years later, these populations have diverged into separate species. But they still share the greatest degree of similarity with their sister species on the other side.

As we will see, it takes a long time for species to form, and the process is not directly observable. But examples like this one provide strong evidence that a species can split in two, thereby generating new species.

9.7 How Do New Species Arise?

Darwin recognized that different populations of the same species often live under very different conditions. He reasoned that these populations would evolve independently, and that each would adapt to the specific environment in which they lived. Ultimately, the populations could become so different that they would be different species.

A number of scientists have studied the process of speciation, and it is clear now that speciation can occur in several ways. But Darwin's view of speciation is a good start. Based on the biological species concept, we have to account for one more fact: if one species splits into two, the two new species must lose the ability to interbreed. How might this happen? Speciation occurs in three stages: genetic isolation, genetic divergence, and secondary contact.

Genetic Isolation

The first stage is genetic isolation (**FIGURE 9.33**). Something must divide populations in such a way that interbreeding is cut off. In the case of the snapping shrimp, the dividing feature was the Isthmus of Panama. The Grand Canyon has divided a number of populations between its north and south rims. Smaller features like a river may divide populations if the animals are not very mobile or cannot swim. In all these cases, some sort of geographical feature arises to physically separate populations.

FIGURE 9.33 The first stage of speciation is genetic isolation: something divides a population in a way that cuts off interbreeding. The division may be caused by a geographical feature such as an isthmus or a river that arises to physically separate the population, or it may be caused by dispersal, whereby a small number of individuals migrate to a new habitat.

Isolation can also occur through dispersal, in which a small number of individuals migrate to a new habitat and form a new population. If the new population is sufficiently distant and the journey between the populations is difficult, then interbreeding cannot occur and the new population is genetically isolated from the old. The dispersal of finches from the mainland to the Galapagos Islands is a good example of this type of isolation.

Genetic Divergence

The second stage is genetic divergence (**FIGURE 9.34**). Founder's effect starts the divergence. The founding members of each population are likely to differ, simply by chance, so the populations start off genetically different from each other.

In each of the separated populations, genetic changes begin to accumulate. Some of these changes are due to the fact that different mutations, and therefore different forms of variation, occur. Natural selection will increase the frequency of some of these mutations and decrease others. This will

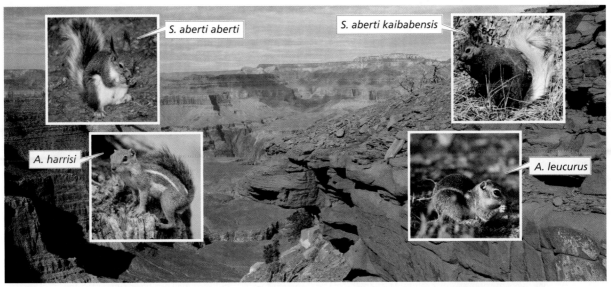

South Rim

North Rim

FIGURE 9.34 The second stage of speciation is genetic divergence: the founding members of each population are likely to differ by chance, and genetic changes begin to accumulate through natural selection. Over time, the populations diverge genetically—squirrels who reside in the different rims of the Grand Canyon are one example of this—and possibly morphologically (structurally) and behaviorally too.

happen differently across populations since the environmental conditions are likely to be different. Over time, the two populations diverge genetically, and maybe morphologically (structurally) and behaviorally, too.

Secondary Contact

In the third stage, the separated populations may come back into contact. Recall that the biological species concept is based on whether individuals can interbreed. Secondary contact provides the opportunity for interbreeding and determining whether speciation has occurred during the time the populations were separated. Whether speciation occurs depends on what happens after contact is reestablished (**FIGURE 9.35**).

It might be the case that in spite of their genetic divergence, the two populations are still able to interbreed successfully. In this case, they probably will do so, and eventually the populations will merge—no speciation will have occurred.

It is also possible that genetic divergence will be so great that members of the populations will not even try to mate. This appears to have occurred with the snapping shrimp. In this case, the two populations satisfy the criterion of the biological species concept, and speciation is complete.

There is a third possibility: "incomplete speciation." Members of the two populations may interbreed and produce hybrid offspring, but the hybrids may not be as successful as nonhybrids. After all, the parents have genes that are adapted to different conditions. The hybrid combines these different genes and may find itself less well adapted to either of its parents' habitats. The hybrids may not be as strong or as fast, and they may die young. Or they might be perfectly viable but less fertile or even sterile, like a mule. In such cases, the parents will end up spending time and energy making and raising their offspring, but they will have either fewer offspring (if hybrid offspring are less viable) or fewer grandchildren (if hybrids are less fertile or sterile).

Natural selection will favor parents that mate with their own kind, and it will work against parents that produce hybrids. Over time, the tendency to produce hybrids becomes rarer and the frequency of hybrids declines.

In spite of the genetic divergence, the populations may still be able to breed successfully; for instance, a coyote and a dog will produce a coydog, which is fertile.

The genetic divergence may be so great that members of the populations will not even try to mate; this is the case with two species of meadowlark whose songs are so different.

Members of the population may interbreed and produce hybrid offspring; but they may not be as successful; for instance, the liger and the zorse are infertile hybrids.

FIGURE 9.35 The third stage of speciation is secondary contact: the separated populations come back into contact. Secondary contact provides an opportunity for interbreeding, and whether speciation occurs depends on what happens after contact is reestablished.

Natural selection that acts to reduce hybrid formation has a special name: **reinforcement**. Reinforcement might select each population to develop different mating rituals, or to mate at different times or in different places, or to develop some type of distinctive difference so that prospective parents don't mistake a member of the other population for a suitable mate. The result of reinforcement is that individuals mate within their own group, interbreeding no longer occurs, and, by the biological species concept, true species have formed.

reinforcement :: natural selection that acts to reduce hybrid formation

9.8 Why Is It So Difficult for the Public to Accept Evolution?

The short answer is this: it's not. Our *Life Application: Public Acceptance of Evolution* feature details public views of evolution from 34 countries. If it came down to a vote, evolution would "win" in 30 of them (though in a few countries, it would be close). But in the United States, as well as Turkey, Cyprus, and Latvia, public acceptance of evolution is low. About 40% of U.S. residents report they have doubts about evolution, 40% accept it, and 20% have not been able to form an opinion. There has been an equal split between those who doubt and those who accept for many years, although the undecideds have increased recently.

While all generalizations have exceptions, the scientific community views evolution as a fact of nature, as well supported as the fact that the Earth revolves around the sun. Many scientists are deeply religious and do not feel their views on evolution and religion conflict. A number of religions have formally adopted the position that evolution is compatible with their spiritual teaching. But not all religions have done so, and the arguments against evolution are largely religious.

The co-option of evolution to support social and political ideology can have terrible consequences. The theory of natural selection suggests that the "natural" struggle for existence leads to the "survival of the fittest." The strong naturally replace the weak. Adopting this view seems to justify aggression, exploitive forms of capitalism, and programs to improve the strength and purity of a human race through selective breeding and pogroms (large-scale massacres of ethnic groups) (**FIGURE 9.36**, p. 277). Some of the concerns about evolution reflect the fear that we would apply its principles to our own society and to the ways that we interact with one another. Most scientists would caution against applying the principles of evolution to our society because human behavior, relationships, and society are too complex to be explained by evolutionary or biological knowledge alone.

Regardless of how we use or misuse it or what its consequences are, evolution is still with us as the unifying theme of life. The significance of evolution lies in its ability to explain what we see in the natural world around us: why some species are more closely related than others, and why each species is adapted to its habitat, for example. There is also a deeper question about how we make decisions about what is true and what is not. Science has presented us with over a century's worth of evidence from every scientific discipline, all of which points to the fact of evolution. Reconciling personal belief and scientific

Public Acceptance of Evolution

A survey of 34 countries revealed that the United States has a significantly lower public acceptance of evolution than any of its European allies or Japan, and the second-lowest level of all the countries surveyed. Surveys like these have been conducted in the United States for 30 years. A June 2012 Gallup poll revealed that only 15% of Americans believe in atheistic evolution, whereas 32% believe that God guides evolution and 46% of Americans believe in creationism.

As they are on so many issues, the people of the United States are split over evolution, in spite of the many changes to science education and in spite of the dramatic advances in the field of evolutionary biology.

One reason for this difference is that the United States has a stronger presence of Christian sects that believe in biblical literalism. Some mainstream Christian religions have reconciled their teachings with evolution, but this is difficult, or perhaps impossible, if one believes the Bible should be taken literally.

Second, evolution has become politicized in the United States as political parties adopt religious positions in the hope of securing votes from like-minded citizens. For example, in a number of recent state-level elections, the Republican platform has called for teaching alternatives to evolution in public schools. When seven Republican presidential candidates were questioned about it during the 2012 election season, five rejected evolution and two accepted it. This shows an even lower level of acceptance among the candidates than in the general public.

The conflict over teaching evolution in public schools began with the Scopes trial in 1925 and continues to this day. Over the past several years, as Kansas State Board of Education members have been voted in and out of office, the board has oscillated between teaching evolution and teaching intelligent design. In 2001, Alabama began inserting stickers into biology textbooks warning that evolution was "only a theory," and Georgia followed suit in 2002. In 2004, the school board of Dover, Pennsylvania, added intelligent design to its K–12 biology curriculum, a move that was struck down in federal court a year later. A large number of homeschool curricula (though by no means all) are structured to teach creationism and intelligent design rather than evolution.

We think that education, particularly of science teachers, can ease these tensions. As more people learn more about science in general and evolution in particular, fewer disagreements will be based on misconceptions. ::

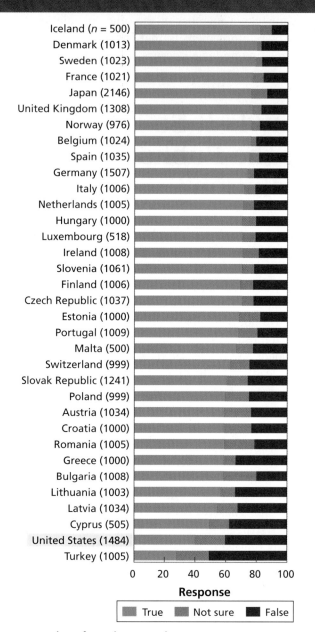

n = number of people surveyed

A survey of 34 countries revealed that the United States has a significantly lower public acceptance of evolution than any of its European allies or Japan, and the second-lowest level of all the countries surveyed. Adults were asked to respond to the following statement: "Human beings, as we know them, developed from earlier species of animals."

FIGURE 9.36 The co-option of evolution to support social and political ideology can have terrible consequences, as it did when the Nazis used the theory of natural selection to justify large-scale massacres of ethnic groups in an effort to "purify" the human race. However, regardless of how we use or misuse it, or what its consequences are, evolution remains with us as the unifying theme of life.

fact has always been a challenging endeavor. But to avoid evolution, one must argue that science is wrong, that evidence doesn't matter, and that personal belief carries more weight than evidence.

Biology in Perspective

How do new species arise and adapt to their environment? Darwin's work provided the first part of the answer, and work by scientists who came later has confirmed and extended his ideas. The evolutionary framework that Darwin and others have developed helps to explain two of the most prominent, observable facts of nature: the diversity of species and the fit of species to their environment.

As a result of evolution, living things have a history. The pattern of geographic variation in humans reflects the way we adapted to local conditions as we dispersed throughout the planet. Some of the puzzling aspects of our own body make sense when we think of them as legacies from earlier ancestors. Understanding why our body is the way it is has medical and social implications, topics we take up later.

For some, the concept of evolution continues the long-standing conflict between science and religion. The consequences of this conflict can be far reaching and can affect education and politics. The consequences also can be polarizing. Education about both religion and evolution is the best way to reconcile belief with science and to defuse the conflict.

Chapter Summary

9.1 How Does Your Body Reflect an Evolutionary History?

Provide three examples demonstrating the effect of evolutionary history on human structure or function.

- The vas deferens does not go directly from the testes to the penis; instead, it takes a circuitous route that is a legacy from earlier ancestors.
- The vast majority of mammals have enzymes that enable them to make their own vitamin C, but humans and other primates inherited a pseudogene that renders the enzyme nonfunctional.
- The vertebrate eye reflects historical constraints; the optic nerve lies in front of the retina, which leaves a "blind spot." .

9.2 What Convinced Darwin of the Fact of Evolution?

Detail the discoveries that led Darwin to change his mind about evolution and describe the steps of natural selection.

- Through biogeography, Darwin found that the rain forest species in South America were more similar to the non–rain forest species of South America than they were to species from other rain forests.
- On the Galapagos Islands, Darwin discovered a number of endemic species, including 14 finches with variations that made them unique; Darwin proposed that "one species had been taken and modified for different ends."
- Darwin's theory of natural selection holds that individuals with certain heritable traits have an increased chance of surviving and producing offspring because of variation, inheritance, and differential survival and reproduction.

9.3 How Do Humans Adapt to Their Environment?

Use specific examples to describe how humans adapt to their environments.

- Humans have adapted to their environment in many ways, including the development of lactase persistence and G6PD deficiency.
  - o Although most adult humans cannot digest lactose, some Africans and persons of Northern European descent can.
  - o The ability to digest lactose in milk is correlated with the origin of domestication of cattle about 10,000 years ago in Northern Europe and Africa.
- In high-malaria areas, a host of adaptations to fight malaria have evolved: one is the sickle cell trait; another is a mutation that puts red blood cells under metabolic stress.

  - o A deficiency in the enzyme glucose-6-phosphate dehydrogenase puts red blood cells under metabolic stress and also prevents malarial parasites from reproducing.

9.4 How Does Natural Selection Produce Adaptations?

Use natural selection to describe the process whereby a population adapts to a specific environment and outline ways that adaptation might be limited.

- Natural selection acts to increase fitness in a population.
- Adaptations increase survivorship and reproduction, which increases the number of genes an organism leaves to the next generation.
- Natural selection has limitations, however: it cannot anticipate or plan for future events, and it can only select from the variation that is present in a population.

9.5 What Are Random Events in Evolution?

Describe the potential role of random processes in evolution and differentiate natural selection from genetic drift.

- Random mutation is the ultimate source of variation required for natural selection to work.
- Virtually every newborn individual will have at least a couple of mutations.
- Genetic drift refers to random changes in the frequency of an allele; in a special case known as founder's effect, a few individuals of a population leave and found a new population. The variation in the starting population limits the potential variations of future generations.

9.6 What Is the Evidence for Speciation?

Summarize the evidence for speciation.

- The biological species concept defines a species as a group of individuals that can breed and have viable young.
- Scientists can extract DNA from different species and compare the sequences for a specific gene on an evolutionary tree.
- In species with few genetic differences, the species are closely related, whereas more genetic differences indicate more distant relatedness.

9.7 How Do New Species Arise?

Outline the stages by which new species arise from a single species.

- Genetic isolation: something divides populations to stop interbreeding.
- Genetic divergence: over time, the populations diverge genetically, structurally, and behaviorally.

- Secondary contact: the separated populations may come back into contact and interbreed successfully (no speciation has occurred), interbreed and produce less-successful hybrids ("incomplete speciation"), or not breed at all (speciation has occurred).

9.8 Why Is It So Difficult for the Public to Accept Evolution?

Assess reasons why some public resistance to evolution still exists despite overwhelming scientific evidence.

- Not all religions have adopted the position that evolution is compatible with their spiritual teaching.
- Evolution has been co-opted in the past to support social and political ideology that justifies selective breeding and large-scale massacres of ethnic groups.
- Some people fear that evolutionary principles will be applied to the ways we interact with one another on individual and societal levels.

Key Terms

adaptation 259
allele frequency 256
biogeography 252
biological species concept 267
Darwin's finches 254
descent with modification 256
endemic 254

evolution 246
evolutionary tree 268
fitness 263
founder's effect 265
genetic drift 265
kin selection 264
modern synthesis 256

mutation 246
natural selection 255
pseudogene 250
reinforcement 275
speciation 267
variation 247

Review Questions

1. Why are most adults unable to eat dairy products, at least not comfortably?

   a. They are missing the lactose gene.

   b. They have lost the ability to break down the sugar lactose.

   c. They have gained the ability to break down the sugar lactose.

   d. They have a gene that increases lactase production.

   e. They have a gene that increases lactose production.

2. List one puzzling feature about the following: human testes, our inability to make vitamin C, our eyes.

3. Briefly give an evolutionary argument for each of the items in question 2.

4. Briefly explain one line of evidence that helped convince Darwin that evolution occurred.

5. On what three observable facts of nature is Darwin's theory of natural selection based?

   a. Individuals in a population vary, variations are inherited, and only the fittest individuals survive.

   b. Individuals in a population vary, individuals in the same species may interbreed, and variations are inherited.

   c. Individuals in a population are all the same species, variations are inherited, and many more offspring are born than survive to adulthood.

   d. Individuals in a population vary, variations are inherited, and in any population, many more offspring are born than survive to adulthood.

   e. Individuals in a population vary, variations are caused by genetic mutations, and more offspring survive into adulthood than die at birth.

6. What was the modern synthesis?

7. Briefly explain how natural selection operated on the medium ground finch during the drought on Daphne Major. How did the birds differ before and after the drought?

8. What is fitness? Explain why fitness involves all your relatives, not just your offspring.

9. Can natural selection fashion perfect structures or adaptations? Explain briefly, with examples.

10. What is genetic drift? What is founder's effect?

    a. Genetic drift is the process by which the frequency of mutations changes randomly over time; founder's effect is a case of genetic drift when a few individuals of a population leave and found a new population.

    b. Genetic drift occurs when a few individuals of a population leave and found a new population; founder's effect is the frequency of random mutation changes over time.

c. Genetic drift is the process by which social behavior becomes easier to explain; founder's effect is a case of genetic drift when a few individuals of a population leave and found a new population.

d. Genetic drift is the process by which inherited mutations cause individuals to be more dissimilar from their descendants over time; founder's effect is a direct consequence of genetic drift.

e. Genetic drift occurs when species become isolated from one another because of geographic changes; founder's effect is when a few individuals of a population leave and found a new population.

11. Define the biological species concept. Provide one example of a case in which the definition does not seem to apply.

12. List the three steps required for speciation to occur.

a. Genetic divergence, secondary contact, and reinforcement

b. Genetic isolation, genetic divergence, and genetic resurgence

c. Genetic mutation, genetic divergence, and genetic resurgence

d. Genetic isolation, genetic convergence, and secondary contact

e. Genetic isolation, genetic divergence, and secondary contact

13. What is reinforcement? Why is reinforcement considered a type of natural selection?

14. In the United States, approximately what percentage of people accept that evolution occurs? Don't accept? Aren't sure? How does this compare to other countries?

15. What is an evolutionary tree? How does it show the degree of relatedness between two species? Briefly explain one way to construct an evolutionary tree.

The Thinking Citizen

1. Select a human trait that varies across different geographic locations. Two good candidates are skin color and the ability to metabolize alcohol. Do some research to make a map of how the trait varies throughout the world. Propose a hypothesis about why the trait varies, and why it is specific to certain geographic regions.

2. Consider a plant species that grows in an area that normally has a lot of rain. However, for several years, there has been a drought and water has been scarce. Describe how the plants could adapt to the drought through the action of natural selection. Be sure to point out the importance of (a) the fact that the plants vary initially in their drought tolerance (or lack thereof), (b) the fact that drought tolerance (or intolerance) can be passed on to offspring, and (c) the fact that many of the young plants produced in a new generation will not survive.

3. A number of bird species have a behavior called "helpers at the nest." Young birds stay with their parents for a number of years, helping to raise their siblings. But the helpers don't have young of their own, at least not in their early years. Are helpers sacrificing their evolutionary fitness by helping? Do helpers get any evolutionary benefit themselves from helping?

4. Lions and tigers are considered separate species. However, they can interbreed. A tigon is the offspring of a male tiger and a female lion. A liger is the result of the opposite cross: a male lion and a female tiger. Male tigons and ligers are probably sterile, while females of both hybrids can have offspring. Do lions and tigers conform to the biological species concept? Which of the three stages of speciation has not occurred in these species? You might want to consult a range map showing where lions and tigers live. If this stage were to occur, what do you predict about the final outcome?

5. Suppose a DNA analysis of four species reveals the following sequences:

Species 1	CCATGAT
Species 2	CCGCGAC
Species 3	CCTTGAT
Species 4	CCCCGAC

Construct an evolutionary tree that shows how these species are related. Recall that to construct an evolutionary tree, you group the sequences into pairs according to how similar their DNA is. Then you connect the pairs, working backward to construct the tree. Which pairs of species are most closely related?

The Informed Citizen

1. Before we became such accomplished global travelers, people tended to be born, grow up, and die in the same locale. Doctors could focus their treatments on diseases and conditions that were specific to their locale. Now that we travel and mix much more, this approach does not work as well. From an evolutionary point of view, why does a person's geographic origin affect the way he or she responds to medical treatment? Should doctors know about these evolutionary differences in order to treat patients better?

2. In 2011, the United States Department of Agriculture (USDA) unveiled MyPlate, which recommends that dairy products, including soy alternatives to cow's milk, be part of a healthy diet (**FIGURE 9.37**). Prior to that, the USDA had recommended a food pyramid expressly promoting the consumption of two to three servings of milk or milk-based products per day. However, approximately 75% of humans are at least somewhat lactose intolerant, and the incidence is even higher in some ethnic groups such as Native Americans, Asian Americans, African Americans, and Latinos. In fact, the group with the lowest incidence of lactose intolerance is that with Northern European ancestry. What role do you think these factors may have played in the government's decision to change its recommendations? What other factors may have influenced the change in recommendations? Given the known high incidence of lactose intolerance, why do you suppose the government waited so long to make the change? What reasons can you think of to explain why milk-based dairy products have been so strongly promoted for so long?

3. A number of grass-roots movements seek to remove the teaching of evolution from science classes in public schools, downplay the importance of evolution, or require teaching alternatives like creationism and intelligent design. How should we decide what to teach? Who

FIGURE 9.37

should decide what we teach? Is it best to allow everyone to believe what they want and teach accordingly? Is it wrong to teach things in science class that scientists agree are not true?

4. We now have some ability to screen for the genes that are implicated in certain diseases such as cystic fibrosis, Tay Sachs disease, and sickle cell anemia. Should we encourage individuals who have these traits to avoid having children? Why or why not? If we were able to identify genes associated with desirable traits and could screen for those, would it be a good idea to encourage individuals who have those genes to have children? Why or why not? Who decides what constitutes a desirable or undesirable trait? What consequences might result if we encouraged or discouraged reproduction in particular individuals? Is it possible to influence the direction of our own evolution? Should we even try?

CHAPTER LEARNING OBJECTIVES

After reading this chapter, you should be able to answer the following questions:

10.1 In What Ways Is Your Body an Ecosystem?
Recognize and provide examples of the different types of species interactions.

10.2 Why Do Diseases Evolve Resistance to Antibiotics?
Describe the process by which some disease organisms increase their resistance to antibiotics.

10.3 Why Are Some Diseases More Deadly Than Others?
Use the trade-off hypothesis to explain and predict the spread and impact of a disease.

10.4 Where Do New Diseases Come From?
Consider the processes of mutation and natural selection and predict likely sources of new disease.

10.5 How Can Evolution Help Us Control Disease?
Propose a strategy to reduce the virulence of a disease using evolutionary medicine.

One reason we get sick is because pathogens, such as bacteria, invade our bodies and do damage. But thousands of species of organisms live in and on our bodies and may even keep us healthy.

CHAPTER 10

The Evolution of Disease

Why Do We Get Sick?

Becoming an Informed Citizen …

Pathogens evolve to overcome our attempts at disease control, and new diseases emerge continuously for which we have no natural immunity, antibiotics, or vaccines. A better understanding of the evolutionary dynamics of disease may provide new approaches to public health.

pathogen :: disease-causing organism, such as bacterium, fungus, or worm, that invades the body and does damage

*hy do we get sick? It's an interesting and surprisingly complicated question. One reason we get sick is because **pathogens**, disease-causing organisms such as bacteria, fungi, or worms, invade our bodies and do damage. Although not considered living organisms, some viruses are also pathogenic. However, as we'll see, thousands of species of organisms, especially bacteria, live in and on our bodies and do not make us sick—they may even help us stay healthy. In this chapter, we discuss how diseases evolve, in some cases to become even more harmful to us, and also what evolution can tell us about how diseases can be controlled.*

First, we will look at the disease that has probably killed more humans than any other: malaria. We began our study of malaria in Chapter 4, but here we address a different question: how has malaria evolved, and how have we evolved in response?

case study

Deadly Malaria

Malaria is a disease suffered by about 300 million people annually, and about 1.5 million die from it each year—almost three people each minute. Malaria has been with us throughout the entire history of our species, perhaps 200,000 years. If you consider all human deaths from all causes, malaria may have killed more than half of us.

There are four species of *Plasmodium* that infect humans (and another 170 species that infect other animals). About 7000 to 10,000 years ago, one of these species, *Plasmodium falciparum*, underwent a massive population explosion. *P. falciparum* is by far the most deadly species, accounting for over 50% of all cases of malaria (more than 75% of those in Africa) and 99% of all malarial deaths worldwide. Why is malaria so deadly?

One reason is that *Plasmodium*, the pathogen that causes malaria, is not as easy to kill as it was in the past. Quinine was first used to treat malaria in 1820. Since then, we have developed a dozen different drugs to treat the disease, but none is effective anymore against all types of malaria. The reason for that is that different types of malaria are caused by different strains of *Plasmodium* (a strain is a type of genetic variation within a species), and some strains of *Plasmodium* have evolved **resistance** to multiple drugs. Resistance refers to the ability of an organism to survive exposure to a drug that had previously been able to kill or disable it. The World Health Organization reports that resistance to antimalarial drugs is now one of the major challenges to our attempts to control the disease.

We humans have not been idle in the evolutionary fight against malaria. Before fighting with drugs and medical practices, we "fought back" the way all organisms do: we evolved resistance. Recall that the malaria parasite causes damage by infecting red blood cells, reproducing inside them, and bursting the cells to release new parasites that continue the damage (**FIGURE 10.1**). All our natural defenses to malaria are mutations that make red blood cells less

resistance :: the ability of an organism to survive exposure to a drug that previously had been able to kill or disable it

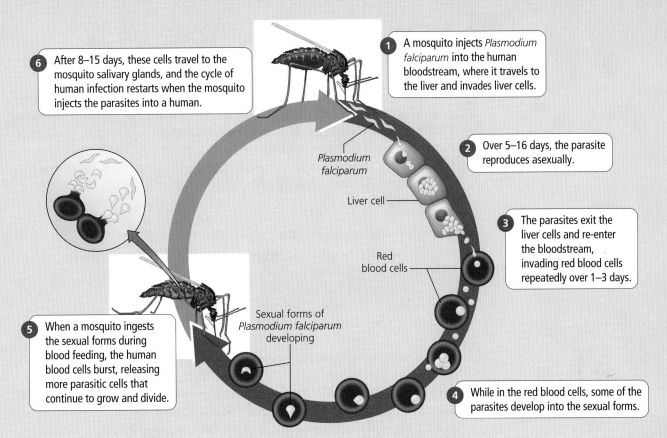

6 After 8–15 days, these cells travel to the mosquito salivary glands, and the cycle of human infection restarts when the mosquito injects the parasites into a human.

1 A mosquito injects *Plasmodium falciparum* into the human bloodstream, where it travels to the liver and invades liver cells.

2 Over 5–16 days, the parasite reproduces asexually.

3 The parasites exit the liver cells and re-enter the bloodstream, invading red blood cells repeatedly over 1–3 days.

5 When a mosquito ingests the sexual forms during blood feeding, the human blood cells burst, releasing more parasitic cells that continue to grow and divide.

4 While in the red blood cells, some of the parasites develop into the sexual forms.

Plasmodium falciparum

Liver cell

Red blood cells

Sexual forms of *Plasmodium falciparum* developing

FIGURE 10.1 The malaria parasite, *Plasmodium falciparum*, causes damage by infecting red blood cells, reproducing inside them, and bursting the cells to release new parasites that continue the damage.

vulnerable to *Plasmodium.* In Chapter 4 you learned how the sickle cell mutation does this. There are actually four different mutations that cause red blood cells to sickle. There are about a dozen other mutations that affect red blood cells in a way that makes them partially resistant to malaria. Like sickle cell, these mutations cause physiological problems in people who inherit two copies of the mutated gene, and they are found only in regions where malaria is common.

Scientists can study the evolutionary history of our resistance mutations by looking at the DNA sequence of the genes. It turns out that these mutations date back to around 7000 to 10,000 years ago—the same time the deadly form of malaria began to spread.

So why is malaria deadly? One of the reasons has to do with the evolution of resistance. Another has to do with 7000-year-old changes in human behavior that allowed a deadly disease to spread. Before we explore these reasons in more depth, however, we'll take a closer look at the many species that inhabit our bodies. Some cause disease, some don't, and some can go either way depending on conditions.

10.1 In What Ways Is Your Body an Ecosystem?

The space your body takes up contains an estimated 100 trillion cells. However, only 10 trillion of them genetically are yours. The other 90 trillion belong to bacteria and other one-celled organisms, fungi such as yeast and mold, and multicellular organisms like mites and a whole host of worms (**FIGURE 10.2**). If you were to somehow remove all "your" cells, these foreign cells would still form a recognizable outline of your body and many of your internal organs.

The Many Species That Live and Evolve in Your Body

Medical researchers are beginning to study the species that live in and on us. Most of these organisms live on our skin and in our mouth and digestive tract, urinary tract, and nose and lungs. The National Institutes of Health have started the Human Microbiome Project to figure out what these species are—the first step toward understanding how they affect our health. What we have learned so far is surprising.

One 2009 study found 182 species of bacteria living on the forearm alone, 30 of which were new to science. When scientists examine other regions of the skin, they discover many hundreds of other species. Furthermore, the species are very different on different parts of your body. Your dry forearm is home to species adapted to a "desert," species adapted to warm, moist underarms and the groin form "tropical" communities, and still other communities live in oily areas like hair follicles and the crease of your nose. Scientists have discovered over 700 species that live in the human mouth and over 800 species that live in the colon. Other parts of your body support similar numbers (see *How Do We Know? The Many Species That Live On You*).

ecology :: the study of the interactions among species and their physical environments

Ecology is the study of the interactions among species and their physical environments. We all have many hundreds of species living with us, but none of us has all the species just mentioned. A few species tend to be found on all of us all the time. But each of us has our own particular set of species. These species form ecological communities that differ throughout our bodies, and the

A *E. coli* bacteria

B *Candida albicans* fungi (mouth, vagina, and skin)

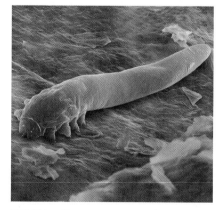

C Follicle mite

FIGURE 10.2 The body is a complex ecosystem. In addition to its own cells, bacteria (A), fungi such as yeast and mold (B), and multicellular organisms like mites (C) live on and within the human body.

How Do We Know?

The Many Species That Live on You

Many biology courses have a lab where students study the species that grow on them. Students run a cotton swab over a certain region of skin, perhaps their armpit or the crease of their nose, and then rub the swab over a petri dish containing culture medium. A few species of bacteria or mold will grow, but it turns out that most of the species that live on your body are highly adapted to the conditions they find there and simply don't grow well or at all on a petri dish. Scientists missed the vast majority of species simply because they couldn't grow them in a lab.

Now scientists are finding other ways to learn about these organisms. A swab from your skin, a small amount of feces, or a cough contains a whole community of microorganisms. Scientists collect the sample, break open all the cells, isolate the ribosomes, and extract the ribosomal RNA (see Chapter 3). Each species has its own specific type, or sequence, of ribosomal RNA. When scientists count how many types of ribosomal RNA they find, they know how many different species there were in the original sample. They can also try to match the RNA to a database of known RNA types, and if they find a match, they can even tell which species were present.

Many biology courses have a lab where students study the species that grow on them. Today, scientists are finding other ways to learn about these organisms without growing them in the lab.

Techniques like these don't tell us much about these microorganisms, how they live, or what they do. But we have learned that there are thousands of species that live on and in us—many more than we expected. This changes our perspective and shows us that our bodies truly are ecosystems. It also changes our notion of what it means to be "clean." ::

species compositions of these communities change over time. Within these communities, individuals survive, reproduce, and compete for resources, just as they do in any biological community. As a result, the species evolve over time, adapting to one another and to the conditions in our bodies. Your body is an ecosystem that provides numerous habitats and supports the evolution of many species.

The Ecology of Our Resident Species

We are in the early stages of understanding the life on us and how it affects our health. Many of our resident species appear to have no effect on us at all. Eyebrow mites, for example, eat dead skin cells and excess oil from hair follicles, but most of the time we don't even know they exist. This ecological interaction is called **commensalism**: one species benefits from the interaction, and the other is neither helped nor harmed.

We do know that our resident species compete with one another for food and space. This competition can be fierce. In the digestive tract, for example, many species secrete toxic compounds that kill other species. Others use up critical nutrients or make the habitat more acidic, interfering with the growth of competing species. Some even form thin films on our intestinal surfaces that prevent other species from sticking there. These free-floating species can be washed away as food moves through us, again removing competitors.

Many species that live with us are **mutualistic**, meaning that both we and the species benefit from the interaction. Some species in our digestive tract help us digest our food, for example (**FIGURE 10.3**). They also eat our food, so technically they compete with us as well. The competitive behavior of some species helps protect us from disease. Pathogens often are invasive species. Before they can infect us, they must survive competition with the species that normally live with us. The competitive behavior our resident species use against one another is often effective against invading pathogens, preventing them from becoming established and infecting us.

commensalism :: an ecological interaction in which one species benefits and the other is neither helped nor harmed

mutualistic :: an ecological interaction in which both species benefit

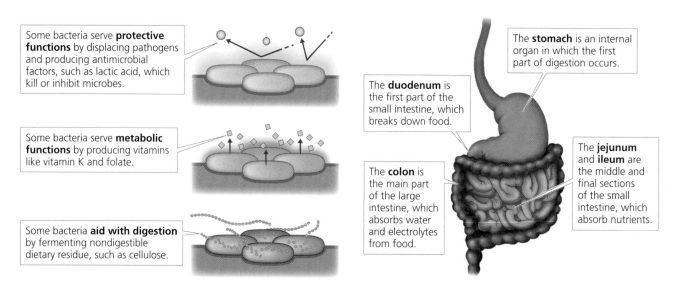

Some bacteria serve **protective functions** by displacing pathogens and producing antimicrobial factors, such as lactic acid, which kill or inhibit microbes.

Some bacteria serve **metabolic functions** by producing vitamins like vitamin K and folate.

Some bacteria **aid with digestion** by fermenting nondigestible dietary residue, such as cellulose.

The **stomach** is an internal organ in which the first part of digestion occurs.

The **duodenum** is the first part of the small intestine, which breaks down food.

The **jejunum** and **ileum** are the middle and final sections of the small intestine, which absorb nutrients.

The **colon** is the main part of the large intestine, which absorbs water and electrolytes from food.

FIGURE 10.3 Many species that live with us are mutualistic, meaning that both we and the species benefit from the interaction. In the gut, for example, bacteria prevent colonization by harmful bacteria, produce vitamins, and help with digestion.

Species That Cause Disease

A species living with us may be commensal or mutualistic, but either way it will still be looking out only for itself. If conditions are right—or from our perspective, wrong—many of these species will change their behavior and attack us. If eyebrow mites experience a population explosion, for example, they can cause inflammation and itching of the skin around our hair follicles. Bacteria living in our digestive tract initially may be helpful. But if their population becomes too large, they may begin to secrete toxins or invade our cells, causing disease.

One example of this behavior is seen in the bacterium *Pseudomonas aeruginosa*. This species lives almost anywhere: in soil and water, on the surface of nonsterile medical equipment, and on your skin. It can live on you and cause no damage. However, *Pseudomonas* can be a serious problem for individuals who are already sick or who have a weakened immune system. In these cases, *Pseudomonas* can cause pneumonia, urinary tract infections, septic shock, and even tissue **necrosis** or death. *Pseudomonas* cells secrete a chemical compound that acts as an alert signal. When enough bacteria are present and the concentration of the chemical becomes large enough, *Pseudomonas* begins to produce a toxin that kills our cells. *Pseudomonas* bides its time until it has a **quorum**: enough cells to overwhelm our body's defenses. Many species of bacteria engage in quorum sensing, waiting until their numbers are large before switching to behaviors that cause disease (**FIGURE 10.4**).

Some species don't have these switches: they always cause disease. Viruses fall into this category. A virus generally can't reproduce without killing the cell it infects. And many species that cause disease are true invasive species. They don't normally live on us, and we "catch" them from other sources: an infected person or contaminated soil or water, for example. These species simply use our bodies as a resource for their own reproduction. We may inhale or swallow the invading species, or they may burrow through our skin or the lining of our digestive tract. They can also get in through wounds in our body's surfaces. Some enter the bloodstream, where they can travel throughout our body to particular locations or organs where they can reproduce. Their reproduction uses our body's energy and can damage our organs and tissues, which causes the disease.

In the rest of this chapter, we will look at how disease organisms evolve, and how their adaptations affect the severity of the disease and the way organisms transmit from host to host.

necrosis :: tissue death

quorum :: a population of bacteria large enough to overcome the body's defenses

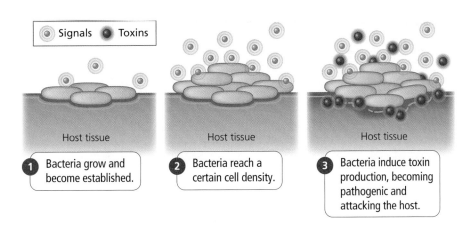

FIGURE 10.4 Whether species that live with us are commensal or mutualistic, many will change their behavior and attack us under certain conditions. Some bacteria engage in quorum sensing, becoming pathogenic only after they have reached a high enough cell density.

10.2 Why Do Diseases Evolve Resistance to Antibiotics?

antibiotic :: drug used to kill a pathogen

We mentioned earlier that some strains of *Plasmodium falciparum* have evolved resistance to at least one of a dozen **antibiotics**, drugs used to kill pathogens. Resistance evolves in every species that confronts antibiotics, including prokaryotic pathogens like bacteria, single-celled eukaryotic pathogens such as *Plasmodium*, pathogenic worms, pathogenic insects, and even pathogenic viruses. And they all evolve resistance for the same reasons. We'll discuss resistance in bacteria, but the principles apply to all species.

Alexander Fleming, a Scottish scientist, ushered in the age of antibiotics. In 1928, Fleming discovered that one of his bacterial cultures had been contaminated by the mold *Penicillium notatum*. Many scientists simply would have discarded the contaminated plate, but Fleming noticed something peculiar: there was a clear ring around the mold where the bacteria did not grow. Fleming performed several experiments and discovered that the mold produced a compound that could kill bacteria. He named this compound penicillin, after the mold that produced it.

Widespread use of penicillin began in 1943, and during World War II it saved the lives of many thousands of soldiers. Since then, we have developed many other antibiotics, and they have continued to cure infections and save lives. One disturbing pattern, however, is that only a few years after a new antibiotic is introduced, new strains of disease evolve that are resistant to it (**TABLE 10.1**).

How Resistance Evolves

Resistance evolves in a very simple manner. When you apply antibiotics to a population of bacteria, most of the cells are vulnerable and they die. However, a few cells will not die. Instead, they will reproduce and form a population of resistant bacterial cells (**FIGURE 10.5**). This is a classic and simple example of natural selection. Resistance evolves like any other trait. Environmental conditions favor some traits over others, and individuals with these traits become more common. And the evolution of resistance is not limited to bacteria. For example, insects develop resistance to insecticides, which presents problems in pest control. For a discussion of how the mosquitoes that carry malaria developed a resistance to DDT, see *Life Application: Malaria and DDT*.

1 Antibiotics are applied to a population of bacteria.

Antibiotic   Bacteria

None   High

Bacterial resistance to antibiotic

2 Most of the cells are vulnerable and they die.

3 A few cells survive the exposure and reproduce, forming a population of resistant bacteria cells.

FIGURE 10.5 Antibiotic resistance evolves like any other natural trait; it is an example of natural selection.

Table 10.1 :: Timeline of Antibiotic Resistance in Bacteria

Antibiotic	Year Introduced	Year Resistance First Observed
Penicillin	1943	1945
Chloramphenicol	1949	1950
Erythromycin	1952	1956
Methicillin	1960	1961
Cephalothin	1964	1966
Vancomycin	1958 (not used much until 1980s)	1986
Cephalosporin (two types)	1979, 1981	1987
Carbapenems	1985	1987
Linezolid	2000	2002

Life Application

Malaria and DDT

If we could kill all the mosquitoes that carry malaria, we would exterminate the disease. DDT (dichlorodiphenyltrichloroethane) is a powerful insecticide that could be used to do the job. During World War II, DDT was used widely to control mosquitoes, lice, and other insects that spread disease among soldiers and civilians in war-torn areas. DDT helped eliminate malaria in many developed countries, although using screened doors and windows, draining swamps, and providing access to better medical care did most of the work.

After the war, DDT was put to a new use: it was sprayed on crops to reduce insect damage. It soon became clear, however, that DDT was linked to many environmental issues. It affected not only insects, but also mammals and birds that ate these insects. Some studies found links between DDT and human birth defects and cancer. In 1962, Rachel Carson published *Silent Spring*, a popular book that chronicled the effects of DDT and touched off the environmental movement. Ten years later, the United States banned DDT, and many countries have followed suit.

Sometimes the science gets lost when issues of public health become entangled with politics. Environmentalists point to the adverse effects of DDT and the damage it can do when it builds up in the environment. Others argue that banning the insecticide is an example of government intrusion, and that millions of children in regions with malaria might be saved if DDT were used to control mosquitoes.

The science behind DDT is clear, however. Spraying DDT will reduce the mosquito population, and this will reduce malaria—initially, at least. But mosquitoes rapidly develop resistance to DDT, as do many insect pest species. DDT persists in the environment long after the insects break free of its effects. It continues to harm mammals and birds and exposes humans to the health risks. Worse, after the mosquitoes become resistant, malaria returns.

Science tells us about the positive and negative consequences of using DDT. But science alone can't help us decide whether the positives outweigh the negatives. In the midst of a particularly severe outbreak of malaria, spraying DDT could save many people. Is its use in such cases justified, in spite of the fact that its benefits are temporary and its disadvantages will persist long term? These questions are hard to answer, but they become even more difficult when political ideology influences the decision-making process. ::

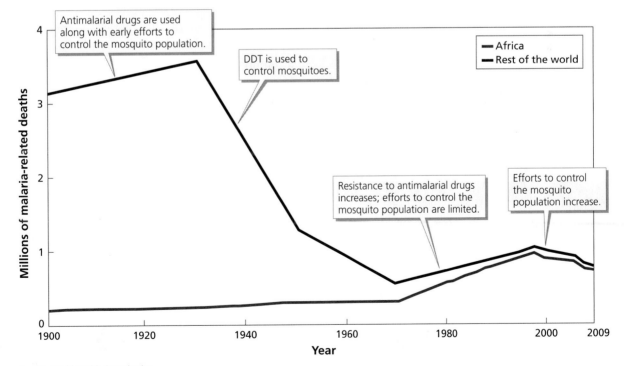

Source: World Health Organization

Although significant DDT spraying led to an initial decrease in malaria-related deaths beginning in the 1930s, mosquito resistance to DDT became more apparent starting in the 1970s.

A *Penicillium* grows on a lemon.

B *Streptomyces* scabs grow on potatoes.

FIGURE 10.6 Most antibiotics are naturally occurring chemicals made by bacteria, molds, and other species. The fungus *Penicillium* excretes penicillin, which kills bacteria in its vicinity (A); the bacterium *Streptomyces* produces streptomycin, which inhibits bacterial growth (B).

resistance gene :: gene that has mutated in such a way that the proteins it makes take on a different function or perform old functions in a new way

But that still leaves several questions about the evolution of resistance. Why are some bacterial cells resistant even before the antibiotics are applied? What characteristics do these cells have that makes them resistant? If resistance is beneficial, why aren't all bacteria resistant? If using antibiotics results in resistant bacteria, why do antibiotics continue to be effective, and how long will this last?

Where Resistant Bacteria Come From

Most antibiotics are naturally occurring chemicals made by bacteria, molds, and other species (**FIGURE 10.6**). Although we can't see them, these microorganisms live in densely populated conditions in the soil, in aquatic habitats, and in you. Competition for food and other resources is very strong. *Penicillium* mold has an advantage: by secreting penicillin, it kills its competitors and keeps more food for itself. Many other species produce antibiotics of their own and get the same advantage.

A species that produces an antibiotic must be resistant to it; otherwise, it would kill itself. This logic is a sort of "proof" that if antibiotics can evolve, resistance must be able to evolve as well. If antibiotic-producing species evolve resistance to their own antibiotics, there is no reason that species that don't produce antibiotics can't do the same.

Resistant bacteria have one or more genes that make them that way. These are known as **resistance genes**. These genes have mutated in such a way that the proteins they make take on a different function or perform old functions in new ways. There are four ways that resistance genes, and the proteins they make, can produce a resistant cell (**FIGURE 10.7**):

1. They can change the cell membrane so the antibiotic cannot get in.
2. They can modify an enzyme so that it breaks down the antibiotic.
3. They can alter the structure of the antibiotic's "target" (what it attacks) so that it is no longer vulnerable to the antibiotic.
4. They can form a "pump" that pumps out the antibiotic before it can cause damage.

Mutation is the reason that every bacterial population has a few resistant cells in it, even when no antibiotics are present. Antibiotics don't cause

FIGURE 10.7 There are four mechanisms that make bacteria resistant to antibiotics.

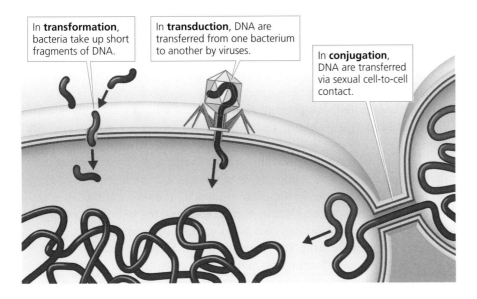

In **transformation**, bacteria take up short fragments of DNA.

In **transduction**, DNA are transferred from one bacterium to another by viruses.

In **conjugation**, DNA are transferred via sexual cell-to-cell contact.

FIGURE 10.8 Horizontal gene transfer spreads antibiotic resistance genes through a bacterial population. Genes enter bacteria through transformation, transduction, or conjugation.

mutations for resistance; they only select for resistance and increase the number of resistant cells. Like all mutations, mutations that make resistance genes are extremely rare and occur only due to chance. However, microbial competitors have been producing antibiotics for millions, maybe billions, of years, so there has been plenty of time for these mutations to occur. Also, bacterial populations are unimaginably large: billions or trillions of cells in each individual host or patch of soil. Every time a cell divides, there is the chance of a mutation. With so many cell divisions over such a long period of time, some mutations for resistance are bound to occur.

Resistance to Multiple Antibiotics

Mutation alone is too slow to account for the rapid appearance of resistant strains. But bacteria have other ways of acquiring resistance genes. Some bacteria can get DNA directly from their neighbors, a process known as **horizontal gene transfer** (**FIGURE 10.8**). DNA can spread through a bacterial population the way news and gossip move through a human community. There are three ways that DNA can be transferred horizontally: transformation, transduction, and conjugation. With **transformation**, DNA released into the environment by bacteria is physically absorbed by other bacteria. **Transduction** involves the transfer of bacterial DNA from one bacterium to another by a virus. In **conjugation**, two bacterial cells physically join together and transfer DNA from one to the other. If one of the bacterial cells has a resistance gene, it can pass it along to other cells, creating many resistant bacteria—and all their descendants will be resistant as well. Worse still, horizontal gene transfer allows many bacteria to become resistant to several antibiotics. With horizontal gene transfer, the spread of resistance occurs much more quickly than by mutation alone.

Not every strain of bacteria is resistant, and not all resistant bacteria are resistant to all antibiotics. Antibiotics continue to be effective against some strains of bacteria, and if one antibiotic doesn't work, you can try another one. But we have already reached the point where it is common for a tried-and-true antibiotic to fail because of resistant strains.

Why Not All Bacteria Are Resistant to Antibiotics

In any environment where antibiotics exist, resistance is a strong evolutionary advantage. But everything in nature, including antibiotic resistance, comes at

horizontal gene transfer :: the process by which bacteria can get DNA directly from a neighbor

transformation :: DNA released into the environment by bacteria is physically absorbed by other bacteria

transduction :: the transfer of bacterial DNA from one bacterium to another by a virus

conjugation :: two bacterial cells physically join together and transfer DNA from one cell to the other

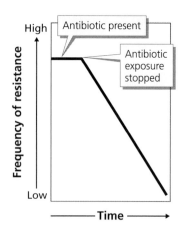

FIGURE 10.9 Bacteria that are resistant to antibiotics become less prevalent in the population when antibiotic exposure ceases.

a cost. It takes energy to break down antibiotics or to make molecules to pump them out. Resistant cells divert this energy from other functions such as growth and reproduction. Also, the modified proteins in a resistant cell may not work as well as the unmodified forms. In short, resistant cells are often less efficient than their vulnerable counterparts.

As we have noted, competition among species within your body is fierce. When antibiotics are absent, resistance is no longer an advantage. And because they are energetically less competitive, resistant strains are outcompeted by more efficient nonresistant strains. Eventually a population of resistant bacteria gives way to a population of nonresistant cells (**FIGURE 10.9**).

This is why it is critically important that you take your antibiotics exactly as prescribed. The specified length of time is calculated to be long enough to kill most of the vulnerable cells but short enough to prevent a population of resistant cells from forming. If you stop your pills early because you start feeling better, you may be leaving too many bacterial cells alive. If you continue treatment for longer than recommended, you may select for a resistant strain.

Antibiotics in the Environment

If resistant bacteria are eliminated in the absence of antibiotics, why is resistance becoming a major global health threat? It is because antibiotics are now available almost everywhere (**FIGURE 10.10**). Besides administering antibiotics in individual patients, hospitals themselves are a huge reservoir of antibiotics.

A high concentration of many kinds of antibiotics is found in hospitals, and as a result hospitals harbor many resistant strains. Over 200,000 patients annually are infected with resistant strains of bacteria, and the Centers for Disease Control and Prevention (CDC) estimates that over 90,000 deaths occur as a result. Since antibiotics are always present, the vulnerable strains of bacteria cannot displace the resistant ones. Hospitals also provide many hosts—some already weakened and sick, and therefore highly susceptible to infection.

Antibiotics are also used in agriculture, which means they can be found in our food, soil, and water. Over half the antibiotics used in the United States are given to livestock, poultry, and crops. Some of these antibiotics are used to treat infection or, in the case of crops, to prevent bacteria from spoiling the food. But farmers often routinely give low doses of antibiotics to healthy livestock and poultry in their feed. These low doses kill some of the bacteria in the digestive tract. This means more food goes to the animal (not to the bacteria that live in it), and it grows larger in a shorter period of time.

FIGURE 10.10 The amount of antibiotics used in agriculture and for human health is staggering. In 2009, almost 29 million pounds of antibiotics were administered to livestock and more than 7 million pounds to humans.

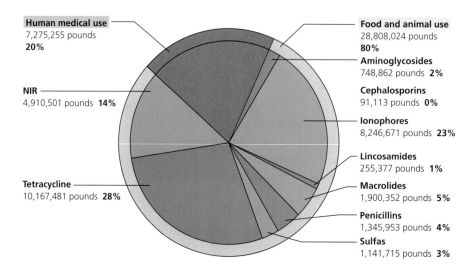

Human medical use
7,275,255 pounds
20%

NIR
4,910,501 pounds **14%**

Tetracycline
10,167,481 pounds **28%**

Food and animal use
28,808,024 pounds
80%

Aminoglycosides
748,862 pounds **2%**

Cephalosporins
91,113 pounds **0%**

Ionophores
8,246,671 pounds **23%**

Lincosamides
255,377 pounds **1%**

Macrolides
1,900,352 pounds **5%**

Penicillins
1,345,953 pounds **4%**

Sulfas
1,141,715 pounds **3%**

As antibiotics become more common in bacterial habitats, resistant strains of bacteria become more common as well. Many medical researchers predict that eventually antibiotics will become ineffective for treating bacterial infection, unless we can find ways around the problem of resistant strains.

10.3 Why Are Some Diseases More Deadly Than Others?

Some infectious diseases can be fatal. Malaria, tuberculosis, and AIDS are responsible for half of the deaths caused by infectious disease each year. Other diseases, like the common cold or chicken pox, cause discomfort but are rarely fatal. Pathogens vary quite a bit in their **virulence**—their ability to cause damage or death. But the one characteristic they have in common is that each pathogen is adapted to its way of life. These adaptations are the result of natural selection acting to maximize the pathogen's survival and reproduction. (See *Scientist Spotlight: Paul W. Ewald* for one scientist's

virulence :: a measure of a pathogen's ability to cause damage or death

Scientist Spotlight

Paul W. Ewald (1953–)

Not many people have redirected their careers because of a case of diarrhea, but that is exactly how Paul Ewald, a professor of biology and director of the Program on Disease Evolution at the University of Louisville, became interested in evolutionary medicine.

Ewald was born in Wilmette, Illinois. His mother was a psychologist and his father was a physicist. Ewald earned his bachelor's degree from the University of California, Irvine, in 1975 and his Ph.D. in 1980 from the University of Washington.

It was in 1977, during his graduate fieldwork studying bird behavior, that Ewald suffered a three-day bout of diarrhea. While he was sick, he wondered about his role as a host for the pathogen that was causing him such uncomfortable physical distress. As Ewald remembers thinking, "There's some organism in there and this diarrhea might be my way of getting rid of the organism—or it might be the organism's way of manipulating my body."

Ewald considered that the organism might be ensuring its success of finding another host to infect by passing efficiently out of his body. He thought about how he should treat his symptoms. "If it's a manipulation and you treat it, you are avoiding damage, but if it's a defense and you treat it, you sabotage the host."

As a result of this experience, Ewald changed his research focus to host-pathogen interactions. He realized that medicine could be enriched by evolutionary theory. In the case of infectious disease, for example, it is essential to look at the disease not simply from the standpoint of the host but also from that of the pathogen.

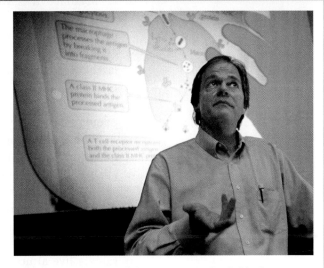

Paul W. Ewald is one of the pioneers in the field of evolutionary medicine.

Ewald is one of the pioneers in the field of evolutionary medicine, which rests on the idea that illness can be understood fully only through inclusion of a Darwinian evolutionary perspective. His ongoing research focuses on the questions of whether some chronic diseases—such as mental illness, cancer, Alzheimer's, and heart disease—are caused by low level, long-term infections and whether pathogens can be made less virulent by altering their behaviors or transmission patterns. ::

Source: http://www.scientificamerican.com/article.cfm?id=a-host-with-infectious-id.

FIGURE 10.11 Evolution selects for pathogens that do minimal harm to their hosts, thereby, prolonging their own life span. The myxoma virus is a deadly pathogen that was introduced to control the rabbit population in Australia; today it kills only about 50% of the rabbits that catch it, and infected rabbits live almost a month.

experience with the tension between the needs of pathogens and hosts.) In this section, we will look at some of the reasons why diseases evolve and adapt in different ways.

Why Some Diseases Become Milder over Time

As the old saying goes, "Never bite the hand that feeds you." But this is exactly what pathogens do. They take energy and nutrients away from their hosts, and when their population size grows, they damage cells and internal organs. This seems shortsighted because pathogens are completely dependent on their host for their very lives. If the host dies, the pathogen dies with it.

Evolutionary biologists recognized this paradox early. They hypothesized that evolution would select for pathogens that did minimal harm to their hosts, thereby prolonging their own life span (or the life span of their descendants). Pathogens that were highly virulent were thought to be out of balance evolutionarily. Most likely these pathogens caused diseases that were new to humans, and evolution had not yet had time to select for milder forms.

There was some experimental evidence to support this view. The best example, which comes from Australia, involves rabbits and a virus called myxoma (**FIGURE 10.11**). In 1859, Australian ranchers imported rabbits from England to hunt for sport. But rabbits, being rabbits, soon reproduced uncontrollably. Within 20 years, rabbits became a major pest, destroying crops and displacing native species.

Myxoma virus, which causes a lethal disease in rabbits, was discovered in Uruguay around 1900. During the 1950s, the Australian government released a plague of myxoma virus that spread throughout the rabbit population. Myxoma virus was virulent and very effective at controlling the rabbit population. It killed 99% of the rabbits that caught it. Their lungs filled with fluids, their eyelids filled with pus and swelled shut, and death occurred within a week. The rabbit population dwindled from around 600 million to 100 million in only a couple of years.

Since then, both the rabbit population and the virus have evolved, adapting to each other. The most virulent strains of the virus were selected against because they killed their hosts before the rabbits could spread the disease to others. The most vulnerable rabbits died off quickly, leaving rabbits that had some resistance to the virus. The net result is that myxoma has evolved to become less virulent, as the theory predicted it would. Today, myxoma kills only about 50% of the rabbits that catch it, and infected rabbits live almost a month—plenty of time for the virus to spread. The rabbit population is estimated to be about a quarter of what it was at its peak. This is still too many rabbits, but exterminating rabbits has proven to be a public relations nightmare for the Australian government, especially around Easter.

The Trade-Off Hypothesis of Reproduction Versus Transmission

In some cases, it is true that pathogens are selected for traits that cause their hosts less harm. But we know now that this is not a general rule. Human malaria, after all, is at least 200,000 years old, and it remains deadly. Interpreting the evolution of disease involves two ideas. First, natural selection selects for increased reproductive success however it occurs. There is no selection for species to be nice to one another unless this happens to increase reproductive success as well. Second, pathogens have a two-pronged challenge for their success. They evolve to compete and reproduce successfully

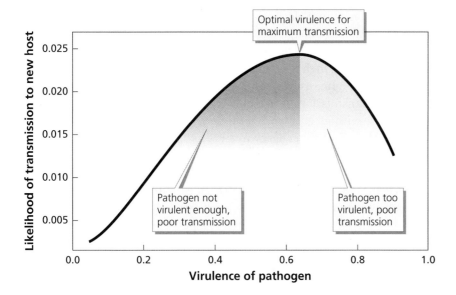

FIGURE 10.12 Pathogens evolve to compete and reproduce successfully within a host, but they must balance this against the need for transmission to a new host. If a pathogen becomes too virulent, the host may die before transmission succeeds.

within a host, but they must balance this against the need for **transmission** to a new host (**FIGURE 10.12**).

There is a trade-off between reproduction within a host and transmission to a new one. If this trade-off favors a milder disease to increase transmission, the selected pathogen will be milder. This occurred with rabbits and myxoma virus. However, if the trade-off favors causing serious damage to the host, then the pathogen evolves to become more severe and deadly. The evolution of disease can take either path. The key is to understand which path is selected for and why. We'll look first at how pathogens evolve inside their host. Then we'll examine the trade-off hypothesis with respect to deadly diseases such as malaria, more mild ailments like the common cold, and sexually transmitted diseases.

How Disease Evolves in Your Body Original infections may involve only a few bacterial cells or viral particles. These invaders reproduce vigorously, producing billions of new bacteria or viruses. Almost all the damage done by disease organisms is related to this population growth. A larger population steals more energy from the host and does more damage to cells and organs. More invaders also mean a stronger response from your body's natural defenses: a higher fever, more mucus, or more inflammation, for example. Sometimes your body's defenses overreact and increase the severity of the disease. The overreaction can even kill you.

As the pathogen population grows, it undergoes evolutionary change. Mutations occur, and genetically different strains of pathogen emerge. Some of these strains may be better at infecting cells or utilizing energy. These strains will reproduce more quickly, causing more damage and outcompeting strains that grow more slowly.

Strains with the largest population size have another advantage: they are most likely to be transmitted to a new host (**FIGURE 10.13**). The most numerous strains will have the largest number of representatives in each cough, tick or mosquito bite, or bowel movement. Therefore, the most numerous strains have the best chance of infecting another host.

So far, it looks like natural selection always will select for the fastest-growing, most virulent strain of pathogen. If this were universally true, diseases would evolve to become very bad indeed. But this is where the trade-off

transmission :: the passage of a pathogen from one host to another

FIGURE 10.13 Strains with the largest population size are most likely to be transmitted to a new host. A person infected with a large population of influenza virus, for example, can easily transmit the pathogen to another person.

hypothesis comes in. Selection for virulence is balanced by the need for the host to survive long enough to pass on the infection.

Testing the Trade-Off Hypothesis Malaria has evolved to be quite deadly, but it has a high transmission rate to compensate. Recall that malaria spreads among humans through mosquito bites. This is a direct and efficient method of transmission. A person with malaria can come in "contact" with many people without ever leaving the bed. If the patient is half paralyzed with fever and chills, with no energy to swat hungry mosquitoes, he might infect hundreds of neighbors in a few days.

Perhaps the trade-off hypothesis also helps to explain the mysterious population explosion of *P. falciparum* 7000 to 10,000 years ago. Before this time, humans lived mainly as small, isolated bands of hunter-gatherers. Under these conditions, what would happen to a pathogen that mutated to a deadly form? It might quickly wipe out an entire band of humans, but there the damage would stop. When the humans all died, this strain of pathogen would die out as well, much like a hot fire that quickly burns out its fuel.

Sometime around 10,000 years ago, humans developed agriculture and our society changed forever. With a more reliable source of food, people began living together in greater numbers. Agriculture also requires irrigation, which leaves open puddles of standing water very close to where people live—a haven for malaria-carrying mosquitoes. Under these conditions, a deadly pathogen has many hosts to infect and may be able to maintain itself in the human population. Certainly we have no proof that agriculture permitted a deadly form of malaria to evolve, but it's a plausible idea.

The trade-off hypothesis predicts that the most severe diseases will be those that spread to people most easily and efficiently. Diseases that spread by mosquitoes, fleas, and contaminated water all fall into this category. The bubonic plague that killed approximately one-third of the people in Europe during the Middle Ages was spread by fleas, and cholera is a classic example of a deadly disease spread by drinking contaminated water.

Diseases like the common cold or "childhood" diseases like mumps, measles, and chicken pox spread by airborne viruses or by contact with droplets of fluids from infected people. These diseases spread when infected people are up and walking around, coming in contact with uninfected people. Strains of pathogens so virulent that infected people quarantine themselves in bed would be at a disadvantage (**FIGURE 10.14**). Many diseases that spread this way are indeed milder than malaria, bubonic plague, and cholera.

While a number of examples and cases support the trade-off hypothesis, it is by no means universally true. Tuberculosis is spread much the same way that colds are spread, yet tuberculosis is a much deadlier disease. Influenza is another example of a sometimes fatal disease spread by direct contact. Still, the general pattern is that the deadliest diseases have the highest rates of transmission, and that in many cases a lower rate of transmission selects for a milder disease.

Sexually Transmitted Disease There is one last class of disease that depends on a "healthy" host for its spread. Imagine a sexually transmitted disease (STD) that was so severe it immobilized you within 24 hours of infection, or caused obvious symptoms like open sores and pustules, and killed you in a couple of days. What do you think about this disease's chances of being spread? For an STD, the ability of the pathogen to spread is tied directly to the host's ability to mate. Any STD that interferes with its host's reproduction will be strongly selected against.

Highly virulent diseases have less opportunity to spread to others.

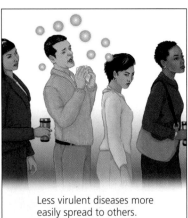

Less virulent diseases more easily spread to others.

FIGURE 10.14 The trade-off hypothesis predicts that the most severe diseases will be those that spread to people most easily and efficiently. Highly virulent diseases do not spread easily unless helped by vectors such as mosquitoes or fleas.

Does this allow us to make predictions about the severity and contagiousness of STDs? The trade-off hypothesis states that like any pathogen, STDs will be selected for competitive ability and fast growth rate within the host. The strain with the largest population has the best chance of transmission to a new host. However, STDs can't cause too much damage too quickly, or their host will be unable to mate often enough. STDs must travel a fine evolutionary line and achieve a balance between these two constraints to be successful. The trade-off hypothesis predicts that STDs will have a slow onset, stay in the host's system for a long period of time, and be fairly mild, at least in the initial stages.

Chlamydia is the most common STD in the United States, which by evolutionary standards means it is the most successful STD (**FIGURE 10.15**). At least 2 million people are infected with chlamydia. But it's hard to get an accurate count because about half the men and three-quarters of the women infected with chlamydia have no symptoms at all. For this reason, chlamydia is sometimes called a "silent" infection. When symptoms do occur, it takes up to three weeks for them to appear—ample time for the disease to be spread among a population of sexually active people. Chlamydia infections can lead to pelvic inflammatory disease, scarring of fallopian tubes, inability to become

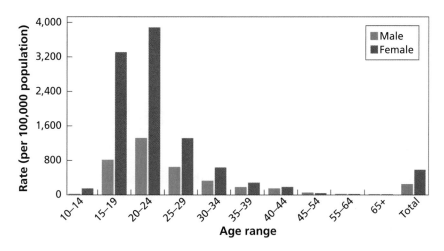

A Chlamydia rates by age and sex, United States, 2011

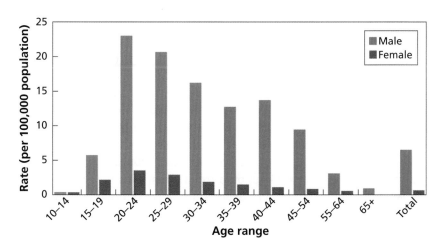

B Syphilis rates by age and sex, United States, 2011

FIGURE 10.15 Chlamydia and syphilis are common STDs, which by evolutionary standards means they are successful. Because few or no symptoms appear early on in the disease, they spread easily throughout the population, with chlamydia being comparatively high in females (A) and syphilis being more prevalent in males (B).

ectopic pregnancy :: a pregnancy in which an embryo develops in the fallopian tube rather than in the uterus; ectopic pregnancies cannot develop successfully to term and if left untreated can lead to rupture of the fallopian tubes

pregnant, or **ectopic pregnancy**, in which an embryo develops in the fallopian tube rather than in the uterus. Ectopic pregnancies cannot develop successfully to term and if left untreated can rupture fallopian tubes.

Syphilis is another STD that may not show symptoms in its early stages. During the first two months after infection, some people develop a series of sores and rashes that begin at the site of sexual contact and later spread to other parts of the body. But many people with syphilis don't develop these sores and rashes, or the breakouts may occur at locations that aren't easily visible. The lack of symptoms results in two major problems. First, infected people won't seek medical treatment, which means that the disease may progress to its final, and lethal, stage. Second, infected people unknowingly spread the disease to their sexual partners.

The final stages of syphilis can take decades to develop. If the disease is not treated, syphilis bacteria eventually invade the tissues of your circulatory system, nervous system, liver, skin, or bones. Depending on where the bacteria invade, syphilis can present a wide array of symptoms in its later stages, all leading to disability and death. But it is possible to be contagious and spread the disease for 20, 30, or even 50 years before symptoms appear. This is bad for infected people and their sexual partners, but good for the pathogen. The absence of early symptoms in some people and the delayed onset of the really bad symptoms is an adaptation of the syphilis bacteria that helps the disease to spread.

Many STDs follow the pattern of symptoms shown by chlamydia and syphilis. One of the most devastating, and most recent, is AIDS.

Where Do New Diseases Come From?

New diseases emerge all the time. Most of these diseases are known only from small outbreaks in localized areas. But there is always concern that some new disease, like HIV/AIDS, will become the next global epidemic. In this section, we will look at the sources of new disease and the conditions they need to become established and spread.

Sources of New Diseases

New diseases come from three sources (**FIGURE 10.16**). First, species that normally live on your body can evolve a new strain that causes disease. Second, you can catch the diseases of other species. Third, species that normally live in soil or water can invade your body. There are over 1400 different species that cause disease in us, and there are many examples of diseases from each of these sources. However, new diseases come overwhelmingly from the second source: pathogens jump to us from other species, mainly other mammals and birds.

We pointed out earlier that your body is an ecosystem, home to thousands of species that are quite well adapted to humans. This is true of all living organisms, especially larger, multicellular ones. Every organism has a set of species living with it, some neutral, some helpful, and others that are parasites and cause disease. Biologically speaking, every large body presents a habitat and resources that other species can adapt to and exploit. Pathogens

A The avian flu virus (normally harmless to humans) can become highly virulent, if combined with a particular human virus.

B Ringworm (shown on this dog's face and chest) can be spread to humans from animals.

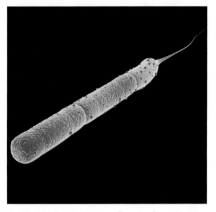

C Cholera bacteria can often be found in water in places with poor water treatment and sanitation.

FIGURE 10.16 New diseases can emerge from three sources: mutation of previously harmless microbes (A), transmission from other species (B), or transmission from the environment (C).

are a universal and natural result of evolution. And pathogens that jump from one species to another are not terribly different from a species that migrates to a new region—a fairly common occurrence. If a species, disease-causing or not, can adapt to the new conditions it encounters, it can thrive in its new home.

New diseases from other species often need some help to make the jump to humans. This help could come in the form of a mutation in the disease organism that allows it to infect humans. Or it could come from a change in human behavior and social practices that allows the disease to become established in human populations. Let's look at how the jump might occur.

Stages of a New Disease

Scientists who study evolutionary medicine have identified four stages that a new infectious disease goes through when it makes the jump to humans (**FIGURE 10.17**). These four stages form a sort of pyramid. At the base—stage 1—we are exposed to many diseases all the time. Only some of these are able to

Exposure:
Something occurs that brings the pathogen into contact with us.

Infection:
A pathogen invades us and reproduces inside.

Transmission:
A small number of microbes that infect us can be transmitted to another person.

Epidemic:
Fewer still microbes can establish an infection that spreads through a large population.

FIGURE 10.17 Scientists who study evolutionary medicine have identified four stages that a new infectious disease goes through when it makes the jump to humans: exposure, infection, transmission, and epidemic.

progress to stage 2 and infect us. Fewer still are able to spread from person to person in stage 3. And, luckily for us, only a small number reach stage 4 and have the potential to cause epidemics.

Exposure The first stage is exposure. Something must occur that brings the pathogen in contact with humans. Contact might come from working with domestic animals or hunting wild ones, being bitten by insects, drinking contaminated water, or coming into contact with contaminated soil. Our behavior can increase our chances of exposure. As we domesticated more and more animals, for example, we dramatically increased our exposure to their diseases. We are exposed to new pathogens all the time, but our immune system does an excellent job of protecting us. Only a few of these pathogens are able to progress to the next stage.

Infection The second stage is infection. Infection occurs when a pathogen is able to invade our body and reproduce inside us. This can be a difficult step. The pathogen is well adapted to the particular conditions inside its original host, and the conditions inside a human are different. The pathogen may not be able to get inside a human. Or, if it does, it may not be able to reproduce, or it may reproduce slowly, giving our natural defenses time to destroy it before it can do damage.

It is unlikely that a nonhuman pathogen will be able to infect a human successfully. However, every time a human is exposed, the pathogen gets another chance to infect. With many opportunities, the chances of successful infection get better. Also, genetic variation among pathogens means that some of them may be a little better than others at infecting humans. Likewise, humans vary genetically, and some humans may be slightly less resistant than average. Eventually, and probably regularly, the right combination occurs: infection by a strain of pathogen that is slightly better at infecting humans, in an individual human who is slightly more susceptible to it. When this occurs, the infection may take hold and the pathogen will be able to reproduce in a human.

It is also worth noting that many animal diseases infect more than one host species. These pathogens already have the genetic flexibility to infect multiple hosts, and for them the jump to humans may not be as difficult.

Transmission to Others Even if a new disease is able to infect a human, transmission to a new human host may prove difficult. First, the pathogen must be able to reproduce many new offspring, perhaps billions in the case of viruses or bacteria. Second, these pathogens must be able to make their way to the appropriate locations and tissues to get out of the body, maybe to the lungs to be coughed out or to the digestive system to be eliminated with feces. Finally, the pathogens must be able to evade our natural defenses long enough for these events to occur.

epidemic :: spread of a disease over a larger area and number of people than expected

Epidemics The final stage in the emergence of a new disease is the ability to spread widely and cause an **epidemic**, in which a greater than expected number of people over a larger area are infected. In a sense, epidemics are the mark of an evolutionarily successful pathogen. An epidemic indicates that the pathogen is well adapted to reproducing in its human host, and also well adapted to spreading among hosts. As the trade-off hypothesis predicts, the disease has "found" the proper balance between these two activities.

One way a disease can reach this stage is through natural selection. Within the host, natural selection selects for the most successful strains of the disease, those that are best at reproduction and spreading to another host.

This process takes time, and there may well be a number of false starts, in which the disease makes it through only one of the first three stages. But with enough tries, enough mutation, and enough natural selection, eventually a strain may succeed in reaching this stage.

A second way to reach this stage is through changes in human behavior and ecology. We've already suggested (though certainly not proven) that the deadly form of malaria may have become possible only after the development of agriculture. Packing more and more people together in cities, sexual practices, and many other changes in human behavior and society—all these can lead to situations that favor the spread of disease epidemics.

HIV/AIDS

Acquired immune deficiency syndrome (AIDS) is the disease caused by the human immunodeficiency virus (HIV) (**FIGURE 10.18**). AIDS was first reported in the United States in 1983, but it has been in the human population for about a century. HIV attacks and kills the blood cells that help to make **antibodies**, compounds that kill pathogens. While your body can make more of these cells, the virus eventually kills them off. Without these cells, you have an "immune deficiency," which makes you more vulnerable to disease. When a person dies from AIDS, it is usually due to other diseases that the patient can't fight, even though these diseases may not be life-threatening to a person with a healthy immune response.

There are two main forms of HIV. HIV-1 is the more virulent form and has the higher mortality rate. HIV-1 jumped to humans from chimpanzees. HIV-2 is less virulent. People infected with HIV-2 live longer and show

antibody :: a molecule that binds specifically to an antigen and kills pathogens

FIGURE 10.18 AIDS was first reported in the United States in 1983, but it has been in the human population for about a century. HIV-1 jumped to humans from chimpanzees (A), while HIV-2 originated in a monkey called the sooty mangabey (B).

FIGURE 10.19 HIV entered the human population when humans ate infected chimpanzees. European colonialism and the ivory trade in Africa played a role in the spread of HIV to a wider area and a greater number of people.

symptoms later, but the disease is still fatal. HIV-2 originated in a monkey called the sooty mangabey.

HIV is a relatively new disease that has progressed to stage 4. It has adapted to humans and has become a worldwide epidemic. So far, it has caused about 25 million deaths, and approximately 40 million people are thought to be infected.

We are not exactly sure how HIV made the jump to humans. We know HIV originated in Africa. Historically, people were probably first exposed when they hunted chimpanzees for food. Butchering and eating chimps in the wild exposed hunters to the blood of chimps infected with their version of the virus. Exposure and "false start" infections may have been going on for thousands of years, but HIV never progressed to stages 3 and 4.

Sometime in the late 1800s or early 1900s, changes occurred that allowed these infections to become established and gave the virus time to adapt to its human host. The list of possibly relevant changes is long. It might have been larger human settlements that allowed the virus to spread more easily. It might have been a result of large work camps set up by European colonial powers to build railroads into the interior of Africa (**FIGURE 10.19**). Over-crowding, physical stress, and malnourishment might have provided the virus with a large pool of vulnerable hosts. Medical practices of the time may also have played a role. In efforts to control tropical diseases, many people were vaccinated. But because syringes were expensive (and at that time not disposable), it was common practice to inject large numbers of people using the same needle, thus helping to spread the virus from person to person. Whatever the actual cause or causes, HIV adapted to humans and has become a global plague.

10.5 How Can Evolution Help Us Control Disease?

Antibiotic resistance shows that we are able to influence the evolution of our diseases. In this particular case, our influence has had a negative impact on our health. Is it possible that we could influence the evolution of diseases in a positive way, one that is beneficial to our health? A number of scientists believe that we can. They believe that integrating evolutionary and medical research can bring about benefits similar to those achieved in biology and chemistry.

This idea is new and somewhat controversial. The medical professions have not adopted all the ideas we'll discuss in this section, and a great deal of testing remains to determine whether these ideas are correct and feasible. But this new field of science shows great promise, and it is an excellent way to illustrate the practical advantages of evolutionary biology.

Antibiotic Resistance

One of the most obvious recommendations is to reduce our use of antibiotics, target their use more narrowly on the most severe diseases, and limit drastically the use of antibiotics for nonhealth-related activities like agriculture. This will not be an easy process. The pharmaceutical industry makes billions of dollars from antibiotics, and ironically (though certainly not by design), antibiotic resistance drives the unending search for newer antibiotics and continued profits. Doctors are under pressure from patients to prescribe antibiotics for relatively minor infections. And the use of antibiotics to protect crops from rotting and to produce larger livestock more quickly has become an economic necessity for many farmers.

A new antibiotic provides an important, but temporary, advantage in biological competition. Antibiotics will likely continue to be important in the way we treat disease. But evolution tells us that antibiotics alone cannot be a long-term strategy for controlling disease.

Vaccinations

The administration of material to provoke the development of immunity to a specific pathogen, or **vaccination**, has been successful in controlling certain diseases, including many childhood diseases like mumps, measles, whooping cough, polio, and most recently chicken pox. The only disease we have driven extinct—smallpox—was eliminated through a massive global vaccination program.

Still, many diseases are difficult or impossible to control by vaccine. Vaccines stimulate your immune system to recognize and attack invading pathogens (**FIGURE 10.20**). But some pathogens, like the influenza virus and HIV, mutate so rapidly that your immune system can't recognize the new strains. This is why you need a flu shot every year, and why scientists so far have been unable to develop a vaccine for HIV.

Evolutionary principles may help us design more effective vaccines (see *Technology Connection: How Vaccines Are Made*). One interesting idea is to use vaccines to target the disease-causing *behavior*, not the *organism* itself.

vaccination :: the administration of material to provoke the development of immunity to a specific pathogen

How Vaccines Are Made

You probably know that there are a number of diseases that you catch only once, like measles or chicken pox. After the disease clears, you are immune for a long time—perhaps your entire life. Vaccines "trick" your immune system into thinking you have a disease. Your body still develops immunity to the disease—but you don't have to go through the trouble of actually being sick.

Your immune system recognizes foreign cells and viruses that invade your body. It does this by recognizing foreign proteins on the cell or viral surface. This is a good system. Foreign protein indicates the presence of foreign DNA, and foreign DNA means that an external invader is present in your body. Once a foreign protein is detected, some of your immune cells make antibodies: proteins that stick, or bind, to the foreign protein and therefore to the foreign cell or virus. Bound antibodies form a "tag" that marks the invading pathogens. Other immune cells seek out tagged pathogens and ingest and digest them. The remarkable thing is that each antibody is specific to a particular protein (i.e., a particular pathogen), and that your immune system "remembers" this protein. Having made the antibody once, the immune system can quickly make more if you are ever infected with the same disease again. This is why immunity lasts so long.

Vaccines are composed of surface proteins from a specific pathogen. In some cases, the vaccine contains the protein alone. In other cases, the vaccine may contain fragments of the pathogen's cells or whole dead cells. When these are injected, your immune system responds to the protein and makes antibodies—and immunity. But the injection does not contain anything that can reproduce and make you sick.

One drawback to these types of vaccines is that the dead or fragmented microbes don't stimulate your immune system quite as well as the real, working pathogen. Of course, injecting you with a working pathogen is not an option. But we can use evolutionary principles to design safe "live" vaccines. The pathogen is grown for many generations in a nonhuman host: maybe another animal, maybe a chicken egg, or perhaps human cells grown in a lab. Through natural selection, the pathogen adapts to these conditions, but in doing so, it becomes less adapted to humans. These pathogens form the vaccine. When injected, they can reproduce slowly in us, and they have the same proteins as the normal disease. They do an excellent job of stimulating the immune system to produce antibodies, but they are not strong enough to make us sick.

Live vaccines can be cheaper to make. They provide better immunity, so booster shots may not be needed or are needed less frequently. And sometimes these vaccines can be given orally or through a nasal spray rather than by an injection. Live vaccines are available for mumps, chicken pox, measles, polio, and other diseases. The nasal form of flu vaccine is a live vaccine. ::

| Vaccines are created from pathogen proteins, fragments of the pathogen's cells, or whole dead cells. | Pathogen is cultured | Proteins or Bacteria or Viruses | The proteins or cells must then be purified and inactivated or weakened. | |

Vaccines are composed of surface proteins from a specific pathogen. The immune system responds to the protein in the vaccine by making antibodies against the protein that results in immunity, but the injection does not contain anything that can reproduce and make you sick.

According to some scientists, the diphtheria vaccine is a good example of this approach. The bacteria that cause diphtheria produce a toxin that kills cells in our upper respiratory tract. These dead cells may release nutrients that the bacteria can use—an advantage to making the toxin. However, the toxin is a protein that takes energy to make—perhaps as much as 5% of the cell's total energy budget. The diphtheria vaccine stimulates the immune system to make antibodies to the *toxin*, not to the bacterial cell. The cells continue to live, but they are harmless.

1 A chemically modified, nontoxic form of the diphtheria toxin, toxoid, is injected into the body.

2 The body makes antibodies that disable the toxoid and remain in the body to provide immunity.

3 If the body encounters the diphtheria toxin, the existing antibodies will fight it off.

Diphtheria bacteria

Toxin

Chemical modifications

Toxoid

Antibodies

FIGURE 10.20 Vaccination stimulates the immune system to make antibodies that disable a pathogen and remain in the body to fight the pathogen, should it appear again. Vaccination for diphtheria targets the diphtheria toxin.

But not all diphtheria cells make toxin. Those that don't have a competitive advantage: they don't spend 5% of their energy making the toxin. In vaccinated people, the cells that make the toxin lose their advantage since the toxin is destroyed. Cells that don't make the toxin save energy, grow faster and reproduce more quickly, and eventually outcompete the cells that do. Instead of trying to kill all cells, only to have them evolve resistance, this approach uses natural selection to select for cells that cause no damage. Just a few scientific studies have shown this effect, but if it turns out that the process works, it opens up new possibilities for vaccine design.

Controlling the Spread of Disease to Select for Milder Forms

The final form of disease control is a direct application of the trade-off hypothesis. Diseases that spread rapidly and easily, and don't require the infected person to be up and about, evolve to be severe and lethal. Diseases that require personal contact of some form to spread are often less life-threatening. If the trade-off hypothesis is valid, then reducing a disease's transmission rate has two important consequences. In the short term, it protects people from becoming infected. But in the long term, reducing the spread selects for milder strains of the disease.

Let's consider how this might play out with AIDS. In a population of intravenous drug users who regularly share needles or a population in which sexual encounters are frequent and partners change frequently, HIV spreads quickly. There is less of an advantage to HIV to keep the host alive for long periods of time. And there is more of an advantage for fast-reproducing strains of HIV that exploit the host severely, increasing their population sizes and therefore their chance of being transmitted to the next host. Under these conditions, the risk is not simply that you'll catch HIV, but that you will catch a particularly virulent strain.

The logic works the other way as well. Behavioral and social practices that make it more difficult for HIV to spread also make it more difficult for virulent strains of HIV to persist. As a personal health choice, it clearly makes sense to practice safe sex and to limit the number of partners you have. An added benefit is that if these behaviors become widespread, we are also changing the evolutionary conditions under which

HIV evolves. We select for milder forms of the disease. There is no way for HIV to develop "resistance" to this approach. And it works for all STDs, not just HIV.

Some scientists make an analogy between evolutionary control of disease and the domestication of animals and plants. Domestication involves taking wild, and possibly dangerous or poisonous, species and selecting them so they evolve into forms that are useful to us. Since we seem to have difficulty driving diseases extinct (out of 1400 infectious diseases, only smallpox has been exterminated), perhaps the idea of "taming" diseases by selecting for milder forms makes more sense.

Biology in Perspective

We are never alone; our bodies are complex ecosystems. Living on or in us are bacteria, fungi, mites, and worms. Most of the time, our coexistence is peaceful—a balance is maintained so that neither we nor our co-inhabitants are harmed. But sometimes a pathogenic organism can overtake the benign microbes or our bodies can malfunction, such that a previously harmless microbe becomes a pathogen. Such events are not rare.

In fact, one-quarter of all human deaths are caused by infectious disease. This figure was much higher in the past. Through better sanitation and medical care, we have made great strides in reducing deaths due to infection. But our relationship with pathogens is a dynamic one because pathogens evolve through natural selection to overcome our attempts at disease control.

Many pathogens have evolved resistance to antibiotics, and some are starting to get around our vaccines as well. New diseases emerge continuously for which we have no natural immunity, antibiotics, or vaccines. As earlier treatments and approaches to disease control become less effective, a better understanding of the evolutionary dynamics of disease may provide new approaches to public health. The field of evolutionary medicine, while still in its early stages, may provide the next big advance in health care.

Chapter Summary

10.1 In What Ways Is Your Body an Ecosystem?
Recognize and provide examples of the different types of species interactions.
- Species that share the same space can compete for resources or interact to the benefit of one or both species through either commensalism or mutualism.
- Competition from nonpathogens can prevent pathogens from infecting us by taking up space and resources so that the pathogens do not have access.

10.2 Why Do Diseases Evolve Resistance to Antibiotics?
Describe the process by which some disease organisms increase their resistance to antibiotics.
- Resistance to antibiotics evolves because of variation.
- Resistant bacteria have mutations that produce proteins that take on different functions or perform old functions in new ways.

- In addition to mutations, some bacteria can acquire resistance genes from their neighbors through horizontal gene transfer.
- Resistance comes at a cost because resistant strains are less efficient.
- Overuse of antibiotics may be the cause of increased resistance to antibiotics.

10.3 Why Are Some Diseases More Deadly Than Others?

Use the trade-off hypothesis to explain and predict the spread and impact of a disease.

- The trade-off hypothesis states that there is a balance between virulence of an organism and its transmission.
- Severe diseases will be those that spread to people most easily.
- Pathogens with lower transmission rates usually have lower virulence.
- Sexually transmitted diseases will not spread if they interfere with the host's reproduction.

10.4 Where Do New Diseases Come From?

Consider the processes of mutation and natural selection and predict likely sources of new disease.

- New diseases come from three sources:

  o Species that normally live on the body can evolve to become virulent.
  o Disease can be caught from another species.
  o Species that normally live in the soil or water can invade the body.
- Most diseases come from other species.
- The stages of a new disease are exposure, infection, transmission, and epidemic.
- HIV/AIDS, an example of a new disease, may have initially occurred from exposure to the blood of chimps in the wild. HIV originated in chimpanzees (HIV-1) and a monkey called the sooty mangabey (HIV-2). Crowding, malnourishment, and physical stress may have enabled initial HIV/AIDS infection with additional transference through reuse of contaminated needles.

10.5 How Can Evolution Help Us Control Disease?

Propose a strategy to reduce the virulence of a disease using evolutionary medicine.

- We should strive to reduce our use of antibiotics when we can.
- We should use vaccines to attack the disease-causing behavior rather than the disease to eradicate quickly mutating pathogens.
- We should change human behavior to reduce the rate of transmission.

Key Terms

antibiotic 290
antibody 303
commensalism 288
conjugation 293
ecology 286
ectopic pregnancy 300
epidemic 302

horizontal gene transfer 293
mutualistic 288
necrosis 289
pathogen 284
quorum 289
resistance 284

resistance gene 292
transduction 293
transformation 293
transmission 297
vaccination 305
virulence 295

Review Questions

1. What is the Human Microbiome Project?

   a. An entity started by the National Institutes of Health to figure out the sources of disease

   b. A study that tracked the process by which a disease becomes an epidemic

   c. An entity started by the National Institutes of Health to figure out all of the species that live in and on us

   d. A study that confirmed there are over 700 species that live in the human mouth

   e. A study that confirmed there are over 800 species that live in the colon

2. What is commensalism?

   a. An ecological interaction in which one species benefits and the other is neither helped nor harmed

   b. An ecological interaction in which neither species benefits

   c. An ecological interaction in which both species benefit

   d. An ecological interaction in which one species benefits and the other is harmed

   e. An ecological interaction in which a parasite harms a host

3. If most of the species that live on the human body can't be grown in a laboratory, how do we know how many species live there?

4. Briefly describe why natural selection results in bacteria that are resistant to antibiotics.

5. Why does the number of resistant strains of bacteria often decrease when antibiotics are not present?

6. Briefly explain the reasoning that suggests diseases should evolve to become milder and less of a threat to their hosts.

7. Briefly outline the trade-off hypothesis. According to this hypothesis, what are the conditions that select for severe or life-threatening diseases? What conditions select for milder diseases?

8. How many new diseases have been reported since 1975?

9. What are the four stages a newly emerging disease goes through?
   a. Exposure, passage, transduction, epidemic
   b. Exposure, commensalism, transmission, epidemic
   c. Exposure, infection, commensalism, epidemic
   d. Exposure, infection, transmission, commensalism
   e. Exposure, infection, transmission, epidemic

10. How does a new disease evolve as it progresses through these four stages?

11. What are the three sources of new disease?
    a. Mutated forms of existing disease, infection, other species
    b. Mutated forms of existing disease, other species, the environment
    c. Infection, other species, the environment
    d. Mutated forms of existing disease, transmission, the environment
    e. Infection, transmission, the environment

12. Which of these sources of new disease supplies the most new diseases?

13. How does a vaccine protect you from a disease?

14. The vaccine for diphtheria has a different "target" than most vaccines. What is the target of the diphtheria vaccine? Why does this difference help to prevent the diphtheria bacteria from evolving resistance?

15. What are the two beneficial, health-related consequences of reducing the rate at which a disease spreads?

The Thinking Citizen

1. In this chapter, you learned that viruses, bacteria, and insects are capable of evolving resistance rapidly to antibiotics or insecticides. Do humans evolve resistance to disease? Discuss some of the evidence that suggests we do. If we do evolve resistance, we evolve it much more slowly than do insects and microorganisms. Why might this be?

2. Go online and look up some facts about several diseases. In particular, find out how they are spread and how life-threatening they are. What sort of support did you find for the trade-off hypothesis described in this chapter? Do the more severe diseases tend to spread by insects or contaminated water, for example?

3. We humans are very good at driving species extinct, even when that isn't what we are trying to do. However, we seem to be very poor at driving diseases extinct, even though we devote considerable effort to it. Of 1400 known infectious diseases, we have exterminated only one: smallpox. List and briefly discuss several reasons why it is difficult to drive a disease extinct.

4. There are a number of diseases, such as rabies and Lyme disease, that infect humans and cause illness but which apparently cannot be spread from one human to another. Is it possible that humans are the main host for such diseases, or are we infected more "by accident"? In terms of emerging diseases, what stage are such diseases at? What are the chances that diseases like these will reach a higher stage?

5. The European conquest of North and South America was aided considerably by disease. Europeans brought with them diseases—such as smallpox, typhus, influenza, diphtheria, and measles—that killed many of the native people of the New World. Why were these diseases common in Europe but absent in the New World before the Europeans came? HINT: consider the differences in social structure of people in both regions.

The Informed Citizen

1. Suppose you are in charge of a poor and underdeveloped state in an African country. A severe outbreak of malaria occurs, threatening many people, especially children. Would you spray DDT to reduce the mosquito population? If so, would you spray as an emergency procedure or as part of a continuing program of mosquito control? What sorts of medical, economic, and ethical considerations would you take into account in making your decision?

2. What should we do about antibiotic use that is not health-related? A lot of antibiotics are used to keep crops from spoiling and to increase the growth of livestock. These practices have important economic benefits to farmers and to consumers who buy their products. But they increase the levels of antibiotics in our environment and lead to resistance. What do you think should be done about using antibiotics this way?

3. Methicillin-resistant *Staphylococcus aureus* (MRSA) is a strain of staph bacteria that is resistant to a number of antibiotics. MRSA is found primarily in hospitals and is becoming a particularly difficult infection to treat. Norway has one of the lowest rates of all infections of any country, and in particular has the fewest people infected with MRSA. Doctors in Norway prescribe fewer antibiotics than in any other developed country. How are these two facts related? Do some research on antibiotic use in Norway and find out the positive and negative effects of reduced usage. Do you think a similar policy of antibiotic usage would work in the United States? Why or why not?

CHAPTER LEARNING OBJECTIVES

After reading this chapter, you should be able to answer the following questions:

11.1 How Do Species Adapt to Their Habitat?
Provide examples of the ways species adapt to the constraints of their habitats.

11.2 Why Do Species Compete?
Summarize how species can coexist without one outcompeting the other.

11.3 How Do Species Exploit One Another?
Describe the ways that species may exploit one another.

11.4 When Can Species Cooperate?
Identify circumstances of cooperation between species.

11.5 How Do Ecological Interactions Affect Us?
Summarize the beneficial and detrimental ecological interactions that we have with other species in the ecosystem.

11.6 What Does a Functioning Ecosystem Do?
Describe the movement of energy and nutrients through an ecosystem.

11.7 What Benefits Do We Get from a Functioning Ecosystem?
Enumerate the benefits we get from a functioning ecosystem.

Ecology is the study of interactions. As species compete, eat one another, and occasionally cooperate, they set up conditions that affect their survival and shape their evolution.

CHAPTER 11

Ecology

How Do We Benefit from a Functional Ecosystem?

Becoming an Informed Citizen . . .

All species, including our own, depend on the benefits ecosystems provide. These benefits may seem free of charge, but if we disrupt them, we must pay for them one way or another. Maintaining ecosystem services is an economic imperative as well as an ethical and cultural need.

s the English poet John Donne wrote, "No man is an island, entire of itself; every man is a piece of the continent, a part of the main" (Meditation XVII, 1624). Donne was reflecting on the social and spiritual interactions that connect people to one another. In reality, all organisms depend on others for survival. Biological interactions connect species to one another and to their habitat. To paraphrase Donne, no species is an island.

ecology :: the study of the interactions among species and their physical environments

Ecology is the study of interactions. As individuals of different species compete, eat one another, and occasionally cooperate, they set up conditions that affect their survival and shape their evolution. Populations also adapt to the challenges of their habitat, and this adaptation is another type of interaction. Interactions also allow an ecosystem to perform "work." Energy and nutrients are distributed throughout the ecosystem as organisms consume one another. Oxygen, carbon dioxide, and nitrogen recycle through the ecosystem, atmosphere, and soil as a result of living metabolism. All species, including our own, depend on the work ecosystems do for their survival.

In this chapter, we will examine how individuals of different species interact, and how their interactions drive an ecosystem's function. We will see how humans interact with other species, and how we have taken advantage of the adaptations produced when other species compete with, exploit, or cooperate with one another. Ultimately a species' ecology, its interactions with its habitat and with other species, determines whether it will survive or become extinct. Consider the fate of Kirtland's warbler.

case study

The Near-Extinction of Kirtland's Warbler

Warblers are small, insect-eating birds known for their colorful plumage and cheery songs. Most North American warblers are successful species and can be found nesting throughout wide regions of the continent. The Kirtland's warbler (scientific name *Dendroica kirtlandii*) (**FIGURE 11.1**) is an exception. It

Presence of nesting Kirtland's warbler

FIGURE 11.1 Kirtland's warbler is extremely rare and nests in only a few sites in northern Michigan.

is extremely rare and nests in only a few sites in northern Michigan. A thorough survey in 1973 detected only 200 singing males (400 birds if the males have mates) in the entire world. What is it about the ecology of these warblers that resulted in their precarious situation? What hypotheses could you develop to account for this near-extinction?

- The warbler's food source is rare.
- Some predator is killing off all the warblers.
- Some disease is killing off all the warblers.
- Human development is crowding the warblers out.
- Suitable habitat and nesting sites for the warblers are rare.

To distinguish among these five hypotheses, you would need to know how the warbler interacts with its habitat, and how it interacts with other species in the area. To make things easier, we'll describe some relevant discoveries about the Kirtland's warbler ecology.

Kirtland's warblers build their nests on the ground. A ground nest is vulnerable to predators and also to environmental problems like flooding during rainstorms. Kirtland's warblers address the predator problem by nesting under the lower branches of a young jack pine tree. These lower branches cover the nest and help to hide it from predators. Also, jack pines grow predominantly in dry sandy soil that drains well, reducing the risk of flooding.

In a young stand of jack pine, the trees are short and relatively spaced out. This allows lots of light into the stand, promoting the growth of grasses and shrubs between the trees. This grassy, shrubby area is home to many kinds of insects, which provide food for the warblers and their young. Young stands of jack pine are the ideal habitat for the Kirtland's warbler, providing nesting sites, shelter and protection, and food.

Jack pines are not rare. They are found in a broad swath across the northern United States and Canada. However, *young* stands of jack pine are rare. As a jack pine stand ages, the trees grow taller and block light from reaching the forest floor. The grassy, shrubby area between trees dies out due to lack of light, and with it goes the warbler's source of food. Tall jack pines also drop their lower branches, which leaves the warblers' nests exposed and vulnerable to predators.

Another ecological threat for the Kirtland's warbler comes from the brown-headed cowbird (scientific name *Molothrus ater*) (**FIGURE 11.2**). Cowbirds are nest parasites: they lay their eggs in the nests of other species—in this case, those of the Kirtland's warbler. The hapless warblers incubate the cowbird eggs along with their own. When the eggs hatch, they feed the young cowbirds along with their own young. Cowbird babies are demanding and aggressive and often succeed in getting so much of the warbler's attention and food that the warbler young starve.

We now have two clues about why Kirtland's warblers are rare: a limited number of young jack pine stands and nest parasitism from cowbirds. Let's now consider how young jack pine stands form. On dry, sandy soils, forest fires are a relatively common event. Jack pines are adapted to withstand fire; indeed, their reproduction depends on it. The cones on a jack pine stay closed, sealed with resin. Only when a fire comes through do the cones fall off the tree, and the fire melts the resin to release the seeds. The fire kills many adult trees, allowing light to reach the ground. Fire also burns away leaf litter and debris, producing a clean bed of soil that promotes seed germination. But fire is decidedly not good for human habitats near jack pines. Our practice of suppressing fire has reduced the number of young jack pine stands, and the warblers have fewer places to live as a result.

FIGURE 11.2 One ecological threat for the Kirtland's warbler comes from the brown-headed cowbird, which lays its eggs in the nests of the Kirtland's warbler. The warblers incubate and then feed the young cowbirds along with their own young; the cowbirds often get so much of the warbler's attention and food that the young warblers starve.

The dry sandy soil where jack pine forests grow also makes an ideal place for human homes (no flooded basements!). In addition, the soil is useful for agriculture: easy to work, and fertile with some addition of organic material. Jack pine forests are cleared to make way for houses and farms. While this change in habitat is detrimental to the Kirtland's warbler, it opens up opportunities for other species. Cowbirds, for example, get their name because they follow herds of bison and other large herbivores. Cowbirds eat soil insects stirred up by these herbivores' hooves. When the herds migrate, so do the cowbirds. This makes nesting difficult, and cowbirds have adapted by laying their eggs in the nests of other species rather than building their own nests. When agriculture opened up in the jack pine forests, perhaps 200 years ago, cowbirds invaded the traditional nesting habitat of Kirtland's warbler. Cowbird nest parasitism has reduced the size of the warbler population.

These ecological facts provide support for two of our hypotheses: *human development* and *limited nesting habitat*. Predators and disease do not seem to be significant, but another hypothesis that we did not consider earlier is *nest parasitism*. A number of other ecological factors play a role as well, including fire, soil conditions, and light penetration through the forest canopy. The Kirtland's warbler is part of a web of ecological interactions that constrain its distribution and abundance, nearly driving it extinct.

To help save Kirtland's warblers from extinction, Michigan has initiated a program of controlled burns and planting jack pines to increase suitable habitats. Cowbirds are being trapped to reduce nest parasitism. Educational programs inform people of the plight of Kirtland's warblers. These efforts appear to be successful. In 2010, there were 1747 breeding pairs, an almost nine-fold increase since 1973.

FIGURE 11.3 Kangaroo rats live in the deserts of North America. To minimize water loss, they forage for food at night and spend their days inside burrows that are humid (due to water evaporation from the kangaroo rats' lungs).

11.1 How Do Species Adapt to Their Habitat?

The *eco* in *ecology* comes from the Greek *oikos*, meaning "home." Ecology is the study of one's home—or, for nonhumans, the habitat. Many physical factors interact to determine whether a habitat is suitable for a particular species: the temperature range, water availability, and nutrients in the soil, for example. The habitat also provides shelter for survival and reproduction.

Species adapt in both their physiology and behavior to meet the challenges of their habitat. In this section, we will examine how species interact with and adapt to their habitat.

Adapting to Physical Conditions

Kangaroo rats (**FIGURE 11.3**) are a group of about 20 species living in the deserts of North America. They have powerful hind legs and hop about on them like tiny kangaroos. Like all animals, kangaroo rats must strike a physiological balance between water intake and water output. Living in a desert habitat makes achieving this balance very difficult.

Kangaroo rats lose water through three channels: urine, feces, and evaporation from respiratory surfaces of the lungs and nose. Compared to other mammals, however, they lose remarkably little water, particularly from their urine and feces. Kangaroo rats have extremely efficient kidneys. Kidneys filter toxic substances from the blood and flush them out in urine. Kangaroo rat kidneys are able to make very concentrated solutions of these toxins and therefore require little water to get rid of them. Their kidneys use only one-quarter to one-sixth as much water as a human would to eliminate a given amount of waste. Their feces are dry as well. That's because the kangaroo rat colon is able to recover almost all the water as it leaves the digestive system.

The inner surface of the lungs must be kept moist in order for oxygen to be absorbed. Nasal passages must be moist to dissolve odors an organism smells. These moist surfaces are in contact with outside air, and water constantly evaporates from them. Kangaroo rats can't eliminate this water loss, but they have adaptations to minimize it.

One of these adaptations is that kangaroo rats are nocturnal: they spend the day inside their burrow. As water evaporates from their lungs, humidity in the burrow increases. This helps to reduce evaporative loss. At night, the rats forage for food, mainly dry seeds and grasses, which they take back to their burrow to eat and store. In the humid burrow, their food absorbs moisture. When the rats eat their food, they actually recover some of the moisture they lost previously from evaporation.

Water intake for kangaroo rats is a puzzle. They have no access to free-standing water, and in fact, in their natural habitat they do not drink. Instead, most of their water comes from metabolizing food. Animals digest food by oxidizing (essentially burning) it. As part of this process, the oxygen an animal breathes is combined with hydrogen atoms in the food to produce water. Depending on the humidity, a kangaroo rat can get between 80% and 100% of its water through this metabolic process. Under conditions typically found in a desert burrow, a kangaroo rat takes in about 40% more water than it loses, without ever taking a drink or eating a succulent plant. This excess compensates for the water loss the rat experiences outside the burrow and enables it to live in the driest of deserts.

Adapting to Limited Shelter

Many species have specific needs for shelter and reproduction, such as suitable sites for burrows or nests. As we saw for the Kirtland's warbler, such sites may be rare finds throughout the habitat. There are several ways species respond to limitations on shelter and reproductive sites.

The yellow-bellied marmot (scientific name *Marmota flaviventris*) (**FIGURE 11.4**) lives high in the Rocky Mountains of Colorado. Marmots get shelter and reproduce in the larger cracks and crevices in the rocks, which they use as dens. Often there are not enough dens for each breeding female to have one of her own. A male marmot will claim and defend a suitable den, and two to five females will breed with this male and share the den. In this case, limitations on suitable sites for reproduction have resulted in a **polygynous** mating system, in which one male mates with several females. Since the sex ratio in marmots is 50:50, if one male monopolizes several females, some males will not have mates in any given year.

White-fronted bee-eaters (scientific name *Merops bullockoides*) (**FIGURE 11.5**) are birds that live in Central Africa and dig cavities in sandy riverbanks for

FIGURE 11.4 Yellow-bellied marmots live high in the Rocky Mountains of Colorado. They use cracks and crevices in the rocks as dens, but the limited number of suitable sites has resulted in a polygynous mating system, in which one male mates with several females.

FIGURE 11.5 White-fronted bee-eaters are birds that live in Central Africa and dig cavities in sandy riverbanks for their nests. Since there is often not enough room for all birds to nest, young bee-eaters frequently stay with their parents and help to raise another brood.

polygynous :: a mating system in which one male mates with several females their nests. Since these sandy areas are limited in size, there is often not enough room for all birds to nest. Young bee-eaters frequently stay with their parents and help them raise another brood of birds by bringing food and helping with nest defense. These "helpers at the nest" provide a large benefit to their parents. Without them, almost half of a nest's hatchlings would die of starvation before they could fly. With them, survival rates can be three or four times higher.

11.2 Why Do Species Compete?

competition :: occurs when two or more species attempt to utilize the same resource; it has a negative effect on both species

exploitation :: an interaction between species in which one species benefits while the other is harmed

mutualism :: an interaction between species in which both species benefit

When individuals of different species come in contact, they interact in a variety of ways. **Competition** occurs when two or more species attempt to use the same resource. It has a negative effect on both species. In **exploitation**, one species benefits while the other is harmed. In **mutualism,** both species benefit. This is analogous to cooperation in humans: everybody wins.

TABLE 11.1 sums up the effects on each of the interacting species. In an ecological sense, benefit and harm are measured by their effects on abundance (population size) and distribution (where it can live). A species is harmed when the interaction reduces its population size or limits its geographic distribution. A species benefits when the interaction increases abundance or distribution.

We often speak of *competition, exploitation,* and *mutualism* quite apart from ecology. We think of competition as a sporting event, exploitation as a crime, and mutualism as active cooperation. The species involved in these interactions don't "think" of them in this way, however. Their interactions are simply the result of how each species makes its living.

The Competitive Exclusion Principle

When species compete, the resource must be in limited supply. All animals use oxygen, for example, but since usually there is enough to go around, this is not really competition. As Table 11.1 shows, competition is detrimental to both species since potentially it reduces the amount of the resource available to either species. With less of an important resource, it is more difficult for organisms to grow, survive, and reproduce, which can result in lower population size.

Since neither species benefits from competition, how can it be avoided? There are two ways this can happen: one competitor can win by driving the other species extinct, at least in the local area; or one or both species can change so that they no longer use the same resource. Competition tends to be an unstable interaction. In the long run, species tend to adapt in order to reduce or eliminate competition.

Table 11.1 :: The Effects of Species Interactions

	EFFECT ON:		
	Species 1	Species 2	**Overall Effect**
Competition	Negative	Negative	Competition has a negative effect on both species.
Exploitation	Negative	Positive	Exploitation benefits one species, while the other is harmed.
Mutualism	Positive	Positive	Mutualism benefits both species.

One of the first scientific studies of competition was performed by F. G. Gause in 1934. Gause worked in his laboratory with two microorganisms: *Paramecium aurelia* and *Paramecium caudatum* (**FIGURE 11.6**). Gause found that both species grew well when they were alone in their test tubes. However, whenever Gause grew them together, he found that *Paramecium caudatum* always died out. Gause reasoned that when grown together, the two species competed for food, and that *Paramecium aurelia* was the better competitor. Gause generalized his results, concluding that when species compete, the best competitor will always win and exclude the other species from the habitat. This conclusion is now known as the **competitive exclusion principle**, which says that two competitors cannot coexist.

The competitive exclusion principle holds up best under laboratory conditions. But even for microorganisms, a test tube is a relatively small world: there's only one food source, conditions in all parts of the tube are the same, and there really is no place to go. It's similar to throwing two fighters into the ring: they have no choice but to compete, and only one can win.

Conditions in nature are much more variable and dynamic, both in time and space. There are more species present, more food sources, and generally more ecological "freedom" for the species. If two species compete for the same food, it may be that one is the better competitor in warmer areas while the other is better in cold areas, for example. If both species were placed together in a constantly warm habitat, the competitive exclusion principle might hold, and one species would be driven extinct. In nature, however, the existence of variable temperatures may allow both species to coexist.

In nature, a species uses a whole suite of resources that promote growth, survival, and reproduction. The particular set of resources a species uses is called its **niche**. A niche includes biological factors like food sources and

competitive exclusion principle :: posits that two competitors cannot coexist

niche :: the particular set of resources a species uses to survive; includes biological factors like food sources and nesting sites and physical factors like temperature

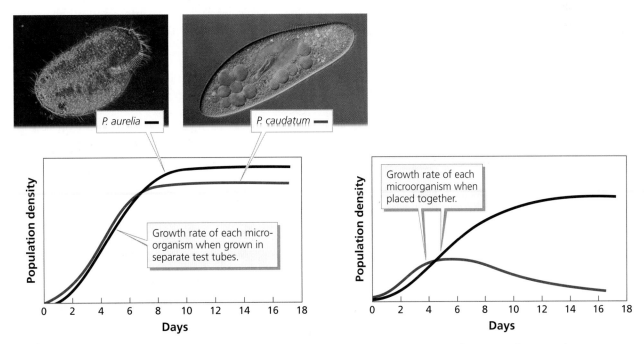

FIGURE 11.6 One of the first scientific studies of competition was F. G. Gause's 1934 research on two microorganisms, *Paramecium aurelia* and *Paramecium caudatum*. Gause discovered that both species grew well separately in test tubes, but when they were placed together, *P. caudatum* always died out.

nesting sites and physical factors like a minimum and maximum temperature. In short, a niche includes everything a species uses to survive. We can generalize the competitive exclusion principle to state that *two species cannot occupy the same niche*. However, they can vary aspects of their niche to reduce competition for some resources, even if they compete for others. In such cases, competitors may coexist. The following example shows how this process has worked on the Galapagos Islands.

Instances When Competitors May Coexist

The Galapagos Islands lie 600 miles off the west coast of Ecuador. They are home to 14 species of finch, made famous by Darwin and studied by many scientists up to the present day (see Chapter 9). Finches are primarily seed eaters, and the size of their beaks determines the size of seeds they can eat most easily. Finches with similar-sized beaks will compete for food.

Two similar species of Galapagos finch are the medium ground finch (*Geospiza fortis*) and the small ground finch (*Geospiza fuliginosa*) (**FIGURE 11.7**). The medium ground finch lives on Daphne Island, and the small ground finch lives on Los Hermanos. Scientists have measured the beak size of species on both islands, and they are about the same: 8–11 millimeters for the small ground finch, 8–12 millimeters for the medium. If these birds lived together, they would compete for the same seeds, but since they live on different islands, no competition occurs.

However, these two species do live together on Santa Cruz Island. Interestingly, on Santa Cruz both species have nonoverlapping beak sizes, which allow them to eat different seeds without competition. When species evolve a change in physical or behavioral characters that allows them to reduce competition, the process is known as **character displacement**. Generally, species tend to divide up resources in their habitat, reducing competition and allowing more species to coexist. See *Scientist Spotlight: Robert Helmer MacArthur*, for a closer look at a scientist who was interested in the problem of how species could coexist in a habitat without competitively excluding each other.

character displacement :: when species evolve a change in physical or behavioral characters that allows them to reduce competition

FIGURE 11.7 The medium ground finch and the small ground finch generally have similar beak sizes and live on different islands (Daphne Island and Los Hermanos, respectively), so they do not compete for food. However, on Santa Cruz Island, the finches live together and have developed different beak sizes so they can eat different seeds.

Robert Helmer MacArthur (1930–1972)

Most scientists find it a challenge to master their own field; it's the rare scientist that can master two. But sometimes it is precisely the blending of two separate disciplines that leads to new insight and new approaches that revolutionize a field. Robert Helmer MacArthur was one of these rare scientists, and his contributions transformed the field of ecology.

MacArthur was born in Toronto in 1930. His father was a professor of genetics who split his time between the University of Toronto and Marlboro College in Vermont. Like many ecologists, MacArthur's early passion was natural history. He particularly loved watching birds. In school, however, MacArthur studied mathematics. At 17, he moved to the United States, earning his bachelor's degree from Marlboro College and his master's in mathematics from Brown University. While pursuing his doctorate at Yale, MacArthur switched his field of study from mathematics to his earlier love, zoology. He earned his doctorate in 1957 and spent an additional year studying ornithology at Oxford University.

MacArthur was interested in the problem of how many species could "fit" into a habitat without competitively excluding one another. One of his earliest studies, now a classic, showed that several species of warblers could inhabit the same spruce tree by dividing up their foraging areas. One species foraged in the treetop, others in the inner branches, some in the outer branches, and others at the bottom.

MacArthur's main contribution, however, was to bring mathematical and experimental rigor to ecology. He developed mathematical models that showed the relationship between species number and habitat size that could be tested experimentally by other ecologists. Working with like-minded colleagues, MacArthur conducted research that provided a theoretical foundation for ecology. He published many papers involving both empirical and theoretical investigations of nature. In 1967, he and E. O. Wilson published *The Theory of Island Biogeography*, which addressed the question of why the number of species varied on islands of different size.

Robert Helmer MacArthur blended his expertise in mathematics and zoology to help revolutionize the field of ecology.

In 1971, MacArthur learned that he had cancer of the kidney and only a year or two left to live. He began working on his last book, *Geographical Ecology*, which summarized much of his life's work and identified open questions that ecologists are still studying. MacArthur died in November 1972. While his professional career lasted only 14 years, he helped to transform the science of ecology. In a memorial written by MacArthur's thesis advisor, G. E. Hutchinson, and colleague E. O. Wilson, they characterize a career cut short: *"Geographical Ecology* is both the reflective memoir of a senior scientist and the prospectus of a young man whose creative effort ended at the point of its steepest trajectory."

11.3

How Do Species Exploit One Another?

Recall that exploitation refers to an interaction in which one species benefits at the expense of another. Predation (in which a predator captures prey to eat), browsing and grazing (when organisms consume plants without killing them), parasitism (in which one organism lives off of another), and disease are all forms of exploitation. In competition, the best interests of both species are the same: each species benefits by avoiding the interaction altogether. In exploitation, the best interests of both species differ. The exploited species benefits from avoiding the interaction, while the exploiter depends on maintaining it.

Exploitation has two effects commonly observed in nature. First, the "population cycles" of both species can be linked: each population increases and decreases regularly over time. Second, an evolutionary "arms race" can develop: the exploited species develops better defenses, and the exploiter then adapts, too, with a better offense.

Population Cycles

Predator–prey systems provide good examples of population cycling, and it's easy to see why. Suppose both predator and prey start off with relatively small populations. In the absence of predators, the prey population can increase rapidly. But this provides more food for the predators, and their population size increases. As the number of predators increases, they begin to overexploit their prey, causing the prey population to decline. As prey decline, the predators begin to starve, and their population declines. This frees the prey to increase again (assuming not all are eaten), and the cycle continues.

One classic example involves the Canadian lynx (*Lynx canadensis*) and the snowshoe hare (*Lepus americanus*) (**FIGURE 11.8**). Both species live in northern North America, and both were extensively hunted for their pelts by fur traders of the Hudson Bay Company. For over a century, the Hudson Bay Company kept good records of how many pelts of each species it traded. Let's assume that the number of pelts taken in any given year is proportional to the size of the lynx and hare populations. We can then estimate the population size of both species over a 100-year period starting in 1845. The population cycles of both species are quite predictable: the lynx population rises just after that of the hare, and it declines just after the hare population begins to crash. These cycles have occurred every 10 years or so for over a century.

Many scientists have extended this example with more recent studies of their own, and as we learn more, the story becomes more complex. For one thing, the hare population cycle is driven not just by the lynx, but by food.

FIGURE 11.8 One classic example of population cycling involves the Canadian lynx and the snowshoe hare. Based on pelt trading records from the Hudson Bay Company, we know that the lynx population rises just after that of the hare and declines just after the hare population begins to crash; these cycles have occurred every 10 years or so for over a century.

As the hare population increases, hares can drastically overgraze the shrubs and twigs that provide winter food. That leads to another set of population cycles, with the *hare* as the exploiter. Also, other species besides the lynx eat snowshoe hares, and lynx have other prey as well. All these factors affect the lynx and hare populations.

An Evolutionary Arms Race

Exploitation can also create an evolutionary "arms race." Over time, the exploited species develops adaptations to prevent or avoid exploitation. The exploiter in turn develops adaptations that allow the exploitation to continue. This adaptation of one species to another is called **coevolution**.

Plants, for example, stay put, and they are loaded with vitamins and carbohydrates. That makes them an attractive target for exploitation by animals. Over long periods of time, however, plants coevolve in response. Many plants employ a chemical defense by making **secondary compounds**, chemicals that are usually poisonous and not needed by the plant for its metabolism. If an animal ingests these compounds, it could die, become sick, or become drugged and therefore vulnerable to its own predators. But some herbivores can adapt, too. Over time, they may develop resistance to the chemical. If other animals do not, they have the plant all to themselves. Of course, the plant may eventually develop a different secondary compound, and the arms race continues.

Milkweeds and monarch butterflies (**FIGURE 11.9**) are an example. Milkweeds are named for their thick, white sap. In many species of milkweed, this sap contains a variety of toxic chemicals known as cardiac glycosides. In high doses, cardiac glycosides speed up an animal's heart rate, sometimes to the point of cardiac arrest. As you might imagine, many animals avoid eating milkweed.

Monarch butterflies, on the other hand, eat nothing *but* milkweed, at least in their larval stage. Adult females lay their eggs only on milkweed plants. The larvae hatch into caterpillars that spend all their time living on—and eating—this milkweed plant. Monarch caterpillars have developed an interesting defense against the cardiac glycosides in their food. These chemicals are gathered up and stored in tiny compartments inside the caterpillar's cells. When the caterpillar metamorphoses into an adult butterfly, the cardiac glycosides remain there and cause no harm to it. However, a bird or other predator that eats a monarch caterpillar or butterfly is in for a rude surprise.

In this case, the monarch not only has won the arms race, it has coopted the milkweed's defensive "weapons" for its own defense. But no defense is perfect. The black-backed oriole of Mexico has developed a tolerance for cardiac glycosides and is one of the few predators that can enjoy a monarch snack.

A second example of the coevolutionary arms race involves nest parasitism, a behavior we introduced in the Kirtland's warbler case study. The European cuckoo is a common nest parasite in England. Unlike the Kirtland's warbler, some species recognize that the cuckoo's eggs are not their own and remove them from the nest. This response, known as **discrimination**, is a defensive adaptation to nest parasitism. The cuckoo, however, has developed a counter-adaptation. Female European cuckoos specialize in parasitizing the nest of only a single species. However, different females will specialize on different host species, so that a population of European cuckoos may parasitize half a dozen or more different hosts. Female cuckoos lay eggs that closely resemble, or mimic, the eggs of their host species (**FIGURE 11.10**). The close physical resemblance of an organism to another or to its surroundings is called **mimicry**. Egg mimicry is a coevolutionary response to discrimination.

FIGURE 11.9 Many plants, such as the milkweed plant, make secondary compounds that are poisonous to predators. The monarch butterfly, which eats only milkweeds, stores these compounds and uses them as a defense against other organisms.

coevolution :: the adaptation of one species to another

secondary compound :: chemical made by plants that is usually poisonous and that is not needed by the plant for its metabolism

discrimination :: when one species recognizes another as not its own

mimicry :: the close resemblance of an organism to another or to its surroundings

FIGURE 11.10 The European cuckoo is a common nest parasite that lays eggs that generally resemble, or mimic, the eggs of its host species. While the host will discriminate against cuckoo eggs, if the cuckoo's eggs are similar to its own, it may leave the eggs in its nest.

Although the host will discriminate against cuckoo eggs, if the cuckoo's eggs are similar to its own, it may sometimes mistakenly leave the cuckoo eggs in its nest.

Coevolution takes time. As mentioned previously, the Kirtland's warbler's lack of discrimination against cowbird eggs is probably due to the fact that the agricultural development that allowed the cowbird into the warbler's range is relatively recent. Kirtland's warbler may not have had time yet to evolve a discriminatory response.

11.4 When Can Species Cooperate?

Cooperation, or mutualism, is a win-win interaction: both species benefit. However, the species engage in mutualism just for that benefit, not to be nice to one another. We can look at mutualism as mutual exploitation. Over time, one species may even become a "cheater," keeping the benefits but no longer providing any to the other species.

Animals may act as mobile agents for plants, as pollinators, seed dispersers, and even defense against predators. Honeybees, for example, are legendary as hardworking pollinators. In return, the plant provides a renewable food source such as nectar, pollen, or fruit, or perhaps provides shelter.

Honeybees benefit plants by transporting pollen from one plant to another. This direct transport is efficient and allows the plant to reproduce without having to make massive amounts of pollen, saving the plant energy. The honeybee uses the plant's nectar and some of the pollen to make honey that feeds the hive's young.

Honeybees can pollinate many different species of plant, but some pollination systems are specific. Perhaps the ultimate example of specialized pollination involves fig trees and the fig wasps that pollinate them. There are 900 species of fig and 900 species of fig wasp. Each fig has its own wasp pollinator, and each wasp pollinates only one species of fig. Furthermore, the wasp reproduces only within the fig flower, and the fig flower is its only food. Neither species can reproduce without the other; the mutualism is obligatory. In these 900 cases, mutualism has become a dependency, and extinction of either pollinator or tree would almost certainly lead to extinction of the other species.

A mated female fig wasp will fly to the **inflorescence** (a plant structure containing many flowers) on her particular species of fig (**FIGURE 11.11**). As she crawls inside the inflorescence, she transfers pollen to the flowers, fertilizing them. She lays eggs on some of the flowers. These flowers develop a gall, a small compartment of plant tissue within which the larva hatches and on which it feeds. The other flowers develop normally and eventually produce

inflorescence :: a plant structure containing many flowers

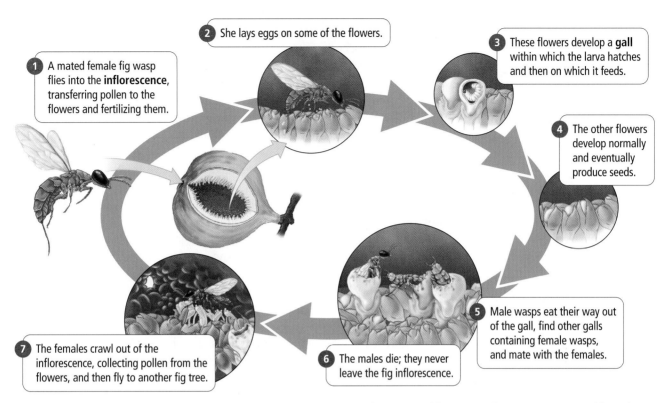

2 She lays eggs on some of the flowers.

1 A mated female fig wasp flies into the **inflorescence**, transferring pollen to the flowers and fertilizing them.

3 These flowers develop a **gall** within which the larva hatches and then on which it feeds.

4 The other flowers develop normally and eventually produce seeds.

5 Male wasps eat their way out of the gall, find other galls containing female wasps, and mate with the females.

6 The males die; they never leave the fig inflorescence.

7 The females crawl out of the inflorescence, collecting pollen from the flowers, and then fly to another fig tree.

FIGURE 11.11 The ultimate example of specialized pollination involves fig trees and fig wasps. There are 900 species of fig and 900 species of fig wasp; each fig has its own wasp pollinator.

seeds. Male wasps eat their way out of the gall, find other galls containing female wasps, and mate with them. Soon the males die; they never leave the fig inflorescence. The females crawl out of the inflorescence, collecting pollen from the flowers as they go. Then they fly to another fig tree of the correct species, and the cycle begins again.

There are a number of cases in which a plant–pollinator mutualism has broken down due to cheating. For example, robber bees chew holes at the base of a flower and extract the nectar without performing any pollination. Plants can be cheaters as well. Some species of orchid have flowers that resemble the females of particular species of bee or wasp (**FIGURE 11.12**). These flowers may also give off **pheromones**, chemical attractants similar to those used by females to attract males of their species. Males seek out these orchids and attempt to copulate with them. In so doing, they get covered with pollen. When they attempt to copulate with the next orchid, they transfer pollen and pick up some more. These orchids provide no nectar or other food for the pollinator, and clearly the "mating" will be unsuccessful as well. The wasp derives no benefit, but the orchid still benefits from pollination.

FIGURE 11.12 Some species of orchid, including *Ophrys speculum* shown here, have flowers that resemble the females of particular species of wasp and may also give off pheromones that attract males. When the males attempt to copulate, they are unsuccessful, but the orchid benefits from pollination.

pheromone :: chemical attractant used by females to attract males of their species

11.5 How Do Ecological Interactions Affect Us?

Humans, of course, are part of the ecological community, too. We engage in interactions with other species, sometimes to our advantage and sometimes to our detriment. We also have learned to use the results of long-standing ecological interactions between other species for our own benefit.

We Compete with Other Species

We are very good competitors. Given our intelligence, our technological arsenal, and our ability to work together, we are able to compete effectively for land and other resources. When we develop large areas for our benefit—by cutting down a forest or draining a wetland, for example—we compete with the species that lived in these areas originally. In many cases like these, we apply the competitive exclusion principle: our competition drives many species to extinction.

When we change the habitat, we have a negative impact on the original resident species. But the new habitat opens up opportunities for other species. Large fields of cotton provide a haven for the boll weevil, for example, and silos full of grain provide good habitat for molds, mice, and seed-eating insects. The species we refer to collectively as pests are actually species that compete with us successfully for food and other resources.

We compete with weeds, insects, molds, birds, and mammals for the food we grow. Weeds are the fiercest competitors, though technically they compete with the crops rather than us, followed by insects and molds. The Environmental Protection Agency (EPA) estimates that $39.4 billion was spent worldwide in 2007 on pesticides and herbicides to combat these competitors—$11.8 billion in the United States alone (**FIGURE 11.13**). And this does not include the money spent each year on research to study crop pests. It does not include government programs that oversee and regulate research and development of pest management. It also does not include the cost of cleaning up pesticide pollution or the health costs related to exposure to pesticides.

In spite of our efforts, our competitors do very well. Estimates of food loss due to pests range from 20% to over 50% of the potential yield. The loss varies depending on the crop, where and how it is grown, and how the harvest is stored. Those of us in developed countries are largely buffered from these losses. We may experience them as an increase in price or decrease in selection at a supermarket. In a world where many people go hungry, however, losing even a small percentage of our food leads to loss of life.

We have historical evidence of another application of the competitive exclusion principle. There used to be several species of humans living at the same time in various parts of the world. Neanderthals (*Homo neanderthalensis*), for example, lived in Europe from about 200,000 years ago until about 28,000 years ago. Our species (*Homo sapiens*) initiated its migration from Africa approximately 80,000 years ago, arriving in Europe around 40,000 years ago. *H. sapiens* and *H. neanderthalensis* coexisted for 10,000 to

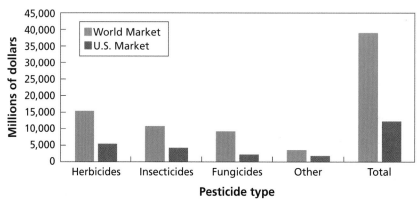

FIGURE 11.13 The Environmental Protection Agency (EPA) estimates that $39.4 billion was spent worldwide in 2007 on pesticides and herbicides—$11.8 billion in the United States alone.

Source: U.S. Environmental Protection Agency

15,000 years, but eventually *H. sapiens* was the only species remaining. There is good evidence that we were somewhat more flexible in our diet and tolerance to changing climatic conditions, and perhaps we were better hunters than the Neanderthals.

We Develop Useful Products and Ideas as a Result of Exploitation Interactions

In our prehistoric hunter-gatherer phase, we engaged directly in predator–prey interactions with many other species and may well have driven many prey species extinct. More recently, we have hunted species for food, pelts, fur, and sport, and we continue to hunt marine organisms such as fish and whales. However, today the bulk of our food comes from domesticated plants and animals. Although our exploitation interactions with natural populations continue, they have become less important to us as a source of food and material.

Exploitation interactions abound in nature, however. As we have seen, these interactions can lead to an evolutionary arms race and result in many adaptations. We have been exceptionally adept at taking advantage of these adaptations, both defensive and offensive, as evidenced by our use of plant secondary compounds.

The cardiac glycosides produced by milkweed to ward off herbivores can have a therapeutic function. They cause the heart to beat more forcefully, and in the proper dose they can be used to treat a weakening heart. Cardiac glycosides have been used to treat heart conditions for over 3000 years.

Young coffee trees produce high concentrations of caffeine. At these high concentrations, caffeine can be lethal to many insects and fungi that would eat parts of the seedling. Caffeine that diffuses into the soil can inhibit seed germination near the seedling, reducing competition from other plants. Caffeine is concentrated in coffee beans, the seed of the coffee plant, to deter herbivores from eating the next generation of coffee plants. Every time you drink your morning java, you reap the benefit of millions of years of defensive evolution in the coffee plant.

The compounds in spices such as pepper, nutmeg, mustard, cloves, cinnamon, mint, and wintergreen evolved as chemical defenses. Many spices have been used for medicinal purposes as well as flavoring. Scientists have isolated many of the chemical compounds responsible for the aroma or taste of spices and have found that they work to repel insects or kill bacteria. Intriguingly, spices are most heavily used in tropical and subtropical regions where food spoilage due to microorganisms is a significant threat.

The religious and recreational use of hallucinogenic drugs from plants and fungi is an ancient practice. For animals living in nature, an altered state of consciousness likely leads to a quick death from predators. This is an unlikely fate for humans, but the use of hallucinogenic and other plant drugs has its own set of physiological and legal consequences. Although developed as an herbivore defense, some of these drugs do have therapeutic value despite their dangers, such as the analgesic properties of morphine from the opium poppy.

Our understanding of exploitation interactions in nature can change our approach to pest control. Our pests have predators and parasites of their own. If we can successfully introduce these predators and parasites, they may take care of our pests for us, a procedure known as **biological control**. This is a promising approach, but it can be difficult to predict *all* of the possible consequences of introducing "pest-controlling" species into an environment. The coevolutionary arms race holds a lesson for us: it is a difficult, if not impossible, race to win.

biological control :: an approach to controlling pests by introducing predators of a pest into an area

Biological Control of Schistosomiasis

When you think of medical technology, you probably don't think about crayfish. Nevertheless, crayfish may be part of the solution to a disease that affects over 200 million people annually: *schistosomiasis*.

Schistosomes are parasitic flatworms, or flukes, about a centimeter long. They burrow into human skin, find a blood vessel, and travel throughout the circulatory system. Males and females mate in the liver, and, depending on the species, a female can produce between 300 and 3000 eggs each day. Many of the eggs pass into the intestines and are eliminated with feces. But many eggs also get trapped in body tissues. The worms themselves cause little damage, but as the body's immune response attacks the eggs, the surrounding tissue can be damaged.

However, this is only half the story. Schistosomes have two hosts: one human and one a freshwater snail. When feces infected with schistosomes contaminate a pond or lake, the schistosomes hatch into swimming larvae that infect snails. Inside the snails, they reproduce asexually to make vast numbers of individuals called cercariae that break free of the snail. The cercariae swim and attach to the skin of humans who play or bathe in the infected water, completing the schistosome life cycle.

This is where the crayfish comes in. The Louisiana swamp red crayfish is a voracious predator of freshwater snails. If the snails are removed, schistosomes can't complete their life cycle, and humans will no longer get infected.

Bruce Hofkin has been studying the interaction of crayfish, snails, and schistosomes since the 1990s, and he has found that lakes and ponds that contain the crayfish rarely contain any snails. In an experiment, Hofkin compared infection rates at two schools. At the experimental school, crayfish were introduced into a nearby pond where students played. At the control school, no crayfish were introduced. Hofkin found that the crayfish established viable populations in the experimental pond. He also saw that the snail populations declined there, and students at the experimental school were significantly less likely to be infected with schistosomes.

Biological control can be a very effective technology, both medically and in agricultural systems to control pests.

A

B

Schistosomes (A) are parasitic flatworms that burrow into human skin and travel throughout the circulatory system. The Louisiana swamp red crayfish (B) is a voracious predator of freshwater snails, and if the snails are removed, Schistosomes can't complete their life cycle.

It is much cheaper than medical treatment or chemical control, and it reduces the need for drug treatment and the amount of pesticides introduced into the ecosystem. A good deal of research is needed before using biological control to make sure the control agents don't become worse than the pests or parasites they control. ::

The arms race between plants and herbivores has been going on for millions of years, and still neither side has won. By using pesticides and antibiotics, we have started an arms race of our own. Will we be able to defeat pests and pathogens this way, or will we simply encourage better defensive adaptations on their part? For another example of how biological control is being used, see *Technology Connection: Biological Control of Schistosomiasis.*

We Capitalize on Mutualisms

We have learned to capitalize on the results of mutualisms, just as we have on the results of the evolutionary arms race. A prime example is the mutualistic origin of fruit.

Technically, a fruit is the structure that contains a plant's seeds (Chapter 7). The fruit develops from the ovary of a flower, where the seeds form. The fruits we are most familiar with—for example, apples, oranges, and mangoes—are sweet and juicy. Fruits like these aid the plant in dispersing its seeds.

Mammals and birds eat the fruit, gaining nutrition and energy. The seeds are indigestible. They pass through the animal, and are deposited away from the parent plant with a dollop of fertilizer to help the seeds get started. Animal seed dispersal is an old plant–animal mutualism that results from the evolution of fruit. We have helped this process along by breeding larger and tastier fruits. But the fruits were already there, the result of a long-standing ecological interaction.

11.6 What Does a Functioning Ecosystem Do?

An ecosystem is like an engine, but how does it get its fuel? Interactions between species distribute the nutrients they need to survive. However, species also recycle the materials of planet Earth on a global scale to keep ecosystems in balance.

Exploitation Interactions Distribute Energy and Nutrients

Ultimately, all the energy in an ecosystem comes from the sun. Plants capture sunlight in the form of energy-rich chemicals, sugars, and starches (Chapter 7). Plants use these chemicals to drive their own metabolism. When an herbivore eats a plant, energy and nutrients are transferred, and when a carnivore eats an herbivore, they are transferred yet again. Ultimately, all living organisms die, and their bodies decompose, returning nutrients to the soil for plants to use. The ecological interaction of predation distributes energy and nutrients throughout the ecosystem, providing it with fuel.

If we make a diagram of "who eats whom (or what)," we get a multilinked structure called a **food web** (**FIGURE 11.14**). We can think of a food web as being divided into levels. At the first level are photosynthetic organisms. These are known as **primary producers** because they are the first link in capturing solar energy. Next we have herbivores, or **primary consumers**. Then several levels of carnivores can exist: these are known as **secondary consumers**, tertiary consumers, and so on. Finally we have **decomposers**, organisms that cause dead bodies to rot, which are the ultimate level of any food web. These various levels that organisms occupy on the food web are known as **trophic levels**. Trophic levels are helpful guidelines, not firm rules. Many animals eat at more than one level; for example, bears and humans eat both fish and berries.

Energy flow through a food web is not 100% efficient, and energy is lost every time one organism eats another. On average, an organism stores only about 10% of the energy content that it eats. Although this figure varies a lot, it means that about 90% of the energy in food is lost as indigestible material and heat. And this loss affects the entire food web: it means that the amount

food web :: a diagram showing a multilinked structure of "who eats whom (or what)"

primary producer :: photosynthetic organism that is the first link in the food web in terms of capturing solar energy

primary consumer :: herbivore

secondary consumer :: carnivore

decomposer :: organism that causes dead bodies to rot

trophic level :: level that organisms occupy on the food web

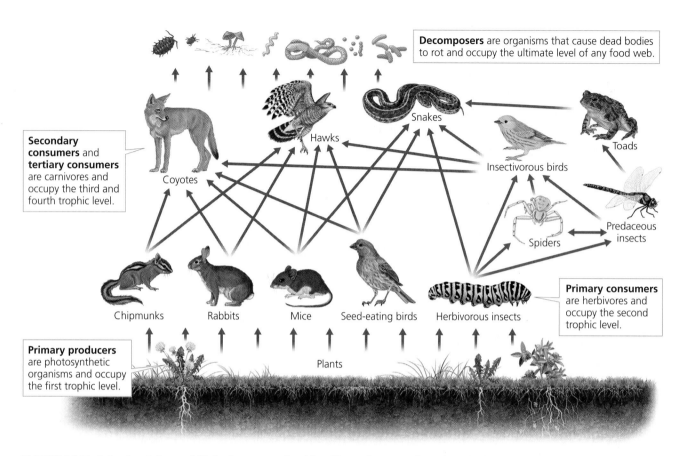

Decomposers are organisms that cause dead bodies to rot and occupy the ultimate level of any food web.

Secondary consumers and tertiary consumers are carnivores and occupy the third and fourth trophic level.

Primary consumers are herbivores and occupy the second trophic level.

Primary producers are photosynthetic organisms and occupy the first trophic level.

Coyotes · Hawks · Snakes · Toads · Insectivorous birds · Spiders · Predaceous insects · Chipmunks · Rabbits · Mice · Seed-eating birds · Herbivorous insects · Plants

FIGURE 11.14 A food web is a multilinked structure that identifies "who eats whom (or what)." Organisms occupy different trophic levels on the food web, but these are not firm, as many animals eat at more than one level.

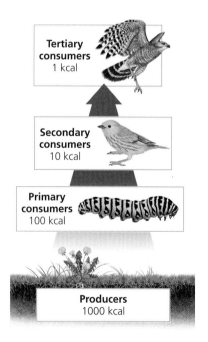

Tertiary consumers 1 kcal

Secondary consumers 10 kcal

Primary consumers 100 kcal

Producers 1000 kcal

FIGURE 11.15 An ecological pyramid shows the change in available energy at the different trophic levels. Population sizes decrease dramatically at higher trophic levels because less energy is available and fewer individuals can be supported.

of energy available decreases dramatically at higher trophic levels. Primary producers capture about 1% of the sunlight that hits their leaves, primary consumers get about 10% of that, secondary consumers get 10% of that, and so on up the trophic levels.

The availability of less energy at higher trophic levels means that fewer individuals can be supported. Population sizes decrease dramatically at higher tropic levels. We can picture the change from level to level with an **ecological pyramid** (**FIGURE 11.15**).

Ecosystems Recycle Material on a Global Scale

All of the matter on Earth is distributed among four reservoirs—the biosphere, atmosphere, hydrosphere, and lithosphere. The biosphere is the global ecosystem and encompasses all life on the planet. The remaining reservoirs are nonliving and refer to the gaseous, liquid, and solid parts of the Earth. **Biogeochemical cycles** are processes that move matter among these four reservoirs. The *bio* in *biogeochemical* tells us that the living organisms in ecosystems are major players in this recycling process. If recycling is disrupted, some things begin to build up while others become depleted, which can cause trouble on a global scale. Almost every chemical important to life is part of a biogeochemical cycle. In this section, we will discuss three of these cycles: carbon, oxygen, and nitrogen. See *How Do We Know? Long-Term Ecological Research* to learn how biogeochemical cycles are studied in nature.

The Carbon Cycle The carbon cycle refers to the movement of carbon throughout the ecosystem (**FIGURE 11.16**, p. 332). The drivers of the carbon

Long-Term Ecological Research

Ecology operates on a moderate time scale. It can take decades for changes to occur in an ecosystem, which makes studying ecosystems a challenge. In 1980, the National Science Foundation (NSF) established the Long Term Ecological Research (LTER) network, a collaboration of almost 2000 researchers at 21 sites throughout the United States and 2 in Antarctica, dedicated to studying ecosystem ecology on a large spatial and temporal scale.

One of the partners in the LTER network is the Hubbard Brook ecosystem study, located in the White Mountains of New Hampshire. The Hubbard Brook watershed is a valley that spans about 8 × 5 kilometers. Hubbard Brook flows east through the valley, and smaller tributaries flow into it from the south or north, carving out a number of smaller valleys that join up with the main valley.

Since the 1960s, researchers at Hubbard Brook have been studying nutrient and mineral cycling in the northeast hardwood forest ecosystem. They measure the input of water and dissolved minerals in rainfall, track their movement through the plants and trees in the forest, and measure the outflow of water and minerals in the forest streams. Because the forest sits on impermeable bedrock, any nutrients or minerals leaving the ecosystem do so in stream outflow. Researchers also have been monitoring the effects of acid rain, since acidified water and soils have a large effect on nutrient and mineral uptake by plants.

In a study designed to test the effect of plants on nutrient and mineral retention, Herbert Bormann of Yale and Gene Likens of Cornell worked with two adjacent valleys. Each valley had a single stream that flowed into Hubbard Brook. One valley was the control. In the experimental valley, they removed all the vegetation—every tree and every plant—and continuously removed all new plant growth during the three-year study, leaving them to decompose. In both valleys,

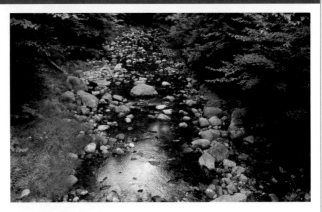

Hubbard Brook White Mountain National Forest in New Hampshire is the original site of the oldest ecological study in the United States and a model for research on watersheds.

they monitored the rainfall and the level of calcium, potassium, nitrogen, and other nutrients. They built a small spillway at the mouth of each stream and measured the volume of water leaving each valley as well as the concentrations of dissolved calcium, potassium, and nitrogen.

Their results were striking. In the control valley, plants took up about 40% of the rain. Virtually all of the calcium and potassium stayed in the forest, and nitrogen levels increased slightly. In the experimental valley, 100% of the rain left via the stream. Mineral and nutrient loss through the stream increased dramatically: a 10-fold increase for calcium, 20-fold increase for potassium, and a 60-fold increase for nitrogen. So much nitrogen was added to the stream that it encouraged the growth of algae, making the water undrinkable.

Long-term studies like the one at Hubbard Brook help to confirm the role forests play in maintaining soil fertility and reducing water pollution caused by runoff. ::

---

cycle are photosynthesis and respiration. Recall that photosynthetic organisms remove carbon dioxide from the atmosphere and convert it into energy-rich carbohydrates (Chapter 7). This is half of the cycle—movement of carbon from the nonliving reservoir of the atmosphere into the living reservoir. The other half of the cycle occurs through respiration. As organisms break down carbon compounds metabolically, they release carbon dioxide back into the atmosphere, completing the cycle.

The Oxygen Cycle The oxygen cycle refers to the movement of oxygen throughout the ecosystem, and it is the mirror image of the carbon cycle (**FIGURE 11.17**). Oxygen is added to the atmosphere by photosynthesis and removed from the atmosphere by respiration. Burning also consumes

ecological pyramid :: a diagram of the changes in trophic level

biogeochemical cycle :: process that moves matter among the four reservoirs: the biosphere, atmosphere, hydrosphere, and lithosphere

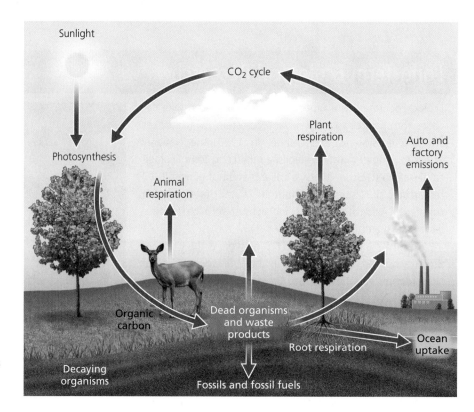

FIGURE 11.16 The carbon cycle refers to the movement of carbon throughout the ecosystem. The drivers of the carbon cycle are photosynthesis and respiration.

atmospheric oxygen, but since oxygen is so much more abundant in the atmosphere than carbon dioxide (oxygen makes up 21% of the atmosphere, carbon dioxide only 0.04%), fires and burning do not deplete oxygen appreciably.

The Nitrogen Cycle The nitrogen cycle refers to the movement of nitrogen throughout the ecosystem (**FIGURE 11.18**). Nitrogen's cycling is more complex than that of carbon or oxygen, and it involves a wider variety of species

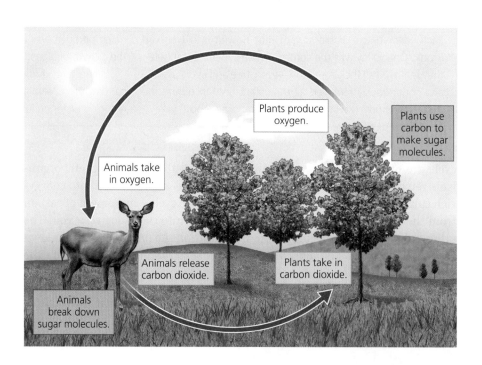

FIGURE 11.17 The oxygen cycle refers to the movement of oxygen throughout the ecosystem. It is the mirror image of the carbon cycle; oxygen is added to the atmosphere by photosynthesis and removed from the atmosphere by respiration.

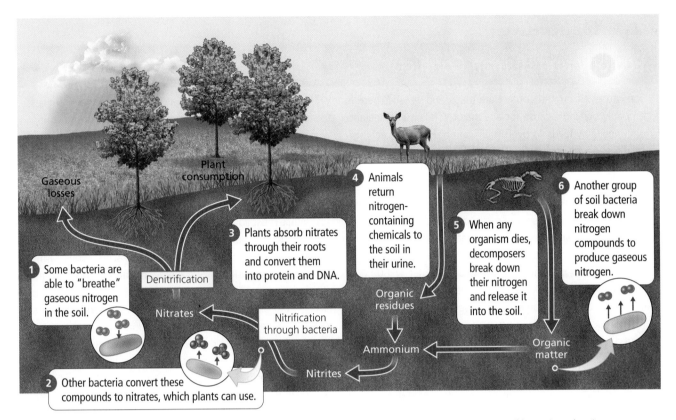

FIGURE 11.18 The nitrogen cycle refers to the movement of nitrogen throughout the ecosystem. Soil bacteria make nitrogen available in a two-step process.

working on a sort of biochemical assembly line. Nitrogen makes up 79% of the atmosphere, making it the most abundant gas. However, no plant or animal can use atmospheric nitrogen. This ability is found only in a number of species of bacteria that live in the soil. Since all living organisms require nitrogen to make protein and DNA, we are all dependent on the activity of these bacteria.

Soil bacteria make nitrogen available in a two-step process. First, some bacteria are able to "breathe" gaseous nitrogen in the soil and capture it in nongaseous chemicals similar to ammonia. Second, other bacteria convert these ammonia-like compounds to nitrates, which plants can use. Plants absorb nitrates through their roots and convert them into protein and DNA. Nitrogen compounds move throughout the food web, providing a usable source of nitrogen for all species. Animals return nitrogen-containing chemicals to the soil in their urine, and when any organism dies, decomposers break down their nitrogen and release it into the soil. To complete the cycle, another group of soil bacteria breaks down nitrogen compounds to produce gaseous nitrogen, which is returned to the atmosphere.

11.7 What Benefits Do We Get from a Functioning Ecosystem?

This is the most significant question for the study of ecology. The natural world provides us with numerous examples of ecology that are interesting to know about and study. Ultimately, however, we face the question of *preserving*

Life Application

Ecology and Human Conflict

The causes of human conflict are many and varied. There is probably no single cause for any major conflict, and seeking one inevitably leads to oversimplification. However, ecological pressure in conjunction with other factors can lead to conflict.

In modern times, it is difficult to find a more disturbing case of conflict than the genocide in Rwanda. This is a complex case, and the causes are many, but Jared Diamond in his book *Collapse* analyzes the ways that a failed ecology contributed to the conflict.

Between April and July 1994, approximately 800,000 Rwandans, about 10% of the population, were murdered, making it the third-largest genocide after the Holocaust. The genocide is usually portrayed as the result of ethnic hatred between the Hutu and the Tutsi, the two major ethnic groups in Rwanda, but that is not the whole story. Diamond reports on a study of the Kamana region of Rwanda, a region that consisted only of Hutu, but where genocide still occurred.

The population density of Rwanda in 1991 was 760 people per square mile, the highest of any African nation. In the Kamana region at the time, the population density was 2040 per square mile, one of the highest in the world. Rwanda has fertile volcanic soil, is at a sufficiently high elevation to receive significant rainfall, and is cool enough to be free of two of Africa's scourges: malaria and tsetse flies. Along with the introduction of modern medical techniques, the habitat supported rapid population growth, leading to the densities just mentioned. However, modern agricultural practices were not introduced. Rwandans continued to farm the way they always had. A combination of deforestation, overgrazing, and unsound techniques for tillage led to a nutrient depletion in the soil and loss of soil due to erosion.

As the land became poorer, the population increased and the average size of a family farm decreased proportionally. In Kamana in 1993, half of the farms spanned less than 0.75 acres, and these were split among family members so that a single family might have only one-tenth of that to farm.

With all the land taken, there was no chance for new families to start out with a farm of their own. Two-thirds of the young women and all of the young men stayed at home on the family farm, further stretching its resources and most likely generating a fair amount of tension and conflict. These small, poor-quality farms simply could not feed the families that owned them.

While most younger families in Kamana had small farms, a few older families had much larger ones. Farms owned by people in their 20s averaged only 0.4 acres, while farms owned by people in their 50s averaged 2 acres. Conflicts and disputes over land increased and became more violent up until the genocide started. When the killing began in Kamana in 1994, at least 5% of the population died. Researchers who studied Kamana concluded that the residents saw this as a chance to settle old scores and conflicts and to redistribute scarce farmlands among the survivors. While no one argues that ecological pressure was the only cause of the killing in Kamana, evidence suggests that it was an important contributing factor. ::

The causes of human conflict are many and varied, but ecological pressure in conjunction with other factors may contribute to conflicts, such as the one that occurred in Rwanda.

ecology. Interest alone does not provide sufficient motivation for the hard and costly work of preservation. In addition, we need to consider the economic benefits of a functioning ecosystem, as well as the costs of a dysfunctional ecosystem. These costs include the possibility of human conflict as we compete for ever-scarcer resources (see *Life Application: Ecology and Human Conflict*).

Ecosystem Services

We can think of the benefits we get from our ecosystem as **ecosystem services**: naturally occurring ecological processes that support our agriculture, technology, culture, and more. These services can be divided into four groups:

- *Provisioning:* these services provide us with food, fresh water, natural fibers, fuel, and biochemical compounds. With modern genetic techniques, even the genes of other organisms provide a useful resource.
- *Regulating:* these services regulate and stabilize the ecosystem, making it more reliable. They include regulating pest and disease control through the checks and balances discussed in this chapter: control of runoff and erosion, pollination, water and air purification, and waste decomposition.
- *Cultural:* these services relate to quality-of-life issues and are uniquely human. They include the recreational, aesthetic, and inspirational value of natural areas and the cultural, spiritual, and religious connections people have to nature. Cultural services also include the joy of learning about nature and the opportunities it provides for scientific discovery.
- *Supporting:* these are services on which all the others depend. They include the capture of solar energy by primary producers, biogeochemical cycling, and the production of soil and oxygen.

Scientists, conservationists, economists, and politicians are working together to measure the benefits we receive from ecosystem services. We already assign an economic value to provisioning services, since they provide products we can use or sell. Historically, though, we have placed little economic value on the other services, implicitly assuming that they are free and will always be there. We realize now that if we disrupt an ecosystem service, it will cost money to repair or to replace it. It may also cost lives. Honest, accurate bookkeeping requires that we take these costs into account in our development projects.

Disruptions to Ecosystem Services

Soil production is a vital ecosystem service, one on which all terrestrial life depends. New soil constantly forms at a rate controlled by the weathering of exposed rocks and the decomposition of dead bodies. Soil lost by erosion enters our streams and lakes as sediment, eventually reaching the ocean, where new rock is formed. The main defense against soil erosion is vegetation. The network of plant roots helps to hold soil in place. However, our agriculture, forestry, and development practices frequently reduce or remove vegetation cover. We have shifted the balance between soil loss and soil erosion dramatically: on a global scale, we lose about 7.5 to 10 tons of soil per acre annually. It is critical that development projects that may increase erosion take the cost of replacing soil into account. If not, the projects will appear to be more economically feasible than they really are.

Our exploitation interactions with other species can also disrupt ecosystem services and the food web. If we overexploit a predator species—by hunting, for example—the herbivores it normally eats can increase their population size. This in turn can have a negative effect on the plant populations

ecosystem service :: a naturally occurring ecological process that supports our agriculture, technology, culture, and more; can be divided into four groups: provisioning, regulating, cultural, and supporting the herbivores consume. Instability of the food web can lead to ecosystem collapse, which in turn reduces the ecosystem's ability to provide its services.

Finally, our activity can alter biogeochemical cycles. We can't do anything to change the total amount of carbon or nitrogen on the planet. But we can have a dramatic effect on where these compounds are stored. Burning forests and fossil fuels (which come from ancient forests) adds more carbon to the atmosphere than photosynthetic organisms can remove. This additional atmospheric carbon then becomes a cause of global climate change.

We also affect the nitrogen cycle. Often there is not enough nitrogen in the soil to support the high rate of plant growth our agriculture requires. Nitrogen is the main component of the fertilizers we apply to our crops. We have developed our own chemical process for capturing nitrogen from the atmosphere, and in fact the majority of nitrogen in our soil now comes from fertilizer rather than soil bacteria. Factories that make nitrogen fertilizer must burn fossil fuels to do so, which increases our impact on the carbon cycle. Excess nitrogen in the soil leaches into the watershed, promoting the growth of algae and altering the aquatic ecosystem.

Biodiversity

biodiversity :: the variety of species living in an ecosystem

There is a common denominator that relates to the health of an ecosystem, its ability to provide services, and the disruption human activity can cause: **biodiversity**. All of the ecological processes discussed in this chapter depend on having a diverse set of species present in the ecosystem. Reducing biodiversity is one of the most common results of development, and maintaining biodiversity may be our greatest ecological challenge. We will take a closer look at biodiversity in the next chapter.

 # Biology in Perspective

We benefit greatly from the ecological interactions among species. Ecological interactions account for many of the adaptations we observe in nature. These adaptations evolve when species develop tolerance for extreme environments. They evolve, too, when species compete, when they get better at catching prey or avoiding capture, and when they cooperate. Many of these adaptations are directly useful to us, providing food or medicine.

We also benefit greatly from the work that functioning ecosystems perform. Food webs fuel the ecosystem and mediate energy and nutrient flow. Biogeochemical cycles move important chemicals between living and nonliving reservoirs. Taken together, interactions among species enable an ecosystem to provide vital services that give us goods, maintain ecosystem health, and preserve the conditions necessary for life.

Our survival depends on the benefits the ecosystem provides. These benefits may seem free of charge, but if we disrupt them, we will pay for them one way or another. Maintaining these services intact is an economic imperative, as well as an ethical and cultural need.

Chapter Summary

11.1 How Do Species Adapt to Their Habitat?

Provide examples of the ways species adapt to the constraints of their habitats.

- Kangaroo rats live in North American deserts and have many physiological adaptations to conserve water, such as being nocturnal and having efficient kidneys.
- Marmots, large rodents in the Rocky Mountains, share dens to adapt to limited shelter.
- White-fronted bee-eaters are birds that nest in limited cavities; year-old offspring are often nest helpers for their parents.

11.2 Why Do Species Compete?

Summarize how species can coexist without one outcompeting the other.

- Competition occurs when two or more species attempt to use the same limited resource: in exploitation, one species benefits and one is harmed; in mutualism, both species benefit.
- The competitive exclusion principle holds that two species cannot share the exact same niche (complete set of physical and biological resources); one will always outcompete the other.
  - o Aspects of niches can be altered to reduce competition so competitors may coexist.
  - o Character displacement occurs when species evolve a change in physical or behavioral characteristics that allows them to reduce competition and coexist.

11.3 How Do Species Exploit One Another?

Describe the ways that species may exploit one another.

- In exploitation (predation, parasitism), one species benefits at the expense of another.
- Predator–prey systems are an example of population cycling: as prey increase, the predator population increases, but then as predators consume prey, the prey population declines; the lynx and snowshoe hare are an example.
- In an evolutionary arms race (coevolution), the exploited species develops adaptations to prevent or avoid exploitation and then the exploiter develops adaptations to counteract these; milkweed plants and cuckoo bird nest parasitism (mimicry) are examples.

11.4 When Can Species Cooperate?

Identify circumstances of cooperation between species.

- In cooperation or mutualism, both species benefit from the interaction.
- Honeybees transport pollen and in return receive nectar.

- Certain species of fig wasp have coevolved with certain species of figs, and neither can reproduce without the other.
- Sometimes species "cheat": for example, orchids attract bees with chemical pheromones, and bees collect their pollen but do not receive any nectar in return.

11.5 How Do Ecological Interactions Affect Us?

Summarize the beneficial and detrimental ecological interactions that we have with other species in the ecosystem.

- Humans are good competitors for land and other resources; we are driving many species to extinction.
- When we alter a habitat, we may drive the original inhabitants out but make it better suited to others.
- There were several species of humans living on Earth, but eventually *Homo sapiens* outcompeted them all.
- Humans don't directly exploit as many organisms now that most of our food is domesticated, but we outcompete many for habitat.
- We take advantage of many chemical adaptations of plants, including cardiac glycosides from milkweed to treat heart problems, caffeine from coffee plants, spices used for medicine and flavoring, and drugs such as opiates and hallucinogens.

11.6 What Does a Functioning Ecosystem Do?

Describe the movement of energy and nutrients through an ecosystem.

- Energy in an ecosystem comes from the sun.
- The food web based on the sun's energy is composed of trophic levels: primary producers or plants, primary consumers or herbivores, secondary consumers or carnivores, and decomposers.
- The ecological pyramid reflects the fact that 90% of energy is lost at each trophic level.
- Biogeochemical cycles recycle the materials in an ecosystem among four reservoirs: the biosphere, the atmosphere, the hydrosphere, and the lithosphere.
  - o During the carbon cycle, photosynthesis removes carbon dioxide from the atmosphere and respiration adds it.
  - o During the oxygen cycle, photosynthesis adds oxygen and respiration removes it from the atmosphere.
  - o During the nitrogen cycle, plants absorb nitrates from the soil, and nitrogen is released back into the soil through decomposition and converted back to gaseous form for release into the atmosphere.

11.7 What Benefits Do We Get from a Functioning Ecosystem?

Enumerate the benefits we get from a functioning eco-system.

- Ecosystem services are naturally occurring ecological processes that support our agriculture, technology, and more;

there are four types: provisioning, regulating, cultural, and supporting.

- We disrupt ecosystem services as we use land for our own purposes or alter biogeochemical cycles.
- Biodiversity is the common denominator that relates to the health of an ecosystem; all ecological processes depend on having a diverse set of species.

Key Terms

biodiversity 336
biogeochemical cycle 331
biological control 327
character displacement 320
coevolution 323
competition 318
competitive exclusion principle 319
decomposer 329
discrimination 323

ecological pyramid 331
ecology 314
ecosystem service 335
exploitation 318
food web 329
inflorescence 324
mimicry 323
mutualism 318
niche 319

pheromone 325
polygynous 317
primary consumer 329
primary producer 329
secondary compound 323
secondary consumer 329
trophic level 329

Review Questions

1. What are the ecological factors that make Kirtland's warbler a rare species?

2. What are the physiological and behavioral adaptations that allow kangaroo rats to live in the desert?

3. What is the competitive exclusion principle?

   a. A conclusion that says two competitors may peacefully coexist

   b. A conclusion that says two competitors cannot coexist

   c. A conclusion that says both species benefit

   d. A conclusion that says one species benefits while the other is harmed

   e. A conclusion that says that if two or more species attempt to utilize the same resource in a habitat, it will have a negative effect on the species that is newer to the habitat

4. What is a niche?

   a. The particular set of resources a species uses

   b. A mating system in which one male mates with several females

   c. An example of population cycling

   d. An evolutionary arms race

   e. A plant structure containing many flowers

5. Pick one of your favorite organisms and list some of the factors that determine its niche.

6. Why does exploitation (particularly predation) often lead to population cycles?

7. What causes a coevolutionary "arms race"?

8. Describe a mutualistic interaction between two (or more) species.

9. What is a food web?

   a. A food chain

   b. Helpful guidelines about which animals are most likely to be prey

   c. Processes that move matter among the biosphere, atmosphere, hydrosphere, and lithosphere

   d. A diagram of "who eats whom (or what)" in a multi-linked structure

   e. An ecological pyramid showing decreases in population size

10. What are trophic levels?

11. What is an ecological pyramid, and why does it narrow toward the top?

12. Draw a labeled diagram of the following: (a) the carbon cycle, (b) the oxygen cycle, and (c) the nitrogen cycle.

13. How is human activity altering the carbon cycle and the nitrogen cycle?

14. What is the original ecological function of spices found in plants like mint and cloves?

15. List the four types of ecosystem services.

    a. Biogeochemical, regulating, provisioning, and supporting

    b. Inflorescence, provisioning, cultural, and supporting

    c. Provisioning, regulating, cultural, and supporting

    d. Provisioning, biogeochemical, cultural, and supporting

    e. Biogeochemical, regulating, cultural, and supporting

The Thinking Citizen

1. Reread the discussion about the small and medium ground finches on the Galapagos Islands. If you assume that beak size controls the size of the seeds these birds eat, are there any islands on which these birds would eat the same seeds? If not, what is the evidence that these birds compete?

2. A typical food web has many pathways through which energy and nutrients flow, and a given species may be involved with more than one pathway. However, no pathway contains many steps. The number of species involved in the path from primary producer to primary consumer to secondary and higher consumers rarely exceeds four or five. What factors limit the length of a trophic pathway? Under what conditions or in what types of habitats would you expect to see the shortest pathways? The longest?

3. In some habitats with few species, the food web may become closer to a food *chain*: only a couple of herbivore species and even fewer carnivores. Compare the effects of losing one species in a food chain to losing one species in a more typical food web. Which is more stable (less likely to crash), a food chain or food web? Generally speaking, how are stability and number of species related, and why does this relationship exist?

4. Discuss the potential effects of the following on the reservoir of atmospheric carbon: (a) cutting down large tracts of rain forest and exporting the lumber; (b) burning fossil fuels on a global scale; and (c) warming of the surface waters of the oceans. (Hint: gases become less soluble in water as its temperature increases.) Think in terms of the input to and output from the atmospheric carbon reservoir: how might these activities change the balance?

5. Select one of the four types of ecosystem services. What ecological interactions among species allow the service to be delivered?

The Informed Citizen

1. Suppose you lived in a rural community that was considering reintroducing wolves into the ecosystem. What benefits and other consequences would you expect? What factors would be most important to you and your neighbors in reaching a decision? What other information would you need to know to make a good decision?

2. Agricultural pests compete with us successfully for our food crops. Pesticides are a major component of our competitive strategy, although they are not a complete or perfect solution. Review the pros and cons of pesticide use, considering biological, economic, and social factors. If your community were charged with designing a strategy for controlling agricultural pests, what role would pesticides play? What other control methods would you consider, and why?

3. There are several ways to assess the value of ecosystem services. One is *replacement cost*: how much would it cost to replace the naturally occurring service with one of our own design? Another is *avoidance cost*: how much money does the service save us by not having to provide it on our own? Select one category of ecosystem service and discuss some of the things we would have to do to replace it if it were disrupted. What are some of the things we *won't* have to do if the service remains intact?

CHAPTER LEARNING OBJECTIVES

After reading this chapter, you should be able to answer the following questions:

Biodiversity refers to the variety of species living in an area. It varies across the planet and has changed over time.

CHAPTER 12

Biodiversity and Human Affairs

How Is the Human Race Like a Meteorite?

Becoming an Informed Citizen . . .

Like all species, we use the resources available to us in our ecosystem for our own benefit. We are extremely good at exploiting our habitat, often at the expense of other species. Preserving biodiversity is critical to our own survival, but it is not easy or cheap, and it requires a number of changes in the way we live and work.

*Y*ou've never seen an ivory-billed woodpecker, a passenger pigeon, or a dodo bird. All three species have gone extinct within recorded human history. But have these extinctions made your life any less enjoyable or any less fulfilled? Why are these extinctions significant?

biodiversity :: the variety of species living in an ecosystem

Saving endangered species is expensive and requires changes in the way we live, so it makes sense to understand why we should even try. In this chapter, we will discuss **biodiversity**, the variety of species living in an area. We will examine the impact humans have on biodiversity—and its impact on us. We will also address the biology behind biodiversity, such as the causes of its decline, how it varies across the globe, how it has changed over time, how it is monitored, and why preservation efforts make sense. What do you think our position on conserving biodiversity should be? How important *is* biodiversity? This chapter will help you decide.

case study

The Discovery of America

FIGURE 12.1 Over 18,000 years ago, Beringia was a large low-lying region connecting continents that later would be named Asia and North America.

Undoubtedly the man had a name, but it was never recorded. If it had been, it would be listed today next to Leif Ericson and Christopher Columbus.

He stared out at the cold waters of the northern Pacific, watching his kin in their small boats catching fish or preparing to hunt seal and walrus. He stood on the subcontinent of Beringia, a large low-lying region connecting continents that later would be named Asia and North America (**FIGURE 12.1**). Legend held that the man's distant ancestors had come from the west, but glaciers a mile thick blocked travel to both the west and east. The relatively mild winds off the Pacific, however, created a pocket of habitable land in central Beringia where a community had prospered for thousands of years.

But it was getting warmer. As the glaciers slowly receded, they exposed habitable land along the coast to the east. Collecting their things, the man and other restless members of his village headed off in their boats in the direction of the rising sun. The weather grew warmer, and the coastal plains widened since the glaciers here began receding much earlier. Eventually, the group reached shore and surveyed their new home. The sea provided ample food, and the wide coastal plains provided a good place to live. The land looked to be a promising source of game. What they did not know is that they had discovered America.

This, or something very similar, took place starting about 18,000 years ago. A small group of humans, perhaps several thousand, was isolated in Beringia. A smaller group of perhaps several hundred colonized the Pacific coast of North America. Their population increased dramatically, and their descendants spread rapidly down the coast. By 15,000 years ago, humans had colonized the western coasts of both North and South America. As the ice age ended, glaciers continued to melt. Sea levels rose and the coastal plains were

FIGURE 12.2 Perhaps 12,000 years ago, humans began moving inland, and they found a continent that was home to many large mammals, including elephants, several species of camels, lions, deer, giant grizzlies, giant armadillos, and dire wolves, to name a few. All these animals are now extinct in North America.

submerged, along with much of the evidence for human habitation on the continents. Perhaps 12,000 years ago, humans began moving inland.

What they found in North America would astound you (**FIGURE 12.2**). The continent was home to many large mammals. There were elephants (mastodons and mammoths), several species of American camel, American lions like the saber-tooth cat, horses, many kinds of deer including four-horned antelopes, giant beavers the size of young grizzly bears, giant ground sloths larger than adult grizzlies, giant armadillos, rhinoceros, cheetahs, and dire wolves, to name but a few. All these animals are now extinct in North America (though the horse was reintroduced).

The extinctions began about 11,000 years ago and took place in a remarkably short period of time, perhaps 400 to 1000 years. Archaeological evidence suggests the extinctions occurred in a wave that moved west to east across the continent. This wave of extinction occurred at the same time as the wave of human expansion throughout North America (**FIGURE 12.3**). Was the timing of extinction and human expansion just a coincidence?

FIGURE 12.3 Archaeological evidence suggests that extinctions 11,000 years ago occurred at the same time as the wave of human expansion throughout North America. Here, an illustration depicts early humans called Hominids attacking a species of saber-toothed cat (Machairdontinae) using wooden branches and a spear.

12.1 What Are the Components of Biodiversity?

If you follow the news, you hear the terms *biodiversity* and *conservation* a lot. In most cases, the reports refer to species that are threatened with extinction. These bear on one important component of biodiversity: the diversity of species alive in an area, or **species richness**. Regions of high species richness allow for more varied interactions among species.

This species-level component is only part of the story, however. Another critical component of biodiversity is **genetic diversity**, the degree of genetic difference among individuals in a species. If a species is genetically diverse,

species richness :: the diversity of species alive in an area

genetic diversity :: the degree of genetic difference among individuals in a species

Life Application

The Importance of Genetic Diversity

How can a jellyfish help stop malaria? The jellyfish *Aequorea victoria* has a gene that makes green fluorescent protein (GFP), which glows green under the proper lighting conditions. Scientists studying a particular gene can insert the GFP gene alongside it, and when the native gene is expressed and makes its protein, GFP is made as well. When scientists observe the green glow, they know the gene they are studying has been turned on.

Scientists have inserted GFP into the larvae of the *Anopheles* mosquito—the mosquito that transmits malaria—next to a gene that is active only in the larval testes. Males have testes that glow green, and the females don't glow at all. This difference allows male and female larvae to be separated easily. In fact, a machine is available that uses laser light to separate 18,000 larvae per hour. Once the males have been isolated, they can be sterilized with gamma irradiation and released. Since female *Anopheles* mate only once, every female that mates with a sterile male will not produce any young. If enough sterile males are released, the number of malaria-carrying mosquitoes can be reduced significantly.

Osamu Shimomura, Martin Chalfie, and Roger Y. Tsien, the three scientists who first isolated the GFP gene from the jellyfish, were awarded the 2008 Nobel Prize in Chemistry for their work. Following their discovery, a whole host of

GFP has been inserted into the larvae of the *Anopheles* **mosquito that transmits malaria.** The GFP causes them to glow green, raising hopes that a gene could be introduced that would make mosquitoes unable to carry the protozoa that causes malaria.

fluorescent proteins have been found, allowing a veritable rainbow of color options for research. The important point for us is that if the jellyfish had gone extinct before this discovery, an entire subfield of molecular biology and chemistry—and maybe even a way to control malaria—would have been lost as well. Preserving biodiversity, even a single species, can be economically and medically valuable.

it is less vulnerable to extinction, which helps to conserve biodiversity. In a genetically diverse species, in which individuals vary in their traits, there is a better chance that some individuals will be resistant to new diseases, more tolerant of changing temperatures, or better able to face other challenges of a changing environment. Species with large populations and a wide geographic distribution are the most genetically diverse. (See *Life Application: The Importance of Genetic Diversity.*)

Genetic diversity is critically important to us. For example, there are only two species of cultivated rice, but we have bred over 120,000 genetically distinct varieties (**FIGURE 12.4**). Some varieties are drought tolerant and others resist disease or have high yield, while others cook more easily or taste better. Because we have bred a lot of genetic diversity into rice, we can grow it under a variety of conditions and plant varieties to meet specific needs. We can also cross varieties to produce hybrid strains that combine the best features of their parents. Nondomesticated species derive the same benefits from genetic diversity. A species spread out over a large area adapts genetically to a variety of different conditions. Some of these different forms may be able to thrive even if conditions change.

The final component of biodiversity is **habitat diversity**. Our planet is home to many different habitats, including coral reefs, rain forests, prairies, marshes, and more. Since different species are adapted to live in different habitats, it makes sense that more types of habitats result in greater species

FIGURE 12.4 If a species is genetically diverse, it is less vulnerable to extinction. While there are only two species of cultivated rice, we have bred over 120,000 genetically distinct varieties.

habitat diversity :: the range of different types of natural homes or environments in which organisms live richness overall. The size of the habitat is important as well. A large forest supports more species than a small one. In fact, a single large forest supports more species than a whole collection of smaller ones, even if the total area is the same.

Biodiversity, then, includes species diversity, genetic diversity, and habitat diversity. Often when we speak of conserving biodiversity, we focus on species diversity alone. We are concerned with species that are on the endangered or threatened list, for example, or species that are close to extinction. Conservation requires that in addition to species diversity, we maintain genetic diversity and protect the diversity of the habitats where species live. Often a species is endangered in the first place because its habitat or genetic diversity has been reduced. Indeed, the small group of humans that moved out from Beringia expanded their numbers but left lower species diversity in their wake.

12.2 What Areas Have the Highest Biodiversity?

Regions of high geographic relief, such as mountains and their surrounding lowlands, often have higher biodiversity than flatter areas. That is because they provide many different types of habitat. A mountain may support a deciduous forest, or leaf-bearing trees, on its lower slopes, but conifers like pine trees higher up. It may become an alpine community above the treeline, where trees cannot grow. Protected valleys and exposed crags exist, along with streams, canyons, and caves. The windward side is likely to be moist, while the lee side is drier, and a grassland or desert may exist beyond the lee slope. These regions provide a diversity of habitats and physical conditions, and as a result many different species can live there.

The Latitudinal Gradient

As you travel north or south from the equator, however, a whole host of *other* physical features change. The temperature decreases, seasons become more pronounced, rainfall decreases, the length of daylight increases or decreases depending on the season, and the intensity of sunlight decreases, to name just a few changes. For a biologist, one of the most striking and consistent changes involves biodiversity, particularly species richness.

The number of species present in an area is highest near the equator and decreases as you travel north or south through the temperate zone into the polar regions. This pattern of species richness is known as the **latitudinal gradient** (**FIGURE 12.5**). The latitudinal gradient applies to almost every species. Compared to polar and temperate regions, tropical regions have more birds, more mammals, more snakes, more amphibians, more insects, more kinds of plants, more deadly disease pathogens—simply more of everything.

TABLE 12.1 compares species richness for several locations. Ecologists doing research in these areas collect data by counting the species they find. At this point, we have almost two centuries of data that confirm the latitudinal gradient. As Table 12.1 shows, in some cases, the tropics have over 100 times as many species as temperate and polar regions of the same size.

We have known about the latitudinal gradient for centuries. When European explorers began to sail across the globe, they brought back specimens

latitudinal gradient :: the decrease in the number of species as one moves away from the equator into the polar regions

Species-poor tundra

Species-rich jungle

Forest with moderate species richness

FIGURE 12.5 The number of species present in an area is highest near the equator and decreases as you travel north or south through the temperate zone into the polar regions. This pattern of species richness is known as the latitudinal gradient.

of the plants and animals they found. These specimens found their way into universities, museums, zoos, and eventually into the public's imagination.

Why the Gradient Exists

Scientists have studied the gradient intensely at least since Darwin's time. We do not yet have a definitive answer as to why the gradient exists. However, a number of hypotheses provide partial answers.

- *The climate hypothesis:* According to this hypothesis, the number of species declines away from the equator because the climate becomes colder, harsher, more variable, and less predictable. Of course, polar bears, walruses, and reindeer do just fine in the north. If these species can adapt, why not others? The answer appears to hinge not on how many species can *adapt* to the ecology of high-latitude environments, but rather on how many species can *fit*. This idea gives rise to the second hypothesis.

Table 12.1 :: Latitudinal Variety of Species

Location	Number of Bird Species
Greenland	56
New York	195
Guatemala	469
Panama	1100
Columbia	1395
Location	**Number of Bat Species**
United States	40
Belize	84
Costa Rica	103
Location	**Number of Tree Species (per 100 m²)**
Eastern deciduous forest	15
Appalachian cove forest	30
Amazonian rain forest	283
Atlantic tropical rain forest	450

FIGURE 12.6 By far, the most diverse places on Earth are coral reefs.

- *The productivity hypothesis:* The tropics are characterized by warm, relatively constant temperatures, and a lot of rainfall and sunlight. These conditions are ideal for plants. And because vegetation captures solar energy, this translates into energy and food to support herbivores— which in turn support several layers of predators and parasites in a food web. Because a rich variety of food is so abundant in the tropics, many species can specialize on particular food items, allowing multiple species to live together because they eat different foods. Higher latitudes may require a species to be more of a generalist, taking advantage of any food source to get enough energy to survive. Under these conditions, competition for food may eliminate many species.

- *The stability hypothesis:* Over millions of years, ice ages come and go, and many species adapted to one climate go extinct when conditions change. It may take a long time for the ecological space to fill up with new species. The stability hypothesis notes that climate in the tropics has been more stable, and so species are not subjected to dramatic changes that could affect their survival. As a result, there may have been time for more species to evolve there.

Thousands of years ago, as humans made their way out of Beringia moving from west to east across the Americas, they encountered varied local environments that supported different levels of biodiversity. But by far the most diverse places on Earth are coral reefs (**FIGURE 12.6**). Interestingly, coral reefs combine the features of mountainous and tropical regions. Reefs stretch from the ocean floor to just below the surface, spanning a complete range of living conditions from cool, dark, deep water to warm, light surface water. Reefs have a complex three-dimensional structure with many tunnels and crevices to hide in and vast surfaces to grow on. And coral reefs are found only in the tropics, so the hypotheses just mentioned may help explain their astonishing diversity.

By some estimates, over 80% of the world's species live in the highly diverse tropics, including rain forests and reefs. These are precisely the regions that are most threatened by human population growth, development, and climate change.

What Can Islands Tell Us About Biodiversity?

12.3

One thing we learn from islands is that larger islands support greater species richness than smaller ones. This pattern has been known for centuries, but it wasn't until 1967 that R. H. MacArthur and E. O. Wilson proposed an explanation for it.

MacArthur and Wilson's **theory of island biogeography** is based on two facts of island life. First, every species present on an island is an immigrant. Each species flew, swam, floated, rafted, was ferried by a bird, or was blown there by a storm. Second, populations frequently go extinct on islands. The populations start off precariously small, perhaps just a single pregnant female. As populations grow on an island, their isolation causes them to reach resource limitations more quickly than a population on a mainland. Competition for these limited resources may be more intense than what occurs on the mainland.

Immigration will increase the number of species on an island, and extinction will reduce it. Large islands and islands closer to the mainland will have a high immigration rate because they are bigger targets. They are more likely to be found, particularly by species that arrive on the island by accident. Larger islands also provide many resources and habitats, which reduces competition and leads to lower extinction rates. With relatively high immigration rates and low extinction rates, large islands support a greater number of species. These conditions are reversed for smaller (or more distant) islands, where low immigration rates and high extinction rates lead to fewer species.

One important result from the theory of island biogeography is that there is a relationship between the area of an island and the number of species it has (**FIGURE 12.7**). This **species-area relationship** has been investigated by

theory of island biogeography :: a proposition that holds that the number of species found on an island is determined by the rates of immigration and colonization by new species and extinction of established species

species-area relationship :: the relationship between the area of an island, or any isolated area, and the number of species it has

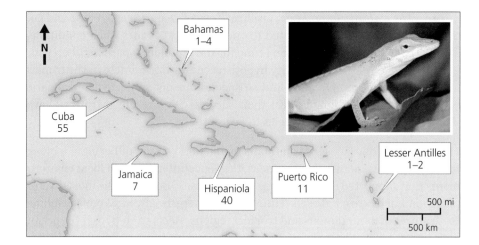

FIGURE 12.7 There is a relationship between the area of an island and the number of species it has. For example, there are about 138 species of *Anolis* lizards among the Caribbean Islands, and the number of species varies in proportion to island size.

FIGURE 12.8 Some ecologists have performed experiments to reveal the species-area relationship. A small island has many more species than a very small island, but a large island and a very large island have about the same number of species.

The hectare (ha) is a unit commonly used by ecologists; it equals 10,000 square meters.

many ecologists, generally by counting the number of species of bird, lizard, or butterfly found on islands of different size. But some ecologists have also performed experiments that show us that the species-area relationship takes the shape of a curve (**FIGURE 12.8**). A small island has many more species than a very small island, but a large island and a very large island have about the same number of species. (See *How Do We Know? Experimental Island Zoogeography* for an account of Wilson and Simberloff's test of the theory.)

How Do We Know?

Experimental Island Zoogeography

Designing a good experiment is challenging under any conditions. Still, working in a climate-controlled lab offers numerous advantages compared to fieldwork. When ecologists experiment outdoors, they must be prepared for inclement weather, nosy observers (people and animals), blood-sucking bugs, and sunburn. Despite these difficulties, ecological theories need to be tested under natural conditions.

In 1967, R. H. MacArthur and E. O. Wilson developed their theory of island biogeography. They proposed that islands

have an equilibrium number of species, determined by the balance between immigration and extinction of species on the island. Their theory has been tested mainly by observation: counting species on various islands.

In 1969, Wilson and Daniel Simberloff designed an experimental test. They studied insects on six mangrove islands off the Florida Keys. These islands are made up of a few to many trees and range from 10 to 50 meters in diameter. They occur within a kilometer of the shore in waist-deep shallows.

Ants, beetles, butterflies, wasps, and many other insects live there.

Simberloff and Wilson hypothesized that if the theory of island biogeography were valid, the number of insect species on each island would be relatively constant. One way to test this theory is to count the insect species on each island, remove them all, and allow them to return. If the numbers of species before and after removal are similar, the theory would be supported.

Counting the species on an island is challenging all by itself. Simberloff and Wilson spent many hours in the Florida sun climbing through mangrove roots and sifting through leaves and hollow twigs to make sure they found them all. Removing all the insect species was even more difficult. Working in the water, they erected scaffolding 20–30 feet high in a cube around the island. They covered the scaffolding with a giant canvas bag and pumped in methyl bromide, the gas used to fumigate pest-infested homes. Once all the insects were dead, Simberloff and Wilson resurveyed the islands regularly for an entire year. They recorded when a species immigrated, when (or if) it went extinct, and how many species existed on each island.

Their results showed that as the islands were recolonized, the number of species on each leveled off at equilibrium.

In many cases, the number of species before and after fumigation was similar, though some variation is always expected in the real world. Considered overall, the findings of the study supported the theory of island biogeography. Interestingly, the identity of species varied quite a bit. Simberloff and Wilson typically found a different set of species before and after, even though the numbers were similar.

Table :: Species Counts Before and After Fumigation

| Island Name | Number of Species | |
	Before Fumigation	After Fumigation
E1	20	11
E2	36	29
E3	22	27
ST2	25	23
E7	23	26
E9	43	29

Source: Daniel S. Simberloff and Edward O. Wilson. "Experimental Zoogeography of Islands: The Colonization of Empty Islands." *Ecology* 50 (1969):278–296.

1 Simberloff and Wilson hypothesized that the number of insect species on six mangrove islands would be relatively constant.

2 They spent many hours counting the insect species on each island.

3 They erected scaffolding 20–30 feet high in a cube around the island.

4 They covered the scaffolding with a giant canvas bag and pumped in methyl bromide.

5 Once all the insects were dead, they resurveyed the islands regularly for an entire year.

Their results showed that the number of species on each island leveled off at equilibrium, but the identity of species varied.

In 1969, Wilson and Simberloff studied insects on six mangrove islands off the Florida Keys. They hypothesized that if the theory of island biogeography were valid, the number of insects on each island would be relatively constant.

Interestingly, the theory of island biogeography applies to any habitat that is like an island ecologically. Any area of good habitat surrounded by a "sea" of uninhabitable area is an ecological island. A marsh and a patch of forest surrounded by houses are good examples. This is important for biodiversity studies because it helps us understand the consequences of human activity. When we cut into a forested area to put up houses, the forest becomes fragmented into "islands." The theory of island biogeography tells us that these small patches will not be able to hold all the species present in the original forest. It is likely that the extinctions that coincided with the migration of humans across the Americas were more prevalent in environments that were ecologically similar to small islands.

Another thing we learn from islands is that many of the species on them are **endemic**—found nowhere else in the world. The finches and giant tortoises on the Galapagos Islands off the coast of Ecuador and the honeycreepers of Hawaii are examples. In fact, over 75% of the native species of Hawaii are endemics, found *only* on the island chain. That makes tropical islands a major source of diversity. It also means that extinction on the island means complete extinction, with no chance for recolonization from the mainland. Human colonization of islands has in fact led to the extinction of many island species, either from human activity or from the domesticated animals (and pests) that humans bring with them.

endemic :: species that are found in only one place in the world

biogeographic realm :: one of eight landmasses of the world with its own distinctive floras and faunas

12.4 Why Do Different Regions Have Different Species?

Rain forests, deserts, grasslands, tundra, freshwater lakes, and coastal intertidal zones—these ecosystems are found all over the globe. Each resident species, or species living in an area, is adapted to the conditions that occur in its ecosystem. It is probably true that a cactus from the Sonoran Desert of the southwestern United States could grow successfully in environmental conditions found in the Kalahari Desert of Africa, or that a monkey living in the Amazon basin would similarly be at home in the type of environment present in the rain forests of Indonesia. But, surprisingly, despite the favorable environments, these species are not found in these other regions of the world (**FIGURE 12.9**).

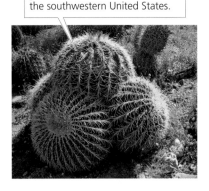

The barrel cactus (*Echinocactus grusonii*) is found in the deserts of the southwestern United States.

The African milk barrel (*Euphorbia horrida*) is from South Africa.

FIGURE 12.9 The barrel cactus and African milk barrel are not related, but they look similar and have developed similar adaptations to desert life, including thick, fleshy parts to store water and spines to deter herbivory. The fact that two species exist where one could suffice increases biodiversity.

Biogeographic Realms

Ecologists have identified eight **biogeographic realms** around the world that correspond roughly to continents (**FIGURE 12.10**). Each realm is home to a large set of species, and each set differs from that of other realms, though many species are found in more than one realm.

A good example of how biogeographic realms differ is found in Southeast Asia. It is difficult to imagine two islands more similar than New Guinea and Borneo. At 786,000 km², New Guinea is the world's second-largest island. Borneo, close behind at 743,000 km², is third. Borneo straddles the equator and is in the Indo-Malay realm, and New Guinea is just south of it and in the Australasia realm. The islands are only about 800 miles apart. Both islands are mountainous and covered with rain forest, and like all tropical areas, their biodiversity is high.

The species found on each island differ dramatically, however. Borneo is home to plants and animals typical of Southeast Asia: tigers, orangutans, rhinoceros, monkeys, starlings, and woodpeckers, for example. New Guinea has none of these. Instead New Guinea is home to honey suckers and cockatoos, and in place of monkeys it has tree kangaroos! Most of the mammals on New Guinea are marsupials that care for their young in pouches like kangaroos and koalas do. Marsupials are absent from Borneo. All told, then, the total number of species in the region is much greater than it would be if they shared many species.

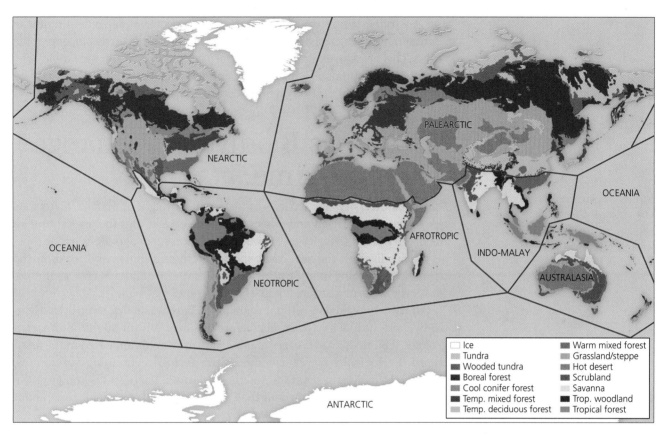

☐ Ice	▨ Warm mixed forest
▨ Tundra	▨ Grassland/steppe
■ Wooded tundra	■ Hot desert
■ Boreal forest	■ Scrubland
■ Cool conifer forest	☐ Savanna
■ Temp. mixed forest	■ Trop. woodland
■ Temp. deciduous forest	■ Tropical forest

FIGURE 12.10 Ecologists have identified eight biogeographic realms around the world that correspond roughly to continents. Each realm is home to a large set of species, which increases global biodiversity.

Wallace's Line

Alfred Russell Wallace was the first to document the differences in species between the two islands. Wallace was a naturalist who in the 1850s traveled throughout the Malay archipelago, collecting specimens for British museums. He kept a map of which specimens were found on which islands in the archipelago. Wallace recognized that his specimens fell into two groups, Asiatic and Australian.

The line between the Asiatic and Australian groups—now known as Wallace's line—runs between the large islands of Borneo and Sulawesi (**FIGURE 12.11**). It continues through the narrow, deep-water strait that separates the small islands of Bali and Lombok. The islands west of this line all share the Asiatic fauna. East of the line the islands share the Australian fauna. It is astonishing that Bali and Lombok, separated by only 20 miles of ocean, could have such completely different resident species. The explanation for these differences can be found in the geological history of the region. The depth of the straits separating Bali and Lombok have always been deep with fast-running currents, hindering the migration of organisms between the islands. Prior to the melting of the glaciers after the last ice age (13,000 years ago), Bali was connected to the Asian mainland. When the glaciers melted, sea levels rose, and shallow water separated Bali from the mainland. Wallace's line is actually the eastern margin of the former Asian continent.

It is not surprising that biogeographic realms tend to be associated with continents. We know now that the continents move or "drift" slowly around the planet. When continents collide or are connected by land bridges, animals and plants can move from one continent to another. But drifting continents can be separated for long periods of time, as was the case with South America

Wallace's Line

FIGURE 12.11 Alfred Russell Wallace was a naturalist who traveled throughout the Malay archipelago collecting specimens in the 1850s. He kept a map of which specimens were found on which islands and recognized that they fell into two groups: Asiatic and Australian; the line between the groups—now known as Wallace's line—runs between the large islands of Borneo and Sulawesi.

and Australia. When this happens, evolution occurs independently on the continent, resulting in a different set of species that make up its biogeographic realm. This helps to explain global patterns of biodiversity: for example, why Australian mammals are marsupials and why North American deserts are home to cactus, but deserts on other continents are not.

12.5 How Does Biodiversity Change Through Time?

Global biodiversity has varied dramatically throughout Earth's history as new species have arisen and others have gone extinct. But the pivotal time in the history of biodiversity was about 550 million years ago. Before that time, all species lived in the sea. Most species were unicellular and lived on the ocean floor, filtering bits of rotting organic matter from the sea for their food. Multicellular animals were small, soft, and slow.

All this changed 550 million years ago. There was a dramatic increase in biodiversity, unlike anything seen before or since. Within 20 million years, a geological blink of an eye, every major group of animals appeared in the fossil record. Ecologically complex communities formed containing vertebrates, arthropods, mollusks, sea stars, and their relatives. This event is known as the **Cambrian explosion**, named for the Cambrian period of geologic time, about 542–490 million years ago (**FIGURE 12.12**).

The cause of the Cambrian explosion is hotly debated. Some scientists argue that increasing oxygen levels allowed for larger and more active animals. Others argue that the explosion simply reflects the evolution of hard body parts, a new ecological "invention" that gave protection from predation. These hard parts fossilize better so we have more fossils of these organisms. Either way, the Cambrian period marks the beginning of global biodiversity.

Cambrian explosion :: the geological period of time from about 542–490 million years ago when every major group of animal appeared in the fossil record

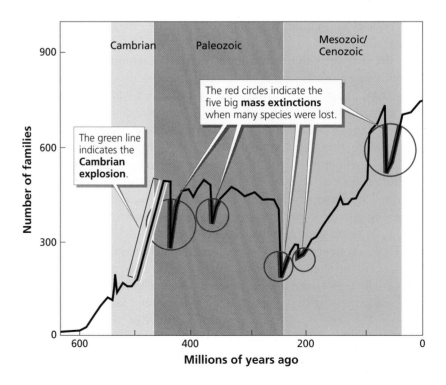

FIGURE 12.12 The Cambrian explosion occurred about 542–490 million years ago. During this explosion, every major group of animals appeared in the fossil record.

12.6

Why Is Biodiversity Needed for a Healthy Ecosystem?

Conservation efforts cost money and can be controversial. That is why it is important to understand the positive, even vital, reasons for preserving biodiversity. First, we look at the biological role of biodiversity, and then we return to the question of why biodiversity is important to us.

Productivity, Stability, and Ecosystem Health

Productivity and stability are two factors associated with ecosystem health. All living organisms require energy. Ultimately, an ecosystem's energy comes from sunlight, but sunlight must be converted into plant material before it can be used (Chapter 7). **Productivity** is an indicator of how effectively plants convert sunlight into food. Productivity is measured by the weight of plant material produced in the ecosystem. More plant material means higher productivity.

Another indicator of ecosystem health is **stability**. All ecosystems exist within a variable environment subject to disturbances: weather changes, drought, and fire are examples. Ecosystem stability refers to an ecosystem's ability to resist these disturbances and to return to its original configuration after them. Because biodiversity contributes to both productivity and stability, it is important to ecosystem health.

Why Biodiversity Increases Productivity

Ecologists have measured the productivity of grasslands, forests, farms, coral reefs, and many other ecosystems. One clear result that emerges from their work is that productivity is correlated with biodiversity. Biologically diverse grasslands produce more plant material than those that are less diverse, and this result holds for other types of ecosystems as well.

There are several reasons why biodiversity increases productivity. One reason is that a diverse collection of species can make more complete use of an ecosystem's resources. Ecosystems vary from place to place. Some areas are wetter, others have deeper soil or different concentrations of soil nutrients, some are shaded, and some are more protected from wind. A single species of plant is unlikely to grow equally well in all these conditions, and where it does grow, it may not use all the available resources. On the other hand, a diverse set of plant species, each adapted to different conditions, can make effective use of *all* the resources and mini-habitats available. More species in this case also means more individual plants, which results in higher productivity.

Some plants help other species to grow. For example, plants in the pea family (or legumes) work with bacteria in their roots to add nitrates to the soil (see Chapter 11). The presence of legumes thus helps species that require high levels of nitrogen. Another example occurs in ecosystems with harsh conditions, such as deserts. Under such conditions, it is hard for new seedlings to survive to adulthood. Woody shrub species provide shade, lower temperatures, and moisture so that seedlings of other species can germinate and grow. A more diverse ecosystem is likely to have more of these helper species, which allows even more species to occur, increasing biodiversity further.

productivity :: an indicator of how effectively plants convert sunlight into food; measured by the weight of plant material produced in the ecosystem

stability :: the ability of an ecosystem to resist environmental disturbances and return to its original configuration after them

Why Biodiversity Increases Stability

Biodiversity buffers an ecosystem from disruption, just as redundancies and backups protect computer systems. In a biologically diverse ecosystem, several different species can take on any given role. If disease or a parasite reduces the population size of one pollinator, for example, another species can step in. If some trees are killed by forest fire, other species, such as pitch pine and jack pine, have adaptations that allow their seeds to germinate and grow quickly.

In short, biodiversity helps ensure that at least some species will do well, whatever happens. This is the natural analogue to holding a diverse portfolio of stocks and bonds. An ecosystem with only a few types of species is a risky portfolio: if conditions become unfavorable, both may crash.

rivet hypothesis :: the idea that ecosystems are like airplanes with rivets; if too many species (or rivets) are lost, the ecosystem will crash quickly (just like an airplane)

Biodiversity promotes ecosystem health by increasing both productivity and stability. When biodiversity is reduced, the health of the ecosystem declines as well. In 1981, Paul and Anne Ehrlich described this process with the **rivet hypothesis**. Ecosystems are like airplanes, and species in the ecosystem are like the rivets that hold the plane together. An ecosystem can tolerate the loss of some species, just as an airplane can tolerate the loss of some rivets. However, if a particularly important species or too many species are lost, the ecosystem will crash quickly, just like a plane that loses too many rivets. Loss of biodiversity can have the same effect on an ecosystem.

How Biodiversity Keeps the Food Web Intact

Why is the world green? This seems like an odd question until you begin to think about it carefully. What this question really asks is why herbivores don't eat all the available vegetation (leaving us with a nongreen world) and then die out. One hypothesis is that plant defenses, like thorns and poisonous secondary compounds (Chapter 11), protect them from herbivores. Another hypothesis is that predators keep herbivores in check, so they don't eat all the plants.

Recently, John Terborgh devised an experiment to test these hypotheses. Terborgh worked in Lake Guri in central Venezuela. Lake Guri is an artificial lake, created in 1986 when a hydroelectric dam flooded 4300 square kilometers in Venezuela's Caroni Valley (**FIGURE 12.13**). What had previously been a valley floor with many rolling hills became a lake with hundreds of islands. Each island had its own set of species, depending on which species happened to be on the hill when the valley flooded. Terborgh and his colleagues monitored biodiversity on 14 sites in Lake Guri. Nine of these sites were free of predators, while five had a complete set of predators similar to that found on the mainland.

Between 1986 and 2002, Terborgh found that the predator-free islands had been virtually stripped of their vegetation. In contrast, the islands with predators had an ecosystem similar to the mainland.

Terborgh's results provide evidence that our world remains green because carnivores regulate the herbivore populations. More generally, biodiversity stabilizes the entire food web. In a diverse food web, if one species becomes too common, its predators will reduce the species' numbers. On the other hand, if a species becomes too rare, its predators can switch to other prey. A diverse food web provides a more diverse "menu," which helps keep population sizes in check.

An outcome of the species extinctions that occurred as humans moved into new habitats across the Americas may have been the collapse of the existing food web, which would have necessitated continued migration. Once gone, this food web would have been replaced with another, perhaps less biodiverse, one composed of different species.

FIGURE 12.13 Keystone species can have a disproportionate effect on an ecosystem's biodiversity. When the sea otter population declined in the 1800s due to hunting (A), the sea urchin population (B) increased and overgrazed primary producers, sending ripples throughout the food web.

Sometimes a single species can have a disproportionate effect on an ecosystem's biodiversity. These are called **keystone species**. An example occurred off the Pacific Northwest coast. Throughout the 1800s, sea otters were hunted for their pelts, the densest fur known among mammals, and at one point the worldwide population of sea otters was as low as 2000.

Freed from predation by sea otters, sea urchin populations increased drastically in size. The sea urchins overgrazed kelp and seaweed beds, sending the populations of these primary producers into decline. The loss of primary producers sent ripples all through the food web, resulting in decreased biodiversity (Chapter 11).

keystone species :: a single species that has a disproportionate effect on an ecosystem's biodiversity

12.7
Why Should We Preserve Biodiversity?

Why, then, should we preserve biodiversity? To help us answer, let's look at an actual case that occurred not that long ago. We shall see that it tells us something about the fate of the ecosystem, and something about our fate as well.

The Spotted Owl Controversy

In 1990, the northern spotted owl (**FIGURE 12.14**) was placed on the U.S. List of Endangered and Threatened Wildlife. At the time, there were an estimated 2000 pairs of spotted owls left. This move came about after years of negotiation and debate among environmental groups, timber companies, and the government.

The northern spotted owl lives exclusively in old-growth coniferous forests of the Pacific Northwest. These forests were over 300 years old, dominated by massive Douglas firs. Trees in the old-growth forest were worth about $4000 per acre, and logging accounted for about half of Oregon's economy and about a quarter of Washington's. To accommodate the spotted owl, over 10 million acres of old-growth forest had to be placed off-limits to logging. That's $40 billion of revenue lost in an economy that was already struggling. It is easy to imagine the controversy surrounding the spotted owl. Families, towns, and industries faced economic hardship to protect a bird most people had never seen.

However, framing the debate as "owls versus people" was a mistake that made it more difficult to understand the real issue. A more insightful way to frame the debate would have been the trade-off between the short-term and long-term benefits of using the forest's resources. Either way it is *people* that benefit from using the forest. It's simply a question of the timescale.

Long-Term Benefits of an Old-Growth Forest

The spotted owl is an **indicator species**. Its presence indicates a healthy, functioning old-growth forest. If we lose the spotted owl, it means we have lost the old-growth forest as well and most if not all of the species it contains. Other organisms may form a new ecosystem eventually, but it will not be as complex or as biodiverse as the old-growth forest it replaced.

The old-growth forests of the Pacific Northwest are a unique ecosystem (**FIGURE 12.15**): a temperate zone rain forest. These forests exhibit very high biodiversity, rivaling that of a tropical rain forest. As in a tropical rain forest, the upper branches of the canopy collect dust, dirt, and organic material and develop a layer of soil. This old-growth canopy supports many plants and animals, a hidden mini-ecosystem hundreds of feet above the forest floor. It takes about 300–1000 years for these forests to form a community of massive trees covered with life from top to bottom.

One valuable resource in these old-growth forests are the genes their species contain. For example, tamoxifen, a drug that was synthesized in the early 1960s and is used to treat breast cancer, originally comes from the Pacific yew tree, a resident of the old-growth forests. In addition to saving lives, tamoxifen provides economic benefit to pharmaceutical companies, accounting for over a billion dollars in sales. There is a good possibility that other useful products eventually can be discovered in this diverse ecosystem.

Many other benefits can be derived from old-growth forests. First, they hold soil well and prevent heavy rains from eroding soil into rivers and out to sea. In contrast, in areas where trees have been clear-cut, sediment from previously forested slopes clouds local streams and has a negative impact on salmon and other fisheries. Second, since water moves more slowly through forested soil, there is more time for the water to be filtered and cleansed of impurities, enhancing the supply of drinking water. Third, old-growth forests absorb a lot of carbon dioxide, helping to slow climate change. And fourth,

FIGURE 12.14 In 1990, the northern spotted owl was placed on the U.S. List of Endangered and Threatened Wildlife. At the time there were an estimated 2000 pairs of spotted owls left.

indicator species :: a species whose presence indicates a healthy, functioning ecosystem

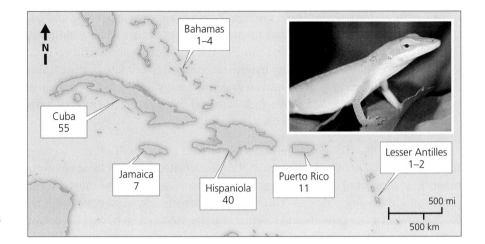

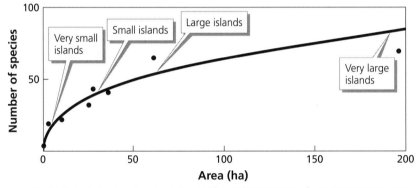

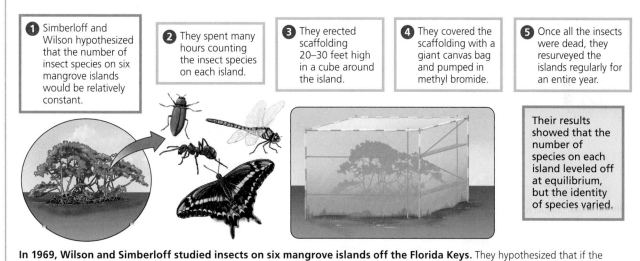

FIGURE 12.15 The old-growth forests of the Pacific Northwest are a unique ecosystem: a temperate zone rain forest. These forests exhibit high biodiversity, rivaling that of a tropical rain forest.

because they are unique and fascinating, old-growth forests support a strong tourism industry.

How Old-Growth Forests Provide Ecosystem Services

In Chapter 11, we introduced ecosystem services, the benefits we get from natural processes that occur in a healthy ecosystem. Let's take another look at these services and see how old-growth forests provide them.

- *Provisioning services:* Old-growth forests provide us with (some) lumber, and their genetic diversity has given us products like tamoxifen.
- *Regulating services*: Old-growth forests retain soil and purify water.
- *Cultural services:* Old-growth forests support a strong ecotourism industry and for many provide a spiritual connection with nature.
- *Supporting services:* Old-growth forests have tremendous productivity, create soil, and remove carbon dioxide from the atmosphere.

Let's return to the spotted owl controversy. In 1990, the rate of logging was so high that most of the old-growth forest would have been cut down in a decade or two, resulting in the same economic hardship. But the people would have been missing critical ecosystem services that the forest provides. The spotted owl legislation did not cause the economic problem; it only forced people to confront it earlier than they would have.

Choosing short-term benefits over long-term is like running up credit card debt. It appears to solve some immediate problems, but ultimately the bill must be paid. When the bill comes in the form of biodiversity loss and loss of ecosystem services, the people who run up the bill—and reap the profits—are not the ones who pay. These bills are paid by future generations. Another concern is that when biodiversity is lost and species go extinct, there is no way to bring them back. It may not be possible to "pay the bill."

The logging industry was likely aware of these issues and would have had plans to plant new, fast-growing trees to replace the ones they cut. This man-made ecosystem would have provided some ecosystem services. But it would have been much less mature and significantly less diverse. As a result, it would have been less productive and less stable, and it would have required our effort and money to maintain it. The most common consequence of our development projects is that we replace highly diverse natural ecosystems with artificial ones that require our care.

We began this section with a question: why preserve biodiversity? One answer is clear. We should preserve biodiversity to prevent long-term economic and ecological catastrophes. There is also an ethical or even spiritual side to preserving biodiversity. We are the only species with any concept of stewardship, and we are the only species capable of affecting biodiversity on a global scale. Those early humans did not understand the potential ecological impact of their actions, but we do. Given these circumstances, preserving biodiversity may be a moral obligation as well.

12.8   How Do We Keep Track of Biodiversity?

Scientists have identified about 1.3 million species. That's a lot, but every time a careful study is done of a new patch of rain forest, an inaccessible mountain peak, an isolated lake, or a deep sea community, many new species are found. Since scientists keep finding new species, it is clear that we have not come close to identifying them all. Using a variety of statistical methods, scientists estimate that our planet could be home to 30 million species or more, in which case we have identified only 4% of them. Most of the missing species are bacteria, other microscopic organisms, or animals like insects and worms. We have a fairly complete survey of mammals, birds, and plants.

The sheer number of species makes it almost impossible to characterize completely the biodiversity of any large area. Nevertheless, monitoring biodiversity allows us to understand how biodiversity changes in response to human activity and other causes.

Recognizing this need, scientists have devised a number of ways to measure and monitor biodiversity. The scale of the study area is important. Working with a smaller ecological community, scientists can be more thorough and more precise, while studying a large ecosystem requires making broader, and less precise, estimates. That includes the broadest ecosystem of all, the planet.

The Species Diversity Index

Recall that species richness refers to the diversity of species alive in an area. Many studies of biodiversity focus on a particular ecological community, such as a stream bed or an individual forest stand. For this type of study, two additional aspects of species diversity play a role. **Species abundance** refers to the number of individuals of each species present in the community.

species abundance :: the number of individuals of each species present in an ecological community

Community 1 has high species richness (four different species of flower) and even species abundance (two of each species).

Species 1: 25%
Species 2: 25%
Species 3: 25%
Species 4: 25%

Community 2 has high species richness (four different species of flower) but low species abundance (dominated by one species).

Species 1: 10%
Species 2: 10%
Species 3: 10%
Species 4: 70%

FIGURE 12.16 Species diversity is highest when species richness is high and species abundance is relatively even (A); if a community is dominated by just a few very abundant species with a smattering of rare ones (B), we would not consider that community to be diverse.

Species evenness refers to the pattern of species abundance. Are most species rare or moderately common, for example?

Species diversity is highest when species richness is high and species abundance is relatively even. If a community is dominated by just a few very abundant species with a smattering of rare ones, we would not consider that community to be diverse. An extreme case of such a community is a cornfield. There may be many species of weeds here and there, but for the most part it's just corn—not very diverse. On the other hand, a prairie dotted with many species of wildflowers would be more diverse, even if its species richness were the same as the cornfield (**FIGURE 12.16**).

Ecologists who study biodiversity at the community level use a species **diversity index** to measure diversity. This mathematical formula takes into account *richness*, *abundance*, and *evenness*. If a community's species diversity index is lower than expected, this indicates adverse conditions that are preventing species from living there. Pollution or drought could eliminate the more sensitive species, for example. Or an invasive species may be outcompeting native species whose populations are limited by the natural checks and balances of ecological interactions. One of the consequences of the expansion of the human population and its migration across the Americas, for example, was a decrease in the species diversity index.

species evenness :: the pattern of species abundance

diversity index :: a mathematical formula that measures species diversity by taking into account richness, abundance, and evenness

Indicator Species and Satellite Images

Given the worldwide scope of human activity, it is important to monitor changes in biodiversity on a global scale. It is not possible to use diversity indexes on a global scale because it is impossible to collect and count all the species in a large geographic region. Instead, ecologists and other scientists use more approximate techniques to monitor biodiversity.

One technique is the use of indicator species, like the spotted owl. Recall that the presence of an indicator species implies that the ecosystem is healthy and will contain its regular set of species. In this case, we need to count only the indicator species. Scientists have learned that birds and flowering plants work well as indicator species of biodiversity, and they are relatively easy to survey. Scientists contribute much of the information, but interested amateurs are important as well. For example, around Christmas every year tens of thousands of bird-watching volunteers spend the day counting every bird they see or hear. The Christmas bird count has been going on since 1900 in the United States, Canada, and 19 other countries. By knitting together published

Technology Connection

Satellite Imagery

How fast is the rain forest being cut down? Satellite imagery can show you photographs of the rain forest taken at different points in time. The deforested areas stand out clearly, and you can use a computer program to compute their area and determine the rate of deforestation.

The U.S. Landsat program has been photographing global land cover since 1972. Landsat 7, which was launched in 1999, has an Enhanced Thematic Mapper (ETM+), a sensor with seven cameras, each of which captures light, but only in a particular band of wavelengths, or "color." Bands 1, 2, and 3 are the familiar blue, green, and red wavelengths of visible light. The remaining four bands are in the infrared zone, longer wavelengths of light that humans cannot see directly. Each camera records an image in its own particular wavelength. The seven resulting images can be combined in various ways to form composite, or "false color," images of the location.

Landsat 7 orbits Earth at an altitude of 705 kilometers (438 miles) and at a speed of 7.5 kilometers per second (16,776 mph). One complete orbit takes 99 minutes. The cameras in the ETM+ photograph a swath of Earth 185 kilometers (115 miles) wide. Each pixel in the camera corresponds to 15 × 15 square meters on Earth. Landsat can distinguish roads, clear-cuts, and larger buildings but not smaller houses, cars, or people.

Landsat 7 orbits from pole to pole, and as it orbits, Earth rotates below it. During each orbit, the satellite hovers over a different region of the globe. Landsat 7 is therefore able to photograph every location on Earth. Because Earth rotates quickly, consecutive swaths are not geographically adjacent. But every spot on Earth is photographed every 16 days.

As Landsat 7 continues to gather information, another orbiting satellite has been added to the sky for additional data collection: Landsat 8 was launched February 11, 2013.

The U.S. Landsat program has been photographing global land cover since 1972. A comparison of pictures taken in Mato Grosso, Brazil, reveals a dramatic loss of rain forest in just 25 years.

reports and encouraging these types of programs, a global system for monitoring the biodiversity of indicator species may emerge.

Satellite images are also used to monitor biodiversity at the ecosystem and global scale (See *Technology Connection: Satellite Imagery*). Landsat and other satellite systems have collected almost 40 years of data about ground cover. By comparing photographs in temporal sequence, scientists can see how glaciers have retreated, how forests have been cut down while deserts have expanded, and how urban centers have grown. If scientists can estimate how many species lived in the original habitat, they can estimate how many species have been lost when the size of that habitat changed.

12.9 Why Might We Be Facing the Sixth Mass Extinction?

About 65 million years ago, a meteorite smashed into the ocean off the coast of Yucatan near what is now the town of Chicxulub. The meteorite measured at least 10 kilometers in radius and was traveling at about 20 kilometers

Table 12.2 :: The Big Five Mass Extinctions

Geologic Period	Millions of Years Ago	Percentage of Species Extinct
Ordovician	439	85 ± 3
Devonian	367	83 ± 4
Permian	245	95 ± 2
Triassic	208	80 ± 4
Cretaceous	65	76 ± 5

Source: D. J. Futuyma. *Evolutionary Biology,* 3rd ed. (Sunderland, Mass.: Sinauer Associates, 1998), Table 25.1, p. 713.

per second. It was 2 million times more powerful than the largest atomic bomb humans have ever exploded. It burned its way through the ocean and deep into the sea floor, sending gigantic clouds of steam into the air and 1000-foot-high tsunamis throughout the Atlantic. The heat of its passage sent a shock wave of superheated air that leveled and burned forests in North and South America. The seabed erupted into a mountain higher than Everest and then quickly collapsed. Millions of tons of ash formed a blanket throughout the atmosphere, cutting out light. Carbon dioxide released by the blast produced a greenhouse effect, and sulfur released into the atmosphere caused global acid rain.

The meteorite completely collapsed global ecosystem services, resulting in the extinction of more than 75% of the species on Earth, including the dinosaurs. This was the Cretaceous extinction, the most recent—and mildest— of the "big five" **mass extinctions (TABLE 12.2)**.

The largest mass extinction so far, the Permian extinction, occurred 245 million years ago. Over 95% of all species on Earth vanished. But even so, hundreds of thousands of species survived. All life on Earth today descends from them.

The biological world that develops following a mass extinction differs greatly from the one that preceded it. Life will continue on Earth no matter what we do. But that doesn't mean we would like it, or even survive, in that future world.

mass extinction :: a period when over 75% of Earth's species disappeared

The Blitzkrieg Hypothesis

Consider again the wave of human expansion into North America and its impact. Is it possible we could have hunted most of the large mammals on this new continent to extinction? Paul Martin thinks so. In 1973 Martin, then at the University of Arizona, wrote a paper whose title we borrowed for our case study, "The Discovery of America." Martin's position, that we are such a powerful force for extinction, has come to be known as the **blitzkrieg hypothesis**.

Donald Grayson at the University of Washington has argued against the hypothesis. Grayson cites limited evidence that humans actually hunted these large animals. Other hypotheses, too, could account for the extinction, such as effects of climate change after the ice age ended. He suggests that some scientists are using the blitzkrieg hypothesis to push a green political agenda that is not based on scientific evidence.

One point in favor of the blitzkrieg hypothesis is that the extinction of large mammals (and large birds) has occurred on *every* continent soon after humans arrived. South America, Europe, and Australia were affected almost as greatly as North America (**FIGURE 12.17**). Only Africa seems to have been immune, perhaps because the large mammals were more familiar with

blitzkrieg hypothesis :: a theory proposed by Paul Martin that humans hunted most of the large mammals on the North American continent to the point of extinction

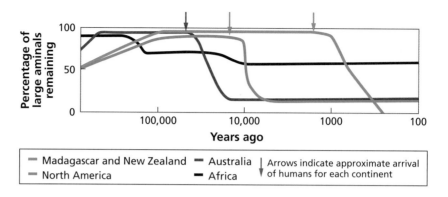

FIGURE 12.17 The blitzkrieg hypothesis proposed by Paul Martin holds that humans hunted most of the large mammals on the North American continent to the point of extinction. The extinction of large mammals (and large birds) has occurred on *every* continent soon after humans arrived, with the exception of Africa.

humans from having lived with them on the continent so long. Not all of these areas experienced climate change when humans arrived.

Humans may or may not have hunted the large mammals of North America to extinction, but there is no doubt that human activity has driven many species extinct—and threatens many more. We appear to be even more of a force for extinction than the meteorite that killed off the dinosaurs 65 million years ago.

Background Extinction

The fossil record shows us that extinction is a naturally occurring, ongoing process. Every species goes extinct sooner or later. The average life span of a species is between 1 million and 5 million years, though some species last much longer. Most of the species that have ever lived on Earth, around 99.9% by some estimates, are extinct.

The extinction rate is the percentage of species that go extinct in a given time period. This is a difficult value to measure. To compute a percentage of species, one needs to know the total number of species. Scientists are unsure of this number for living species. For species that lived in the past, the number is even more uncertain, since the fossil record has not revealed them all.

Paleontologists employ statistical techniques to make "best-guess" estimates of extinction rates. Marine invertebrates tend to last longer than terrestrial mammals, for example. Extinction rates also vary throughout time. A mass extinction is a period of very high extinction rate, whereas most of the time extinction rates are considerably lower. The **background extinction rate** is defined as the extinction rate during one of these "typical" times, not during a mass extinction. There is no exact value for the background extinction rate, but a commonly used estimate is one extinction per million species per year.

Current extinction rates are between 100 and 1000 times the typical rate of background extinction. This is in the same class as the "big five" extinctions in Earth's geologic history, and many have concluded that we are currently in the middle of the sixth mass extinction. According to the International Union for Conservation of Nature, 13% of all bird species are threatened with extinction. The threat is even more serious for other organisms: 25% of mammals, 40% of amphibians, 28% of reptiles, 27% of fishes, and 71% of flowering plant species are in danger of extinction. The causes of this ongoing mass extinction may not be as dramatic as a meteorite impact, but they are just as serious and inevitable. *If these extinction rates are accurate, 60% of all species alive today will be extinct within 100 years.*

background extinction rate :: the extinction rate during "typical" times, not during a mass extinction

Human Activity Threatens Biodiversity

Island species most vulnerable to extinction have already gone extinct. However, many of these extinctions happened long before our population began to increase significantly and before we developed technology. Initially, the population of humans that moved across the Americas was quite small. Yet their impact on biodiversity was dramatic. In the modern world, our burgeoning population requires space for farms, houses, and commercial ventures. We require resources and food that we harvest from the global ecosystem. We produce waste products and climate-changing gases. And piggybacking on our global commerce, many species are transported, intentionally or not, to new regions. All these activities have a negative impact on biodiversity.

Let's consider the ways that human activity causes biodiversity decline. In his book *The Diversity of Life*, E. O. Wilson compiled the percentages of endangered and threatened species from each activity (see *Scientist Spotlight: E. O. Wilson*):

Habitat destruction	73%
Displacement by invasive species	68%
Chemical pollutants	38%
Overharvesting	15%

Scientist Spotlight

E. O. Wilson (1929–)

What would a scientist be doing on a list that includes Oprah Winfrey, Jerry Seinfeld, Al Gore, Robert Redford, Sandra Day O'Connor, and Courtney Love? The list was *Time* magazine's 1995 list of the 25 most influential Americans. The scientist was Edward Osborne Wilson. Let's take a look at why Wilson made the list.

Wilson was born in 1929 in Birmingham, Alabama. He lived with his father and stepmother and moved around a lot as a child. Wilson loved nature and spent as much time as possible outdoors, hunting, fishing, and observing birds and insects. Insects, in particular ants, were to become his passion. As Wilson puts it, "Every boy goes through a bug stage. I just never grew out of it."

At age 13, Wilson discovered a colony of ants he'd never seen before down by the docks in Mobile, Alabama. This was the first colony of invasive, aggressive fire ants found in the United States. Wilson's first scientific paper, published in 1949 at the age of 20, documented the spread of fire ants throughout Alabama. He has studied ants ever since, publishing hundreds of papers and many books about them. His work has won recognition and many honors, including two Pulitzer Prizes, election to the National Academy of Sciences in 1969, and the National Medal of Science in 1976. Wilson earned his doctorate at Harvard and went on to join the Harvard faculty.

Impressive as these achievements are, they are not the reason Wilson was included on *Time*'s list of influential people. Building on a lifelong career in science, Wilson has become a philosopher, thinker, and writer in several areas.

In 1995, E. O. Wilson made *Time* magazine's list of the 25 most influential Americans. Wilson has been recognized as an important scientist and advocate of the conservation of biodiversity.

(Continued)

E. O. Wilson (1929–) (*Continued*)

He has also become passionate about the urgent need to preserve biodiversity. Wilson's work has taken him all over the globe and he has witnessed firsthand the effects of habitat destruction on biodiversity. He also understands the economic, cultural, and ethical benefits of preserving biodiversity. Wilson writes elegantly about the need to conserve and serves on many national and international organizations dedicated to preserving biodiversity.

In 2006, Wilson published *The Creation: An Appeal to Save Life on Earth*, a book framed as an open letter to the religious community. It is an appeal to put aside differences over evolution and the origin of life so that science and religion can work together on the urgent issue of preserving biodiversity. Even if motivations differ, he argues, the preservation of nature is both a scientific and moral goal. "Science and religion are two of the most potent forces on Earth and they should come together to save the creation." Despite the damage done so far, Wilson remains optimistic that biodiversity can be preserved.

Many species are threatened by more than one activity, so the total adds up to more than 100%. We will take a closer look at two of these: habitat destruction and overharvesting.

Habitat Destruction Habitat destruction is most obvious, and most harmful, in the deforestation of our planet's rain forests. Tropical forests account for about 7% of Earth's surface, but they are home to over 50% of Earth's species. Satellite images allow us to document the pattern of tropical deforestation (**FIGURE 12.18**). The process begins with roads into the forest, which provide access for logging operations. Selective logging occurs first, where trees of particular species and sizes are harvested for lumber. Villages grow up along the roads to support this industry, and the road system becomes more extensive. Next farmers and ranchers move in and clear-cut the forest for agricultural purposes. The villages grow into towns or small cities as the area is settled.

About 1.1% of the rain forests in Southeast Asia and about 0.7% of the rain forests in Africa and South America are cut each year. In the Amazon basin, this translates into 25,000 square kilometers—about the size of Massachusetts—and the rate of deforestation is increasing. Even low rates of logging can clear the entire forest in a surprisingly short period of time. Wherever it occurs, habitat destruction attacks all three components of biodiversity: species, genetic diversity, and habitat.

Overharvesting When individuals are taken from a population faster than they can grow back, this is known as overharvesting. It results in a negative

FIGURE 12.18 Satellite images are used to document the pattern of tropical deforestation. For instance, by comparing photographs in temporal sequence of Rondonia, Brazil, scientists can see how rain forests have been cut down.

June 1975—a single road is cut into the rainforest to provide access to loggers.

June 1992—side roads spring up in a "herringbone" pattern, along which selective logging occurs.

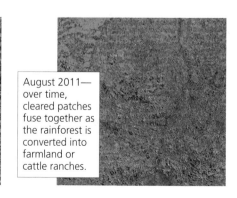

August 2011—over time, cleared patches fuse together as the rainforest is converted into farmland or cattle ranches.

growth rate that if continued will drive the population extinct. While habitat destruction targets all resident species, overharvesting targets particular species that are valuable to us. But as we have seen, removing even one species can destabilize an ecosystem and reduce the services it delivers, particularly if it is a keystone species.

Marine ecosystems have experienced a good deal of overharvesting. Whales, seals, sea otters, many types of fish, and crustaceans like crab and lobster have all experienced population declines due to overharvesting.

In marine ecosystems, it is not easy to monitor population size or to determine whether overharvesting is occurring in the early stages. The first symptoms may be a population crash, followed by the reduction in the catch and possibly the collapse of the harvesting industry. While many species are threatened by overharvesting, conservation programs have brought many of them back from the brink of extinction. Learning more about the population size, growth rates, and demography of the marine species we harvest will promote the development of more sustainable harvesting procedures.

12.10 How Can We Preserve Biodiversity?

There is no question that biodiversity is threatened on a global scale. The threat is a cause for concern, but not for pessimism. Preserving biodiversity is not easy or cheap, and it requires a number of changes in the way we live and work. But it can be done.

Become educated about the long-term benefits. Biodiversity enhances the productivity and stability of ecosystems. This in turn allows the ecosystem to provide more reliable services.

Develop economic policies that take advantage of long-term biodiversity benefits. Costa Rica has set aside over one-quarter of its land area for national parks and biological preserves. As a result, ecotourism in Costa Rica is a multibillion-dollar industry and has become one of the country's best performing economic sectors. Another way to capitalize on biodiversity is to "mine" diverse areas for new products, including medicines and genes.

Develop a cooperative approach. We have focused largely on the scientific issues surrounding biodiversity, but maintaining biodiversity requires the cooperation of scientists, politicians, economists, conservation organizations, organized religion, and of course informed citizens. International cooperation is required as well. Most of the regions of highest biodiversity are in developing countries. These countries will most likely require help if they are to forgo short-term exploitation of their biodiversity resources.

Curtail human population growth. As we will discuss in Chapter 13, global population growth shows evidence of slowing down, but it will still take decades before growth levels off. During this time, we will continue to require more space, more food, and more resources. And biodiversity will continue to decline as a result. It is clearly not possible to halt population growth quickly, but increased awareness of the need to reduce population growth can give us a long-term goal to work toward.

Design effective methods of conservation. Island biogeography has helped us understand the relationship between area and biodiversity. Continued research on particular species helps us understand their ecological needs.

This research can help us design parks and preserves more effectively and ensure that our conservation efforts help the species we are most concerned about.

None of these steps will be easy. But awareness of the benefits of biodiversity, and the costs of losing it, is becoming more widespread. Many governments, organizations, and individuals are working to curb biodiversity loss. There are reasons for cautious optimism, but we have a lot of work ahead of us.

 # Biology in Perspective

How is the human race like a meteorite? Like a meteorite, our activities are global in scale, and we have the ability to drive extinct a large percentage of Earth's species. Education about biodiversity and its benefits is the first step in developing approaches to preserve biodiversity.

Biodiversity comes from species diversity, genetic diversity, and habitat diversity. Biodiversity is highest in the tropics and in regions of high geographic relief. It has varied dramatically over time but has generally increased throughout Earth's history despite five episodes of mass extinction.

Like all species, we use the resources available to us in our ecosystem for our own benefit. With our intelligence and technology, we are extremely good at exploiting our habitat, often at the expense of other species.

Preserving biodiversity is critical to our own survival. We depend on the services that our ecosystem provides, and the ability of an ecosystem to deliver these services depends on its biodiversity. Life itself also has intrinsic value, and some argue that this is a sufficient reason to preserve species. Preserving biodiversity will not be easy and will require a number of political, economic, and social changes. In particular, preservation requires recognizing the long-term benefit of a biologically diverse ecosystem.

Chapter Summary

12.1 What Are the Components of Biodiversity?

Distinguish between species diversity, genetic diversity, and habitat diversity and describe how they relate to biodiversity.

- Biodiversity of a habitat depends on species richness (diversity of species alive in an area), genetic diversity (degree of genetic difference among individuals of one species), and habitat diversity (variety of types and sizes of habitats, each with different species).
- Conservation requires that we maintain all these factors.

12.2 What Areas Have the Highest Biodiversity?

Explain the relationships between geographic relief, latitude, and biodiversity and the reasons why the latitudinal gradient exists.

- Biodiversity depends on location.
- The latitudinal gradient is the pattern whereby species richness is highest at the equator and decreases as you move away; there are three hypotheses for why it exists: climate, productivity, and stability.

12.3 What Can Islands Tell Us About Biodiversity?

Describe the species-area relationship and relate it to the theory of island biogeography.

- Larger islands support greater species richness; the theory of island biogeography explains this:
  - Immigration increases the number of species on an island, while extinctions reduce them.

o Larger islands have higher immigration and more resources, while smaller or more distant islands have fewer immigrants and fewer resources.
- Species on islands are often endemic.

12.4 Why Do Different Regions Have Different Species?

Explain the relationship between Wallace's line and the distribution of species.

- Ecologists have identified eight biogeographic realms globally, roughly corresponding to the continents.
- The naturalist Alfred Russell Wallace described "Wallace's line" dividing neighboring South Pacific islands into either the Asiatic or Australian realm; continental movement explains why each realm has unique species.

12.5 How Does Biodiversity Change Through Time?

Describe, in general, the variation in global biodiversity over the course of Earth's history.

- Over Earth's history, new species arise and others go extinct; numbers vary dramatically.
- The Cambrian explosion, 550 million years ago, was a dramatic increase in species diversity, from unicellular or soft, simple multicellular organisms to every major group here on Earth now.

12.6 Why Is Biodiversity Needed for a Healthy Ecosystem?

Detail how biodiversity enhances the productivity and stability of an ecosystem.

- The overall health of an ecosystem depends on productivity and stability.
- The rivet hypothesis compares a healthy ecosystem with many species to an airplane with many rivets: a few rivets can be knocked out and the airplane can still fly, but if too many are lost, the plane will crash.
- If removal of a single species from an ecosystem causes its collapse, that species is the keystone species.

12.7 Why Should We Preserve Biodiversity?

Evaluate the importance of old-growth forests for providing ecosystem services.

- Northern spotted owls are an indicator species.
- In 1990, spotted owls were declared endangered, and over 10 million acres of old-growth forest were put off-limits for logging.

- Old-growth forests are a unique, highly diverse, and productive ecosystem that provide a number of ecosystem services, including provisioning in the form of lumber and medicines; regulating in the form of soil retention and purification; cultural in the form of ecotourism; and supporting in the form of carbon dioxide removal from the atmosphere.

12.8 How Do We Keep Track of Biodiversity?

Describe how biodiversity is monitored.

- The species diversity index allows us to monitor biodiversity.
- High richness and relatively even abundance equals the highest biodiversity.
- An unexpectedly low diversity index suggests adverse conditions.
- Monitoring indicator species or observing changes in ground cover from satellites are ways to monitor biodiversity globally over time.

12.9 Why Might We Be Facing the Sixth Mass Extinction?

Compare the background and current rates of extinction and explain how habitat destruction and overharvesting impact biodiversity.

- Over the history of life on Earth, there have been five mass extinction events, in which 75–95% of species living at the time disappeared.
- The blitzkrieg hypothesis holds that humans are a powerful force for extinction.
- Aside from mass extinctions, most species last between 1 million and 5 million years; the current extinction rate is 100–1000 times "normal."
- Habitat destruction by humans is responsible for about 73% of threatened and endangered species.
- Overharvesting is when humans remove individuals from populations faster than they can be replaced.

12.10 How Can We Preserve Biodiversity?

Examine the steps that can be taken to preserve biodiversity.

- Educate people about long-term benefits of stable ecosystems.
- Enact economic policies that take advantage of those long-term benefits.
- Develop a cooperative approach among scientists, politicians, economists, conservation and religious organizations, and informed citizens.
- Reduce human population growth.
- Develop effective methods of conservation.

Key Terms

background extinction rate 364
biodiversity 342
biogeographic realm 352
blitzkrieg hypothesis 363
Cambrian explosion 354
diversity index 361
endemic 352

genetic diversity 344
habitat diversity 345
indicator species 358
keystone species 357
latitudinal gradient 346
mass extinction 363
productivity 355

rivet hypothesis 356
species abundance 360
species evenness 361
species-area relationship 349
species richness 355
stability 355
theory of island biogeography 349

Review Questions

1. What is the "blitzkrieg hypothesis"? List one alternative hypothesis that could account for the same phenomenon.

2. Define the following terms: species richness, species abundance, and species evenness.

3. What are the three main components of biodiversity?

    a. Species diversity, species productivity, and species stability

    b. Species diversity, genetic diversity, and habitat diversity

    c. Species diversity, species-area relationship, and species evenness

    d. Genetic diversity, species abundance, and species evenness

    e. Species richness, species abundance, and species evenness

4. What is a biogeographic realm? List eight of them. Why do biogeographic realms increase biodiversity at the global level?

5. What was the Cambrian explosion, and approximately when did it occur?

6. What was the Permian extinction, and approximately when did it occur?

7. What three factors does a species diversity index take into account?

    a. Species abundance, species stability, and species richness

    b. Species productivity, species stability, and species richness

    c. Species richness, species diversity, and species productivity

    d. Species richness, species abundance, and species evenness

    e. Species evenness, species abundance, and species stability

8. How does an indicator species simplify the task of monitoring biodiversity?

9. What have ecologists learned about the relationship between biodiversity and productivity of an ecosystem?

10. What have ecologists learned about the relationship between biodiversity and the stability of an ecosystem?

11. Briefly describe one line of evidence that supports the hypothesis that predators prevent herbivores from eating all the plants, rather than the plants' defense systems keeping herbivores at bay.

12. What is one possible effect of removing a keystone species from an ecosystem?

    a. Increased species richness

    b. Increased biodiversity

    c. Decreased biodiversity

    d. Increased species stability

    e. Increased species evenness

13. What are the four main human activities that cause a loss of biodiversity?

    a. Preservation, habitat destruction, overharvesting, and use of chemical pollutants

    b. Conservation, preservation, use of chemical pollutants, and overharvesting

    c. Overharvesting, conservation, use of chemical pollutants, and invasive species displacement

    d. Habitat destruction, overharvesting, conservation, and use of chemical pollutants

    e. Habitat destruction, invasive species displacement, use of chemical pollutants, and overharvesting

The Thinking Citizen

1. How can you be sure that a species has gone extinct? This question refers to species that have been alive within human memory, not long-dead species like dinosaurs. What are some of the difficulties involved with assigning a cause to the extinction?

2. Consider the three hypotheses about why species diversity is high in the tropics. Probably all three play a role. But which of the three do you think offers the most compelling argument? Briefly explain your thinking.

3. The theory of island biogeography states that the number of species on an island is in equilibrium, reflecting the difference between the number of immigrant species and the number that have gone extinct. In the text, we point out that extinction rates are higher on smaller islands, which shifts the equilibrium toward fewer species. How does distance from the mainland affect an island's immigration rate? Consider a series of islands with the same area but at different distances from the mainland. How would distance affect the equilibrium number of species?

4. Many people have argued that losing one species here or one species there will not have a significant effect on anything important to humans. Briefly explain the Ehrlichs' "rivet theory." Does the rivet theory support this argument, refute it, or possibly support and refute it depending on circumstances? Explain your reasoning.

The Informed Citizen

1. Along with the spotted owl, the snail darter (a small freshwater fish) is the poster child for the conflict that frequently arises between conservation and development. Research the snail darter case, which began in 1973 and involved the Tellico Dam in Tennessee. What was the controversy? What were the arguments on both sides? How was the case ultimately decided? If you were in a position to make decisions about cases like this one, what set of guidelines would you use? Briefly explain your reasoning.

2. One of the biggest difficulties in arguing to preserve biodiversity is the trade-off between current and future benefits. A Brazilian farmer could choose to cut down his portion of rain forest now and use the land as a farm to feed his family. Or he could choose to preserve the rain forest because at some unspecified time in the future, a scientist may discover a plant that contains the next wonder cure for a disease. In most cases, the decision to farm would not be a difficult one. What are your suggestions for ways to preserve biodiversity in cases like these?

3. Another factor in the spotted owl controversy involved the barred owl. Over the past 30 years or so, the barred owl has expanded its range into the old-growth forests of the Pacific Northwest. The barred owl is larger and more aggressive than the spotted owl, and as its range expands, the spotted owl population continues to decline. If the barred owl drives the spotted owl extinct, did we waste millions of dollars to protect it? Were the lumber companies right in wanting to cut timber for profit and to provide jobs? Should we feel differently about this type of extinction than we do about extinction due to habitat destruction?

CHAPTER LEARNING OBJECTIVES

After reading this chapter, you should be able to answer the following questions:

13.1 How Can Populations Grow So Fast?

Distinguish between linear and exponential growth and explain how the growth rate is determined.

13.2 Why Don't Populations Grow Forever?

Describe the effects of population density and carrying capacity on the principle of logistic growth.

13.3 How Is Population Growth Influenced by Age and Sex?

Explain what age pyramids can tell us about a country's conditions and social challenges.

13.4 Why Do Developing and Developed Countries Grow Differently?

Compare and contrast the factors that contribute to growth in developing and developed countries.

13.5 How Do We Use Information About Population Growth?

Demonstrate how data on population growth can be used to plan for population shifts and resource depletion.

The continued growth of the human population is not sustainable. Fortunately, because we are capable of recognizing the problem, we are in a better position to find ways of solving it.

CHAPTER 13

Human Population Growth

How Many People Can a Single Planet Hold?

Becoming an Informed Citizen . . .

Our ability to reproduce successfully is programmed in our genes, the result of millions of years of evolution. However, the continued growth of the human population is not sustainable. To make the tough choices ahead, we need to first understand the principles behind population growth.

ne thing that all species are good at is reproduction. If they weren't, they would be extinct. Our ability to reproduce successfully is the result of millions of years of evolution. But will this ability be a source of our downfall? What happens if we reproduce without limit?

As a population gets larger, its habitat fills up, and overcrowding leads to death by starvation, disease, and predation. Advances in agriculture, medicine, engineering, and technology have supported healthy population growth for most of our history. However, the continued growth of the human population is not sustainable. As we become more numerous, we also deplete resources and produce more waste. Overcrowding can lead to political and economic instability as well.

We have an advantage over other animal populations: we can recognize the problem and discover ways to solve it. But we can do this only if we understand the principles behind population growth. In this chapter, we learn what affects population growth, and we compare patterns of growth in different regions of the globe. We begin by showing how fast and dramatic population growth can be—and how easy it is to underestimate.

case study

A Story About Bacteria

Bacteria have the simplest form of population growth possible. A bacterial cell divides in two, each of these divides to produce a total of four cells, then eight, and so forth. But even this simple form of growth can cause trouble. Let's consider the problems faced by one group of fast-dividing bacteria. Although we take some liberties with reality, the principles and consequences of our story are very real.

This particular species of bacteria divides once every minute (much faster than actual bacteria) and lives in test tubes, eating the culture medium in the tubes. If one bacterial cell is placed in a test tube at 11:00 a.m., at 11:01 there will be two cells, at 11:02 four cells, and this growth will continue until the tube is completely full and all the food is gone. This particular tube contained just enough culture medium so that the tube would be full, and the food all gone, at noon. When will the tube be *half* full?

Let's pretend these fictional bacteria, unlike real bacteria, are intelligent and capable of conscious thought and planning. They are fully aware of their finite, test tube world and their 12:00 deadline. They know they have to make plans for their long-term survival, but in the minutes just after 11:00 there are only a few bacteria and lots of food and room to go around. They decide to enjoy their current bounty and postpone making plans until their tube is half full.

As you probably expect, this is a bad decision. Their test tube becomes half full at 11:59, and because the bacterial population doubles every minute, it will take just one minute to be completely full. This gives the bacteria only one minute to plan for the complete ecological collapse of their habitat. Recognizing that they don't have enough time, intrepid bacterial explorers leave their tube and discover a remarkable resource: 15 more test tubes, each the size of their current home. Here's your second question: how long will it take

Starting with one cell, it took the imaginary bacteria from the case study 59 minutes to fill half of the first test tube, but only 5 more minutes to fill that tube and 15 more.

the bacteria to fill up these 15 tubes (assuming transportation to tubes is not a problem)?

At noon, the bacteria have 16 tubes (the original full tube plus the 15 new ones). You might think the extra tubes are a reprieve for the overcrowded bacteria, but they don't help much. If half of the bacteria in the first tube move to a second tube, both will be filled by 12:01. Then if half of the bacteria move from each of these tubes to two additional tubes, all four tubes will be filled by 12:02. Continuing this process, the bacteria will fill eight tubes by 12:03, and one minute later—12:04—they will have filled all 16 tubes. Starting with one cell, it took the bacteria 59 minutes to fill half of the first test tube, but only five more minutes to fill that tube and 15 more (**FIGURE 13.1**).

The parallels between us and our imaginary bacteria are clear. We, too, live in a finite world. We, too, need to be concerned about the finite nature of our resources and the Earth's ability to absorb waste and pollution. Underestimating population growth and how quickly we will use up natural resources leaves us vulnerable to the same mistake the bacteria made: not leaving enough time to prepare for its consequences. Understanding population growth requires a different type of reasoning than we use in our daily lives—the type of reasoning we explore in this chapter.

FIGURE 13.1 Even though bacteria have the simplest form of population growth possible, it is possible to draw parallels between their growth and human population growth. If we underestimate how quickly we will use up natural resources, we run the risk of not leaving enough time to prepare for its consequences.

13.1 How Can Populations Grow So Fast?

We begin by considering the pattern of human population growth. **FIGURE 13.2** shows our growth between 1000 AD and 1500 AD. If you lived in 1500, how would you project population growth throughout the rest of the millennium? Based on the data, you might decide to draw a straight line starting at 1500 AD and continuing through to 2010 AD. This is a logical approach. You would be assuming that the second 500 years would be like the first. Mathematically, you are assuming **linear growth**.

linear growth :: growth that proceeds at the same rate over a given time frame

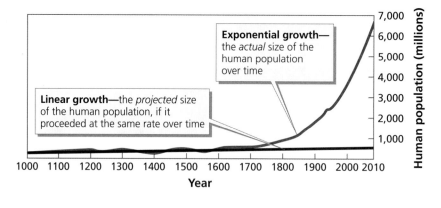

FIGURE 13.2 Linear growth proceeds at the same rate over time, while exponential growth proceeds at an increasing rate. Exponential growth is a fundamental property of all populations, including the human population.

exponential growth :: growth that proceeds at an increasing rate over a given time frame

compounding :: a principle that underlies exponential growth; occurs if new entities are themselves capable of creating new entities

growth rate :: how fast a population grows

The Difference Between Linear and Exponential Growth

But populations don't experience linear growth. The curve line in **FIGURE 13.2** shows that during the second half of the millennium, population growth actually followed a rapidly rising curve. This is known as **exponential growth**, and it is a fundamental property of all populations. Note that by the year 2000, linear growth predicts 500 million people, but the actual figure is 6 billion—the linear projection is off by 5.5 billion people. Starting in the year 1000, when there were fewer than a half billion people on Earth, it took us 960 years to reach 3 billion people, but only 40 more years to double to 6 billion.

The principle that underlies exponential growth is **compounding**. In any growth process, new entities are added to the population. Compounding occurs if the new entities are themselves capable of making still more new entities. Any compounding growth process can grow exponentially.

Without compounding, new entities contribute to growth just by being there, but they do not themselves make more new entities. For example, trees in the temperate zone add one new ring each year, but the rings themselves don't make new rings; the growth in the number of rings is linear. That is why the age of a tree equals the number of rings it has (**FIGURE 13.3**).

On the other hand, when a tree makes branches, each new branch can make several new branches. This is an example of compounding, and the number of branches on a tree grows exponentially (**FIGURE 13.4**). As we saw for bacteria, cell division is a compounding process, too, since each single cell can make two new cells.

Every biological population grows by adding new individuals. Since these new individuals can reproduce and make even more new individuals, this is a compound growth process as well. All biological populations therefore have the potential to grow exponentially.

Defining Growth Rate

Two questions are important for analyzing population growth. How fast does it grow? How big will it get? We'll consider the first question here and address the second in *How Do We Know? Modeling Population Growth*, p. 378.

A population grows when new members are added. Let's consider a small village of 100 people. Suppose five babies are born in the village and two people die during a particular year. Our village of 100 people grew larger by three people. This allows us to determine the **growth rate**: 3/100, or 3%.

FIGURE 13.3 Trees in the temperate zone add one new ring each year, but the rings themselves don't make new rings; the growth in the number of rings is linear. That is why the age of a tree equals the number of rings it has.

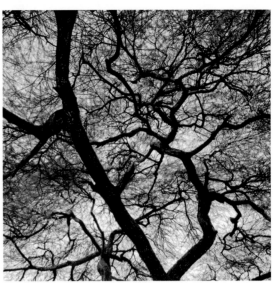

FIGURE 13.4 When a tree makes branches, each new branch can make several new ones. This is an example of compounding, and the number of branches on a tree grows exponentially.

The growth rate tells us how fast a population grows. It also tells us how many people are added to the population during a year (or other appropriate period of time). To figure out how many people are added, just multiply the population size by its growth rate. In this example, a population of 100 growing at 3% adds three people: $0.03 \times 100 = 3$.

These calculations illustrate the most important property of exponential growth: *the number of people added to the population is proportional to the number already present.* This means that when a population is small, few people are added and the population grows slowly. However, as the population gets larger, the number of people added each year increases as well. Population growth accelerates, starting off slowly but eventually increasing dramatically. When you graph exponential population growth over time, you get a characteristic *j*-shaped curve. Human population growth so far appears to follow an exponential curve, as do many animal populations in nature (**FIGURE 13.5**).

FIGURE 13.5 Just as human population growth so far appears to follow an exponential curve, so do many other animal populations in nature, including that of elephants. The graph is based on data from a 1900 to 1967 study of elephants conducted at South Africa's Kruger National Park.

Modeling Population Growth

Throughout this chapter, you will read statements like "The world's population will reach 9 billion by 2050." How do we make projections like this?

The first step is to figure out how fast a population is growing. We've already seen how to do this with exponential growth. The number added to the population is the growth rate multiplied by the population size. Mathematically we write it this way:

$$\text{number added} = rN$$

where r is the growth rate and N is the population size. To get the new population size, we make N bigger by the number of people that were added.

But this tells us only how much the population grew in a single year. One way to figure out how big the population will be in 50 years is to determine how many people are added in year 1, year 2, year 3, and so on for 50 years and then add them all together.

But this involves a lot of work, so mathematicians have devised a way to do all these additions in one step. To determine how big a population will be at some time in the future, you can use this equation:

$$\text{size} = N_0 e^{rt}$$

where N_0 is the starting population size, r is growth rate, and t is the number of years in the future you're considering; e is a mathematical constant, like π. Raising e to a power is called the *exponential function*, which is where *exponential growth* gets its name. You probably have a button labeled

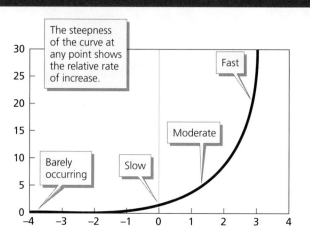

To predict population size, scientists calculate growth using *e*, the exponential function.

"e^x" on your calculator, which enables you to calculate this equation.

Let's work through an example. Suppose a population starts with 1000 individuals (N_0) and grows at 5% per year ($r = 0.05$). How big will the population be in 50 years (t)? Multiply r by t, raise e to this power, and multiply by N_0. The answer you get is 12,182 individuals, a 12-fold increase in 50 years.

This equation represents the exponential model of population growth, which works well for the early stages of a population's growth. But there are many other models that take into account factors that are missing in the exponential model. What happens when the population grows too large and starts to run out of food? How does a population's age structure affect its growth? Throughout this chapter, we will describe how these factors affect population growth.

The shape of the exponential growth curve helps to explain why we often fail to plan correctly for processes that grow exponentially. The early phase of slow growth lulls us into a false sense of security. Then we are caught unawares by the dramatic increase at the end.

Determining Growth Rate

When we speak of growth rate we really mean net growth rate. As we saw earlier, the number of babies born in a population doesn't determine growth rate by itself. You need to subtract the number of deaths in a population to arrive at the net change. We can also clarify the difference between numbers and rates. Rate refers to per capita (or per person) changes. If the *number* of individuals added to a population of 500 is 20, then the *rate* of change is 20/500: 0.04, or 4%.

In a population, birth rate is the number of babies born per capita, and death rate is the number of deaths per capita. The (net) growth rate is the birth rate minus the death rate. Consider our earlier example:

$$\text{growth rate} = (\text{birth rate} - \text{death rate})$$
$$= (5/100 - 2/100)$$
$$= 0.03$$

This value is sometimes referred to as the natural growth rate, since it is based on natural processes like birth and death. When ecologists study the population growth of animals and plants they often use this natural growth rate. For human populations, however, we need to introduce another consideration. Since we are a globally mobile species, migration can also affect a country's population growth. Like birth and death rates, migration breaks down into two components: immigration (those moving in) and emigration (those leaving). Like the birth rate, immigration adds to a population, and like the death rate, emigration subtracts from it. If we take migration into account, we get our final formula for the growth rate:

$$\text{growth rate} = (\text{birth rate} - \text{death rate}) + (\text{immigration rate} - \text{emigration rate})$$

TABLE 13.1 lists these rates for the United States during the period 2010–2012. The country's growth was under 1%, and you can see how the growth rate was determined by the other four rates. When working with human populations, these rates are not expressed as percentages or fractions. As Table 13.1 shows, these values are expressed as number per 1000 individuals in a population (number per 100 would be a percentage). Although percentages are more familiar, we will adopt this standard for the remainder of the chapter and report rates as number per 1000.

Equilibrium

You can think of growth rate as having two positive components (birth and immigration) and two negative components (death and emigration). The growth rate is the "balance" among these four components. If the positive components outweigh the negative, the growth rate is positive and the population grows exponentially. However, if the negative components outweigh the positive, population growth is negative and the population gets smaller.

Table 13.1 :: U.S. Growth Rate Components: April 1, 2010–July 1, 2012

	Value	
	Number per 1000	**Per Capita Rate**
Birth rate	14.4	0.0144
Death rate	9.1	0.0091
Immigration rate	3.0	0.0030
Emigration rate	0	0
Growth rate	8.3	0.0083
Population size	308,745,538 (April 1, 2010)	

Source: United States Census Bureau, http://www.census.gov/popest/data/state/totals/2012/index.html.

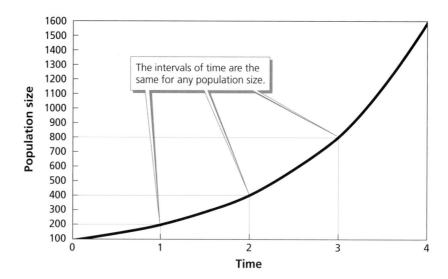

FIGURE 13.6 Doubling time is the amount of time required for a population to double in size. It depends only on the growth rate, not on the population size.

equilibrium :: a state when the population's size does not change

If birth and immigration exactly equal death and emigration, the growth rate is zero. When this occurs, the population is in **equilibrium** and its size does not change. As we will see later in the chapter, countries show a variety of growth rates. Many developing countries have large positive growth rates, while many European countries have negative growth rates.

Doubling Time

Exponential growth has an interesting and useful property: the amount of time required for a population to double in size is always the same (**FIGURE 13.6**). This **doubling time** depends only on the growth rate, not on the population size. To calculate doubling time, divide 0.69 by the growth rate (0.69 is the natural log of 2 and is a numerical constant used in calculations of exponential growth):

doubling time :: the amount of time required for a population to double in size

doubling time = 0.69/growth rate

Doubling time often gives a clearer and more intuitive interpretation of population growth than does the growth rate. For example, Table 13.1 shows that in 2010, the growth rate for the United States was 8.3 per 1000 individuals, or 0.0083—neither number is easily visualized. The doubling time makes more sense. At this rate of growth, the U.S. population would double in 0.69/0.0083, or 83 years. For comparison, Kuwait's population doubles in only 19 years. Doubling times can be meaningful for social planning purposes, such as deciding how many schools an area needs and how quickly they must be built.

 # Why Don't Populations Grow Forever?

Exponential growth as just described continues forever, with the population growing continuously larger and ever faster. If any natural population grew like this, we would surely notice. Before long, we'd be swamped with that species. The bacteria from our case study would blanket the entire planet in less than two hours. Clearly, exponential growth is at best temporary. But what makes it stop?

The Effects of Population Density

As a population grows larger, it begins to have a negative impact on its environment and on its members. The amount of food and space available to each individual decreases, and waste products and pollution increase. Diseases and parasites spread more easily. Close contact among individuals may lead to aggressive conflicts. Factors like these are called **density dependent**, because they grow worse with density, or crowding. Density-dependent factors make it harder for individuals to reproduce and less likely that individuals will survive as population size increases (**FIGURE 13.7**).

Is our population affected by density-dependent factors? The late 18th-century British scholar Thomas Malthus was among the first to argue that it is. He observed that the human population grows exponentially, but the human food supply grows only linearly: each newly cultivated acre produces only 1 acre of increased yield. Malthus predicted that human population growth would soon outstrip its food supply, and dramatically so. He feared that starvation and social collapse would surely follow.

Malthus's warnings have come true in many areas hit by famine, drought, or other causes of crop failure. Globally, food production has kept up with population growth because we have developed new varieties of high-yield crops and productive, technology-dependent agriculture. While some scientists believe that technological advances will allow the human population to grow without limit, most recognize that exponential growth is not possible if resources are finite. See *Scientist Spotlight: Donella Meadows* to learn about one of the first and most influential wake-up calls regarding sustainable living.

density dependent :: a factor responsive to population crowding

FIGURE 13.7 Density-dependent factors, such as limited food or space, grow worse as populations become larger and make it harder for individuals to survive and reproduce. Shown here, Pakistani women struggle as they try to order food outside of a subsidized food store on the outskirts of Rawalpindi, Pakistan.

Scientist Spotlight

Donella Meadows (1941–2001)

Donella Meadows was a true environmental visionary in the tradition of John Muir, Aldo Leopold, and Rachel Carson. Meadows shared their sensitivity to the environment, and she lived her personal life in a way that reflected her environmental ethic. She brought a particular skill to her professional work: a strong scientific background that allowed her to examine environmental problems analytically and quantitatively. Her work presented a formidable challenge to conventional wisdom about the environment and the economy, and the issue of population growth took center stage in her analysis.

Meadows grew up on the outskirts of Chicago. After earning her doctorate from Harvard, she became a research fellow at MIT, working with Jay Forrester. Forrester's research involved what he called "system dynamics," using a computer to simulate change in a complex system. They were able to ask how a whole host of factors like resource use, medical advances, and pollution affect population growth—and how population growth in turn affects them.

In 1972 Meadows, Forrester, and others published a seminal book, *The Limits to Growth*, in which they explain exponential and logistic growth. They simulated the growth of almost every important aspect of the world economy using system dynamics and reached a startling conclusion:

> If the present growth trends in world population, industrialization, pollution, food production, and resource depletion continue unchanged, the limits to growth on this planet will be reached sometime within the next hundred years. The most probable result will be a rather sudden and uncontrollable decline in both population and industrial capacity.

It is difficult to imagine the stir this caused. The very thought of limits collided with an economic system based entirely on growth. Meadows's work posed a severe threat to conventional thinking and legitimized the budding environmental movement.

Limits to Growth argued for sustainable living, rather than one based on growth. Meadows embraced this idea in her personal life. She taught environmental systems and ethics at Dartmouth for 29 years. In 1997, she founded the Sustainability Institute, which she referred to as a "think-do tank," rather than a think tank. For 16 years, Meadows wrote a syndicated column titled "The Global Citizen," in which she discussed many issues of environmental and social concern. She lived simply on an organic farm in New Hampshire. Meadows died in 2001 at the early age of 59, after a two-week battle with bacterial meningitis.

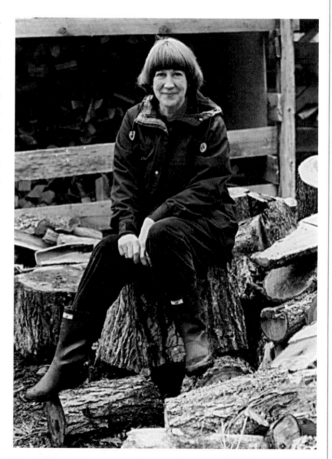

Donella Meadows was a true environmental visionary. She brought to the world's attention the fact that we would not survive unless we started living in a sustainable manner.

Logistic Growth

The reason why exponential growth assumes a population grows forever is that density-dependent factors are not taken into account. Leaving out density dependence makes sense in the early parts of a population's growth: when a population is small, there is plenty of food and space available, and for a while its growth *is* exponential.

But as density-dependent factors come into play, the population's growth rate slows down. Recall that growth rate is the difference between birth rate and death rate (we're ignoring migration here). Before density dependence comes into play, birth rate is high and death rate is low. But density-dependent factors make reproduction more difficult and dying more likely. As the population grows, birth rate declines and death rate increases.

One result of these changing rates is that the growth rate decreases. The population starts off growing exponentially, but eventually growth slows as density-dependent factors reduce the growth rate. A second result is that as birth rates go down and death rates go up, there must be a point at which they are equal. When the growth rate is zero, the population is in equilibrium and its size does not change. This unchanging population size has a name—the **carrying capacity**. You can think of carrying capacity as the maximum number of individuals the environment can "carry," or support.

When we take density dependence into account, we get a new pattern of population growth called **logistic growth**. Logistic growth starts off fast, almost exponentially, when the population is small. But growth slows down as the population increases. And growth stops altogether when the population reaches its carrying capacity. Logistic growth produces an s-shaped pattern of growth (**FIGURE 13.8**). Logistic growth is a more realistic description of how populations grow in nature, since density-dependent factors are a fact of life for all biological populations.

What happens when a population exceeds its carrying capacity? Individuals will starve and die until the population shrinks back to a sustainable level. The birth rate will decline, and the death rate will increase until the growth rate is negative.

carrying capacity :: the maximum number of individuals the environment can support

logistic growth :: a pattern of population growth that takes density dependence into account; starts off exponential and stops when the population reaches carrying capacity

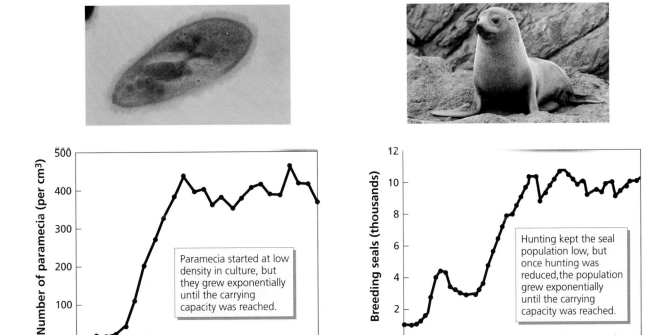

Paramecia started at low density in culture, but they grew exponentially until the carrying capacity was reached.

Hunting kept the seal population low, but once hunting was reduced, the population grew exponentially until the carrying capacity was reached.

FIGURE 13.8 Logistic growth takes density-dependent factors into account: it starts off fast when the population is small, but then slows down, eventually leveling off, as the population gets larger. Logistic growth produces an s-shaped curve, as shown in population studies of paramecium and fur seals.

FIGURE 13.2 (p. 376) provides good evidence that our population is growing exponentially, at least for now—our population doubled between 1960 and 2000. If it continues to double every 40 years, we will reach 12 billion by 2040 and 24 billion by 2080. Most experts agree that our food production will not increase nearly as fast. Neither will our ability to provide clean, fresh water. In fact, according to a 2012 United Nations report, 783 million people do not have dependable access to water suitable for drinking. This represents 11% of the world population. Similarly, our abilities to remove waste and pollution, or to provide health care and education, lag behind our population's needs. If we are unable to provide basic services for the population, the impact on political and economic stability will be dire.

Fortunately, data show that we are approaching a logistic phase in our population growth. Many organizations—including the United Nations, the United States Census Bureau, and even the Central Intelligence Agency—monitor the global population. They have found that global population growth is slowing down. During the last half of the 20th century, global population growth was estimated at around 2% per year. Currently, growth is near 1.3% annually, and it is projected to decline to 0.5% by 2050. If these estimates are correct, our population size will reach 9 billion by 2050 and level off near 10 billion by 2200 (**FIGURE 13.9**).

Our Carrying Capacity

What is the carrying capacity for the human race? After more than 300 years of study, there still is no definitive answer. The carrying capacity depends on things that are harder to predict, from technological breakthroughs to subjective judgments regarding the quality of life. The global carrying capacity will be lower if every individual expects the quality of life we have in the United States, for example.

Despite the uncertainty, many scientists have taken a crack at assessing the carrying capacity. Estimates run as low as 2 or 3 billion, in which case we are already significantly above our carrying capacity. Most estimates, however, are in the range of 8–15 billion.

For now, the projected decline in growth rate is encouraging. Still, a 50% increase in population size by 2050 is significant. And the decline in growth rate is not universal. Population continues to increase in many developing countries, including regions that have difficulty sustaining the population they already have.

FIGURE 13.9 Many organizations that monitor population have found that global population growth is slowing down and we are approaching a logistic phase.

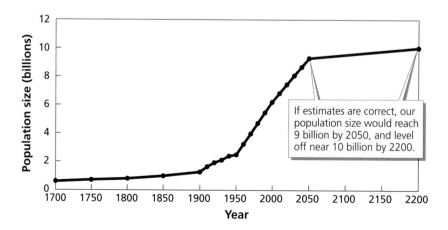

If estimates are correct, our population size would reach 9 billion by 2050, and level off near 10 billion by 2200.

How Is Population Growth Influenced by Age and Sex?

Our story about the bacteria at the start of this chapter had a hidden assumption: that all bacterial cells in the population were the same. In other words, all individuals contributed equally to population growth. The exponential and logistic growth models alike make this assumption about populations. In reality, people do not contribute equally to population growth. Their contribution varies depending on their age and sex, which affect their fertility and their chances of survival.

Age, Sex, and Population Growth

Females have a greater impact on population growth than males. A population can't grow faster than the rate at which females produce children, so adding (or subtracting) females more directly affects how fast a population grows than does adding or subtracting males. In many animal populations, for example, only a few of the males actually mate—sometimes only one. Adding more males won't increase population growth, and removing males won't decrease it.

Not all females contribute equally to population growth, however, mainly due to age. Women go through three life-cycle stages: prepubescent, reproductive, and menopausal. A woman may not contribute at all to the birth rate for more than half her life. Even in their reproductive stage, women vary in their fertility. Typically, females in the middle of their reproductive years have more children than younger and older females.

Age also affects growth rate for another reason: the chances of survival. Diseases often affect people of different ages differently. A disease of the elderly has little effect on population growth. A disease that affects children has little effect on current population growth but may reduce population growth in the future. A disease that affects women at the peak of their reproductive years can have a dramatic effect on growth. In short, the age distribution of the population affects both reproduction and mortality.

Age Pyramids

Demography is the science that analyzes population growth by examining a population's age structure and how age affects fertility, survival, and population growth. An **age pyramid** (**FIGURE 13.10**) is a snapshot of the population at a given point in time. The age pyramid has separate sides for males and females, and each bar on the pyramid shows the number of individuals in a given five-year age class. Thus, an age pyramid shows both the sex ratio and age structure of a population.

Age pyramids fall into three shape categories. Typically, the age pyramid is widest at the bottom, since children usually are the largest age group, and narrows toward the top due to mortality. Other shapes are possible, however, including reasonably vertical sides for the non-elderly age classes, as well as a tapering bottom that is narrower than the top (**FIGURE 13.11**). Properly interpreted, the shape of an age pyramid tells us a great deal about conditions in a country and the social challenges it is likely to face.

From the end of World War II in 1945 until 1964, the United States experienced a growth spurt—the baby boom. In 1955, children from this growth

demography :: the science that analyzes population growth

age pyramid :: a snapshot of a population at a given point in time; displays the number of individuals in age categories

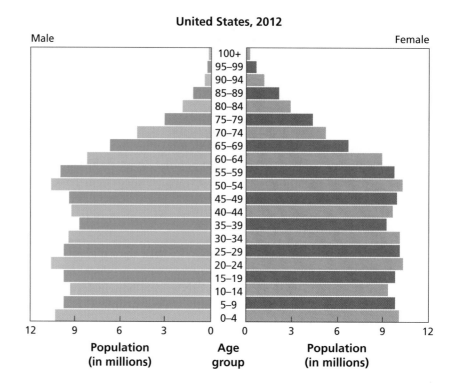

FIGURE 13.10 An age pyramid (such as the one shown for the United States in 2012) is a snapshot of the population at a given point in time. It shows both the sex ratio and age structure of a population.

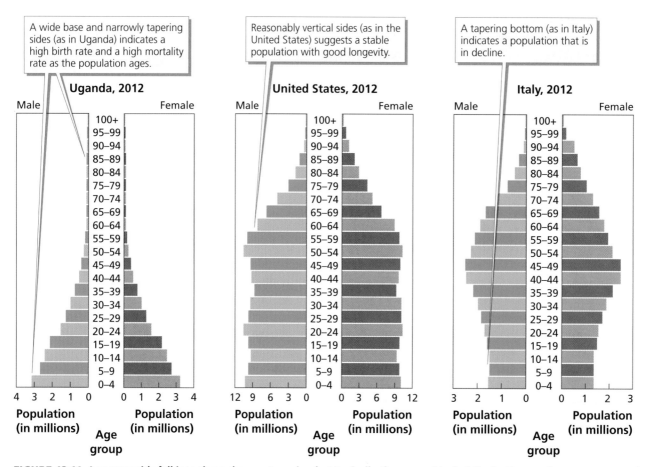

FIGURE 13.11 Age pyramids fall into three shape categories, but typically they are widest at the bottom and narrower toward the top. Properly interpreted, the shape of an age pyramid tells us a great deal about conditions in a country and the social challenges it is likely to face.

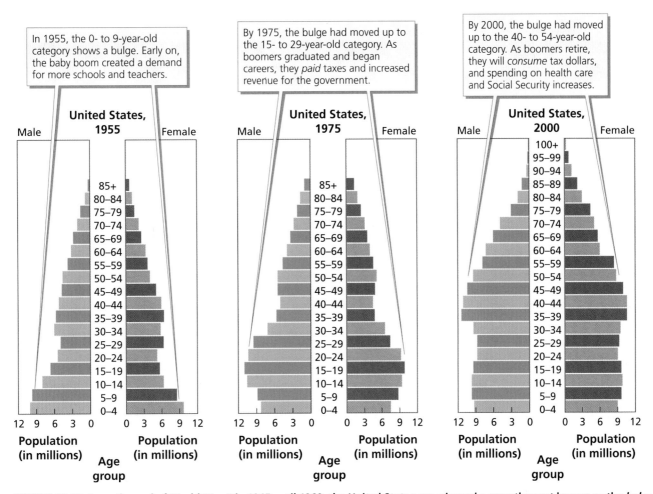

In 1955, the 0- to 9-year-old category shows a bulge. Early on, the baby boom created a demand for more schools and teachers.

By 1975, the bulge had moved up to the 15- to 29-year-old category. As boomers graduated and began careers, they *paid* taxes and increased revenue for the government.

By 2000, the bulge had moved up to the 40- to 54-year-old category. As boomers retire, they will *consume* tax dollars, and spending on health care and Social Security increases.

FIGURE 13.12 **From the end of World War II in 1945 until 1960, the United States experienced a growth spurt known as the *baby boom*.** The movement of the baby boom through the age pyramid has important consequences.

spurt were between 0 and 10 years old. The 1955 age pyramid (**FIGURE 13.12**) shows a bulge in these age classes. You can follow this bulge up the age pyramid as the baby boomers age. In 1975, for example, children born between 1945 and 1960 were between 15 and 30 years of age. In 2000, the boomers were between 40 and 55 years of age, and the bulge has moved up the pyramid accordingly.

The movement of the baby boom through the age pyramid has important consequences. Early on, the baby boom created a demand for more schools and teachers. As the boomers graduated and began careers, they paid taxes and increased revenue for the government. As boomers retire, however, they *consume* tax dollars, as spending on health care and Social Security will likely increase. At the same time, the number of individuals of working age will decline, along with the tax revenue they produce. Of course, policy decisions can compensate for the shifting age structure of the population. Tax rates and spending can both change. Nevertheless, it is easy to see how age pyramids help us decide what changes are needed.

The 2000 age pyramid shows a small population bulge about 25 years behind the baby boomers, the so-called baby boomlet (**FIGURE 13.12**). These are the children of the boomers, and you can predict that other and ever-smaller wavelets of growth may occur at 20- to 30-year intervals.

If you look closely at the numbers of baby boomers (the width of the bars), you will note that they don't change in size much. This is because survivorship

Life Application

The Demographics of China

No country has felt population pressure as strongly as China, which has almost a fifth of the world's total inhabitants. Its population control methods have been successful, but also controversial.

China's age pyramid for 1990 (the earliest data available on the Census Bureau site) shows the characteristic wide base and inward sloping sides of a growing population. By 1979, however, China had already announced its one-child policy: each couple was permitted only a single child, and financial penalties were imposed for violations. Children born between 1976 and 1980 would have been 10–14 in 1990. The "notch" in the age pyramid for 10- to 14-year-olds shows the result of the one-child policy. And as you might imagine, the rule wasn't enforced equally everywhere. For instance, it did not apply to rural areas, where children were needed for farm labor. Even so, one child per couple in principle results in a 50% decrease in population size, as each couple is only half replaced—a drastic step.

By some measures, China's efforts have been successful. Its population is predicted to peak at 1.46 billion around 2030 and decline thereafter. Perhaps as many as 300 million people will not have been born as a result. In 1990, China had 21 births per 1000 individuals, and couples had 2.2 children on average, still a positive growth rate. By 2050, the births may drop by 50%, and the aging population will actually decline. We can picture this prediction in an age pyramid for 2050.

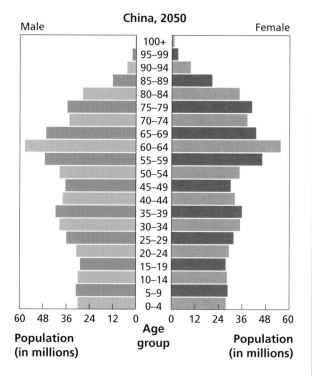

	1990	2050
Births/1000	21	10
Deaths/1000	7	13
Growth rate (%)	1.4	−0.3
Total fertility	2.2	1.7

China, home to nearly 20% of the world's population, has implemented dramatic population control measures. In 1990, China had 21 births per 1000 individuals, but by 2050 the births are expected to drop by 50%.

For all their success, government-imposed restrictions on reproduction have been very unpopular. Critics argue that the one-child policy has led to widespread abortion and even infanticide. It has definitely led to a preference for boys over girls, resulting as of 2009 in 1.13 male boys for every girl. That means 17 million more boys than girls. Among adults there are 27 million more men than women.

All countries must eventually address population growth. Will that happen by choice, by natural causes, or by government regulation? It is an open question.

is high, and a relatively small fraction of the boomers die. The small increase in numbers is due to immigration of people in the same age classes as the baby boomers.

The age pyramids also show that longevity has increased. The 1955 and 1975 age pyramids show only 1–1.5 million over age 85. In 2000, there were over 4 million people in this age class. For younger people, the sex ratio is 50:50. But at 75 years and older, the number of males begins to decline more than the number of females. Men still live shorter lives than women, though both sexes live longer.

Finally, age pyramids are also useful for comparing growth of different countries. We will return to analyze age pyramids for Uganda, the United States, and Italy later in the chapter. *Life Application: The Demography of China* details the patterns of population growth and consequences of government measures instituted to control this growth.

13.4 Why Do Developing and Developed Countries Grow Differently?

Two additional demographic factors affect population growth: total fertility and the age of a woman at the time she has her first child. Developing and developed countries differ strikingly when it comes to these two factors.

Total Fertility and Age at First Reproduction

Total fertility is the number of children a woman *could* have, on average, if she lived at least until the end of her reproductive years. Birth rate is the average number of children a woman *does* have (**FIGURE 13.13**). Total fertility gives a higher estimate of growth than does birth rate, because some women will die before having all the children they could. On the other hand, total fertility gives a more accurate picture of a population's potential to grow if female longevity were not a factor.

As we have seen, population growth rates can be negative, zero, or positive. Total fertility provides similar information. If total fertility is 2, then each couple exactly replaces itself with children. We can call this zero population growth (ZPG). If total fertility is more than 2, each couple more than replaces itself, and population size increases. If total fertility is less than 2 the population size decreases.

total fertility :: the number of children a woman could have, on average, if she lived at least until the end of her reproductive years

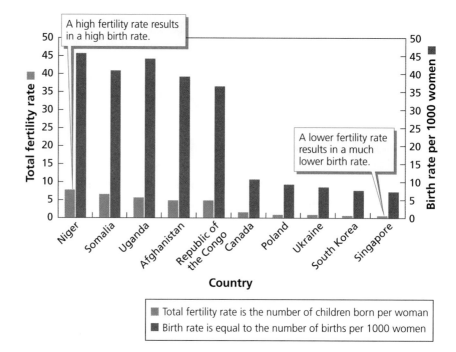

FIGURE 13.13 **Total fertility is the number of children a woman *could* have, on average, if she lived at least until the end of her reproductive years.** Birth rate is the average number of children a woman *does* have and is equal to the number of births per 1000.

A second important demographic factor is age at first reproduction (**FIGURE 13.14**). This refers to when women actually give birth to their first child, not the age of puberty. Women in developed countries tend to bear children later in life. This works to reduce population growth in several ways. First, a woman who postpones having children may not live to have as many. Second, postponing childbirth reduces the number of generations alive at any one time. If people live to see their great-grandchildren, then four generations are alive at once.

Fertility and Mortality Differences

Developing countries have a much higher total fertility rate than developed ones. **TABLE 13.2** shows that Uganda has a total fertility rate of 6.06 children per female, the fourth highest in the world. This compares to 2.06 and 1.41 for the United States and Italy, respectively. If we consider developing countries as a group, their total fertility is notably higher than that of developed countries.

Why is there such a big difference in total fertility? Clearly couples choose their family size based on a large number of economic, religious, and social considerations. Part of the explanation comes from infant mortality (**FIGURE 13.15**).

FIGURE 13.14 **Age at first reproduction refers to when women actually bear their first child.** Starting younger allows a woman to have more babies, as the likelihood of conception decreases as she ages.

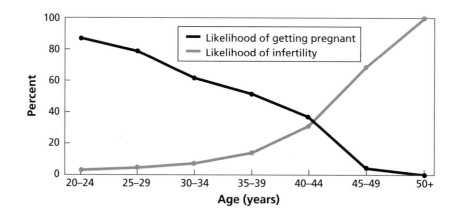

Table 13.2 :: Demographic Data for Uganda, United States, and Italy (2012)

Demographic Factor	Uganda	United States	Italy
Growth rate	3.3% (4th)	0.9% (124th)	0.38% (157th)
Birth rate/1000	45.8 (3rd)	13.7 (147th)	9.06 (208th)
Death rate/1000	11.6 (33rd)	8.4 (84th)	9.93 (54th)
Migration rate/1000	−0.02 (115th)	3.62 (26th)	4.67 (22nd)
Total fertility (births/female)	6.06 (4th)	2.06 (121st)	1.41 (203rd)
Infant mortality/1000	64.2 (22nd)	6.0 (173rd)	3.36 (215th)
Life expectancy (years)	53.45 (205th)	78.49 (51st)	81.86 (10th)

Source: CIA World Factbook, https://www.cia.gov/library/publications/the-world-factbook/.

In developing countries, about 48 of each 1000 infants die before they reach their first birthday, while in developed countries, 6 infants out of 1000 on average suffer this fate. In developing countries, bearing more children is a safety net to ensure that at least some of your offspring survive.

There are many causes of infant mortality in developing countries, including malnutrition, disease, lack of sanitation, and limited medical care. These causes are density dependent, which means they have a greater impact when population density is high. Overall, the population density of developed countries is about 24 individuals per square kilometer, while the density of developing countries is almost three times greater.

Family Planning Differences

The decision about when to start a family also varies greatly from individual to individual and nation to nation. Typically, women in developed countries begin having children later in life than women in developing countries.

FIGURE 13.15 In developing countries, about 48 of each 1000 infants die before they reach their first birthday, while in developed countries, 6 infants out of 1000 on average suffer this fate.

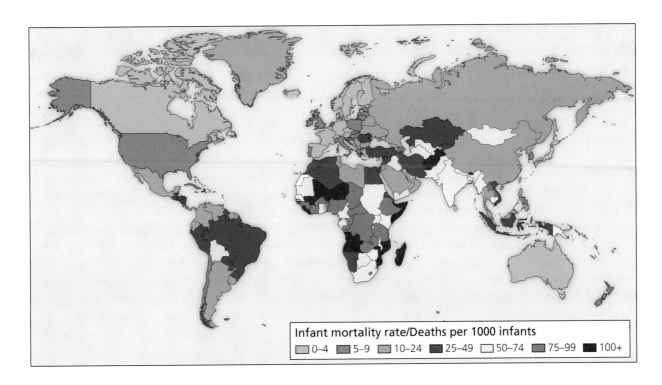

Infant mortality rate/Deaths per 1000 infants
☐ 0–4 ■ 5–9 ■ 10–24 ■ 25–49 ☐ 50–74 ■ 75–99 ■ 100+

Biologically, many girls can become pregnant before they are 13. In some developing countries, cultural practices consider the onset of puberty to be the age of consent for marriage. Childbirth can follow quickly thereafter. In countries where life expectancy for parents is short, early marriage can provide a daughter with security for the future. This practice also results in fewer mouths for the parents to feed.

For example, girls in Uganda become sexually active at 16 on average, and in 2006 girls under 18 accounted for almost one-third of all pregnancies. Early marriage tends to increase total fertility and lower the age of first reproduction, both of which increase population growth. Sadly, growth is mitigated a bit because young girls can die of complications from pregnancy, as 6000 teenagers do annually in Uganda.

In contrast, women in developed countries often have greater access to family planning and may be more likely to postpone childbearing because of education or a career. In the United States, for example, about half of women marry by age 22, but the other half marry substantially later or not at all. In addition, in industrialized developed countries, children are not viewed as an important source of labor (as they are in agrarian developing countries), so there is less urgency to have children early. The result of these and other relevant factors is that total fertility is reduced, the age of first reproduction is raised, and there are fewer years devoted to childbearing, which reduces population growth. To learn more about research on the development of new methods of regulating reproduction, see *Technology Connection: Male Contraception*.

When it comes to population growth, perhaps the single most important difference between developing and developed countries is women's rights. In some countries, economic or religious conditions encourage women to marry young. For example, the average age for marriage is 20 or younger in Chad, India, Uganda, Bangladesh, and Niger. Some of these young women are denied an

Technology Connection

Male Contraception

Who would have guessed that scientists studying deafness in Iranians would discover a clue that could lead to an effective male contraceptive? But that is exactly what happened. Previously, males had only three options for contraception: condoms, vasectomies, and abstinence. Contraception has often been regarded as a woman's responsibility, but several lines of research into new forms of male contraception are beginning to show promise. Given the importance of contraception to limiting population growth, these discoveries are timely.

Testosterone is a hormone most people associate with males, although females produce it as well. Not surprisingly, testosterone plays a big role in sperm production, and low testosterone is one of the causes of male infertility. What is surprising is that adding *extra* testosterone actually *reduces* sperm production—at least in experimental animals like rams, mice, and monkeys.

Why this happens is not completely clear. It's possible that the brain uses testosterone level as its "fuel gauge" for several hormones that are involved with male fertility. If the testosterone level is too high, the brain may shut off production of *all* these hormones. Testosterone level remains high (because testosterone is being added), but when the other hormones shut down, sperm production drops. In one experiment with rams, external testosterone reduced the weight of the testis to 24% of normal. It also reduced sperm production to only 3% of normal. After scientists stopped adding extra testosterone, the rams returned to normal testis size and sperm production.

In a different experiment, scientists were performing a genetic analysis of Iranian families that had a high incidence of deafness. They performed a pedigree analysis (see Chapter 4) to determine how deafness was inherited. They also found that in two families, male infertility seemed to be inherited. The scientists analyzed DNA from these subjects and discovered that they had a mutation in a gene called *CATSPER1*. The *CATSPER1* gene makes a protein that is found on the flagellar "tail" of sperm cells. When a sperm cell encounters an egg, its tail begins to wiggle violently, a process known as hyperactivation. Hyperactivation is required for the sperm to get through the egg's outer membranes for fertilization. The *CATSPER1* mutation prevented the sperm cells from hyperactivating, so they could not fertilize the egg.

The *CATSPER1* mutation also suggests a method of contraception. If it were possible to knock out *CATSPER1* in males, they would not be able to father children. One way to do this is to inject antibodies that attack the protein, much in the same way that your immune system makes antibodies to attack the proteins in disease organisms. This approach is known as immunocontraception. The process would be reversible, just by stopping the injections. It would also be very specific. Unlike testosterone, which acts in many different ways throughout the body, the protein produced by *CATSPER1* is found only in sperm cells, so these would be the only cells affected.

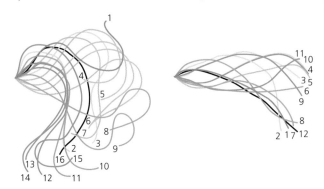

A Normal sperm   **B** Mutant sperm

Normal sperm exhibit hyperactivation swimming behavior (A), whereas CATSPER mutant sperm do not (B). The numbers refer to the positions of the sperm tail as it moved and then was captured in freeze-frame, at one beat per cycle and 16 frames in total.

education and the right to vote, are kept subservient to their husbands with little access to contraception, and have limited financial and social independence.

Education of women has a particularly powerful influence on reproduction. In Kenya, for example, adding one more year of schooling decreases by 10% the probability of giving birth while a teenager. Advancing women's rights, especially the right to an education, is a critical ethical goal, but it is also one of the most effective ways to control population growth.

Demographic Transition

Over time, developing countries become developed countries. Industrialization occurs, urbanization increases, and the number of people working on farms decreases. Education becomes more important as children become a less important part of the labor force. Demographic factors also change, a process known as **demographic transition**.

Demographic transition divides a country's history into several stages: pre-development, transitional, and developed (**FIGURE 13.16**). During the pre-development stage, the country experiences very high birth and death rates. These populations live close to their carrying capacity, and overall population growth is limited.

The transition begins when the country starts to develop due to improved agriculture and increased interaction with developed countries. This interaction can lead to better medical practices, more sanitation, a cleaner water supply, more technology, and better education, which increases the carrying capacity. The death rate drops rapidly, as more children survive and life expectancy increases. Since birth rates at first remain constant, population growth increases in an exponential fashion.

demographic transition :: the change in population growth patterns seen over time as developing countries become developed countries

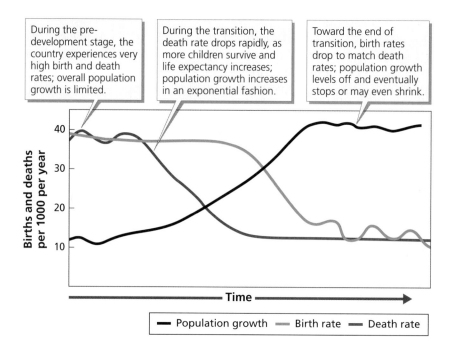

During the pre-development stage, the country experiences very high birth and death rates; overall population growth is limited.

During the transition, the death rate drops rapidly, as more children survive and life expectancy increases; population growth increases in an exponential fashion.

Toward the end of transition, birth rates drop to match death rates; population growth levels off and eventually stops or may even shrink.

Births and deaths per 1000 per year

Time

— Population growth — Birth rate — Death rate

FIGURE 13.16 Demographic transition refers to the changes in population growth patterns seen over time as developing countries become developed countries.

Toward the end of the transition, birth rates drop to match death rates. Urbanization increases, the status of women improves, access to contraception reduces the number of births, and women tend to postpone childbirth well past puberty. In time, population growth levels off and eventually stops. In some cases, the birth rate may drop below the death rate, causing the population to shrink.

Most of the world's developing countries are in the middle of their demographic transition. Death rates have dropped, but reproductive practices have not yet resulted in lowered birth rates. As a result, these countries are growing exponentially. They will account for the majority of population growth over the next century. We all stand to benefit if population growth slows, so economic, medical, and technological aid to developing countries is a worthwhile investment.

Demographic transition provides a uniquely human explanation for logistic growth. Ecologists studying animal and plant populations find that biological factors alone affect population growth. These species have been selected to maximize their lifetime reproduction, and their population growth is limited exclusively by starvation, disease, predation, bad weather, and similar factors. Over our evolutionary history, we have been selected in much the same way. But now we have intelligence, technology, and the ability to make choices about our reproduction. We should, therefore, be able to find better solutions to the challenges of population growth.

13.5 How Do We Use Information About Population Growth?

Every government keeps track of population data, and many nongovernment organizations do so as well. Insurance companies use population data to make predictions about life spans, and market researchers keep track of demographics when figuring out what to sell and to whom. The CIA keeps tabs on population data from every country, because it sees population pressure as a potential security threat.

Why is demographic information so important? It is an integral part of our representative government. Anticipating demographic shifts is vital to planning for almost all segments of society. And the demographic analysis of population growth also helps us analyze our use of critical resources.

The Constitution and the Census

The drafters of the U.S. Constitution recognized an important fact: a representative government requires detailed knowledge of how many people are being represented. A state's number of representatives in Congress and votes in the Electoral College depend on its population. Article 1, Section 2 of the Constitution, therefore, mandates a census at least every 10 years. The first national census was taken in 1790, and the Census Bureau was established as an official department in 1903.

Today, in addition to counting individuals, the Census Bureau keeps records of age, sex, ethnicity, family size, marriage, divorce, employment status, household income, and more. These data are used to allocate funds appropriated by Congress for public utilities, upkeep of roads and bridges, and school construction. About $300 billion is distributed annually, and accurate information on population size and growth is critical to making the distribution as equitable as possible.

Planning for Population Shifts

As we have seen, the shape of a country's age pyramid provides information about its future. By exploring a country's history, demographers can project what the country's pyramid will look like many years from now.

Recall the sharply narrowing sides of Uganda's age pyramid (Figure 13.11, p. 386), which reveal low rates of survival in all age classes, due in part to political upheaval. Following independence from British colonial rule in 1962, Uganda experienced the purges of a dictator (Idi Amin, 1971–1979) and the ravages of civil war (led by Milton Obote, 1980–1985). Up to half a million people were killed directly, and many more died from the loss of food, medical care, and other social services. Uganda was also hit early and hard by the AIDS pandemic. Ten years after the first case was diagnosed in 1983, the percentage of people infected reached 29% in the capital, Kampala, and 14% elsewhere.

The wide base of Uganda's age pyramid reflects a high birth rate and the early stages of demographic transition. Death rates are declining for several reasons. There is more political stability, and Uganda is often credited as being a model for sharply reducing the incidence of AIDS. The percentage of adults currently infected with HIV was reported to be 7.3% in 2012, a drop from 1993. Because births still exceed deaths, the population of Uganda is in an exponential phase of growth.

FIGURE 13.17 shows a projection of Uganda's age pyramid in 2050. At first glance, the age pyramids from 2012 and 2050 look similar. The sides of the 2050 pyramid narrow slightly less steeply, indicating improved survivorship. If you look at the horizontal axis, however, you'll see that the 2050 age pyramid predicts a much larger population size. During this 40-year period, Uganda is expected to double in size twice, from its current population of about 32 million to almost 138 million in 2050—half of them still children.

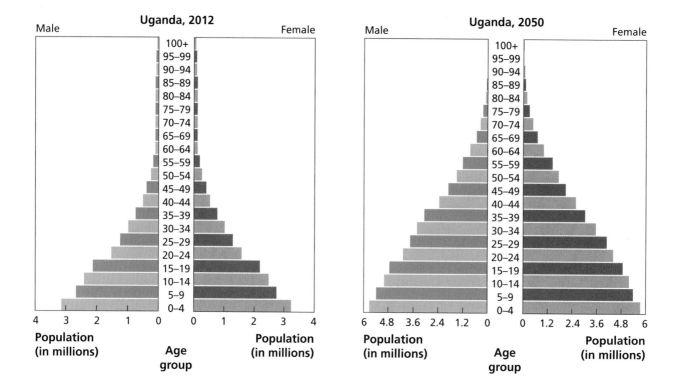

FIGURE 13.17 Uganda's age pyramids for 2012 and 2050 have a similar shape, with a wide base and sharply narrowing sides. However, the 2050 pyramid represents a population that is over four times larger.

Death rates in Uganda are likely to continue to drop. An infusion of agricultural techniques, improved medical care, and increased foreign aid will likely make higher survivorship possible. However, Uganda's projected age pyramid points to the challenges the country will have to address, especially the needs of childhood: improving nutrition and education, controlling childhood diseases, and eventually providing jobs.

While Uganda is a country of children, Italy is a country with an aging population, and its population is projected to decline. Italy's projected 2050 age pyramid (**FIGURE 13.18**) is similar in shape to that of 2012, though smaller and with the bulge of older individuals moved upward 40 years. As these individuals retire, the country's workforce will shrink. This could mean fewer workers to support the aging population. Italy's future needs are likely to be the opposite of Uganda's. Italy likely will be concerned with caring for the elderly: more focus on addressing cancer, heart disease, and palliative end-of-life care, and also on means of increasing worker productivity.

Resource Depletion

One lesson we learned from our story about bacteria is that population growth and resource depletion are two sides of the same coin: exponential growth of bacterial cells could also be viewed as exponential growth in the use of food or test tube space. The principles of growth are not limited to biology. The same ideas can help plan our use of finite resources, such as oil. Besides its obvious utility as a source of heat and combustion for the transportation of vehicles, oil is also used to manufacture a variety of other products that we use, including ink, floor wax, clothes, tires, car dashboards, deodorant, paint, soft contact lenses, and so on.

For the last 30 years, the United States has used about one-quarter of the world's annual oil production. **FIGURE 13.19** shows U.S. oil use in the period between 1860 and 1980. Despite some ups and downs, the data generally follow exponential growth.

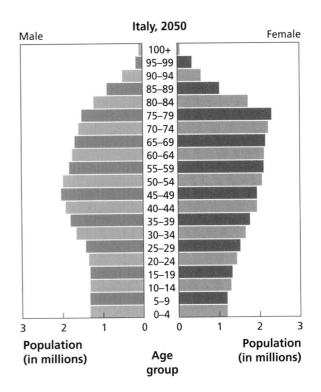

FIGURE 13.18 Italy's age pyramids for 2012 and 2050 show a declining population, with fewer workers and more retirees.

Developing countries currently consume relatively little oil. But as their development proceeds, they will consume more. In China between 1980 and 2009, the amount of oil consumed and the rates of consumption were strikingly similar to those in the United States between 1920 and 1960 (**FIGURE 13.20**). Currently, China uses only about 9% of the world's oil. But China's oil consumption is growing at 6% annually and is projected to double every 12 years, and other Asian countries are following a similar pattern.

There is a great deal of debate about how much oil is left, how long it will last, and when (or if) we will hit "peak oil": the point at which oil production begins to drop. It's a complicated debate, but we should not make the mistake of thinking about oil use as a linear process. It may have taken us 150 years to use about half our oil, but that doesn't mean the second half will last another 150 years. If our oil consumption continues to grow exponentially, once it is half gone, we will consume the second half in only one doubling period—about a decade or two.

Of course, we have another choice: reduce our oil consumption below the level of exponential growth. The cost of producing oil increases as some

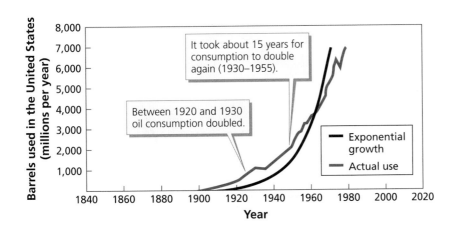

FIGURE 13.19 This graph shows that the use of petroleum in the United States from 1850 to 1980 increased exponentially. This curve has a growth rate of 7.4% and a doubling time of 9.4 years.

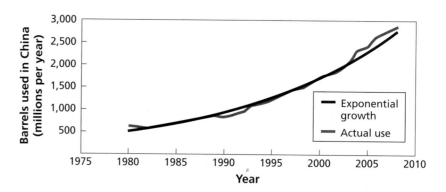

FIGURE 13.20 This graph shows that between 1980 and 2009, both the amount of oil consumed and the rates of oil consumption in China were strikingly similar to those in the United States between 1920 and 1960.

reserves run dry and more difficult reserves are tapped. As the price of oil goes up, consumption may come down. And when we understand that oil is finite, we can take steps to conserve it and to reduce consumption. Oil consumption in the United States declined during the oil crises of the 1970s. And since 1980, oil consumption has fallen way below the exponential increase that occurred previously (**FIGURE 13.21**).

The Limits to Growth

Population size, resource consumption, and pollution: all of these are growing, some linearly, some exponentially. But our planet is finite. There are experts who contend that our technology will overcome any ecological limitation and that growth can continue indefinitely. Most scientists argue, however, that there are limits to growth and the consequences of reaching those limits will be severe. Our future will depend on the choices we make now.

But since we do have choices, there is cause for optimism. Donella Meadows put it this way in her 1972 book *The Limits to Growth*:

> *If there is cause for deep concern, there is also cause for hope. Deliberately limiting growth would be difficult, but not impossible. The way to proceed is clear, and the necessary steps, though new ones for human society, are well within human capabilities. . . . The two missing ingredients are a realistic, long-term goal that can guide mankind to the equilibrium society and the human will to achieve that goal. Without such a goal and a commitment to it, short-term concerns will generate the exponential growth that drives the world system toward the limits of the earth and ultimate collapse. With that goal and commitment, mankind would be ready now to begin a controlled, orderly transition from growth to global equilibrium. (p. 183)*

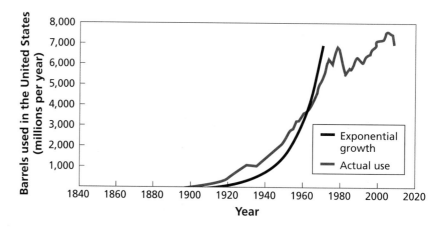

FIGURE 13.21 Since 1980, oil consumption in the United States has fallen below the exponential increase that occurred from 1850 to 1980.

Biology in Perspective

How many people can the Earth support? Our best current prediction is that our population will reach 10 billion people by 2200, so we can hope that it will hold at least that many. From the 1700s until the present, the human population has grown exponentially. We may be at a transition point where our population growth slows and becomes logistic, leading to the predicted leveling off around 2200.

To make the tough choices ahead, we need to understand population growth and how it differs in developing and developed countries. A country's age structure helps demographers make predictions about population trends that can guide future planning and resource allocation.

Demographic changes act as a brake on population growth in developing countries. Industrialization results in a decrease in birth and death rates. However, developed countries also deplete essential resources. If we don't take steps to limit our growth, density-dependent factors like disease pandemics, resource depletion, and insufficient food eventually may reduce our growth rate. If we acknowledge the problems of overpopulation, there is a chance that we can make choices that limit our growth in more acceptable ways.

Chapter Summary

13.1 How Can Populations Grow So Fast?

Distinguish between linear and exponential growth and explain how the growth rate is determined.

- Populations grow when new individuals are added; if this is the result of migration, the growth rate is linear, but if the individuals added reproduce, the growth is compounded, and the population grows exponentially.
- Growth rate addresses two questions: How fast will it grow? How big will it get?
- Population growth rate is calculated by adding births and immigrations and subtracting deaths and emigrations; a population is in equilibrium when the increase equals the decrease.
- Doubling time is the amount of time required for a population to double in size.

13.2 Why Don't Populations Grow Forever?

Describe the effects of population density and carrying capacity on the principle of logistic growth.

- Density-dependent factors—insufficient resources, waste production, disease, and aggression—eventually inhibit exponential growth.

- The density-dependent factors result in logistic growth for most populations; growth starts out fast, slows as the population becomes larger, and stops as it reaches carrying capacity (*s*-curve on a graph).

13.3 How Is Population Growth Influenced by Age and Sex?

Explain what age pyramids can tell us about a country's conditions and social challenges.

- Demography is the study of population growth based on its age structure.
- Females have a greater impact on population growth than males, and it varies based on their age.
- An age pyramid shows the demographics of a population, both the sex ratio and age structure.
- There are three possible shapes to age pyramids that describe conditions in a country and allow comparisons between countries:
  o Widest at the bottom shows that children are the largest group, and the number of individuals declines due to mortality associated with aging.

o A tapering bottom, narrower than the top, indicates an aging population with a low birth rate.

o Comparatively vertical sides suggest a stable growth rate and good longevity.

13.4   Why Do Developing and Developed Countries Grow Differently?

Compare and contrast the factors that contribute to growth in developing and developed countries.

- Total fertility is the number of children a woman could have, on average; birth rate is the average number she does have.

- A total fertility of two would just replace the parents and result in zero population growth (ZPG).

- Age at first reproduction is when women first reproduce, and it tends to be later in life in more developed countries.

- Developing countries tend to have both higher total fertility rates and lower ages of first reproduction, resulting in faster population growth.

- Countries undergo demographic transition as they develop.

13.5   How Do We Use Information About Population Growth?

Demonstrate how data on population growth can be used to plan for population shifts and resource depletion.

- Demographic information is necessary for anticipating shifts and planning government and resource allocation accordingly.

- The U.S. Census Bureau provides detailed information about the demographics and number of people living in the United States.

- The age pyramid at any given time allows predictions to be made about demographic shifts in the future.

  o Uganda has a lot of children, so its population will increase and the country will need to provide care for children and, eventually, jobs.

  o Italy's population is much older, so they will need to plan to support a retiring workforce with fewer younger workers.

- As countries develop, they increase their use of limited resources, so it's important to plan accordingly by developing new resources or minimizing consumption.

Key Terms

age pyramid 385
carrying capacity 383
compounding 376
demographic transition 393
demography 385

density dependent 381
doubling time 380
equilibrium 380
exponential growth 376
growth rate 376

linear growth 375
logistic growth 383
total fertility 389

Review Questions

1. Draw a graph with "population size" on the vertical axis and "time" on the horizontal axis. Sketch three curves that illustrate linear, exponential, and logistic growth.

2. What is compounding?

   a. A principle that underlines linear growth

   b. A principle that underlies exponential growth and that occurs when new entities are themselves capable of making more new entities

   c. A principle that underlies demographic transition

   d. A principle that underlies population density

   e. A principle that underlies logistic growth

3. Give an example of a process that exhibits compounding.

4. What are the four factors that determine a population's growth rate?

   a. Birth rate, death rate, immigration rate, and emigration rate

   b. Birth rate, death rate, compounding, and immigration rate

   c. Birth rate, death rate, doubling time, and compounding

   d. Birth rate, death rate, doubling time, and carrying capacity

   e. Birth rate, death rate, carrying capacity and emigration rate

5. If a population of 5000 is growing at a rate of 8% (0.08) per year, how long will it take for the population to double in size? Is it necessary to know how big the population is to determine the answer? Briefly explain.

6. What are density-dependent factors? Give examples of potential density-dependent factors for a top predator, like a wolf, and a hibernating creature, like a bear.

7. What is carrying capacity?

   a. Logistic growth that starts off fast but slows down as the population gets larger

b. Exponential growth

c. The maximum number of individuals an environment can support

d. Growth rate

e. The minimum number of individuals an environment can support

8. Why does population size level off as it approaches carrying capacity?

9. What information is shown in an age pyramid?

a. Exponential growth and logistic growth

b. Carrying capacity and sex ratio

c. Logistic growth and age structure

d. Demography and sex ratio of a population

e. Sex ratio and age structure of a population

10. Briefly describe the overall shape of the age pyramid for a country whose population is: (a) increasing, (b) relatively stable, and (c) shrinking.

11. How do birth and death rates change during demographic transition?

12. How does a country's population size change in the early stages of demographic transition? How does it change toward the end of demographic transition?

13. List and briefly explain two reasons why a government would want an accurate census of its population.

The Thinking Citizen

1. Birth rates decrease eventually during demographic transition, and women tend to have their children later as the country develops and the transition proceeds. What are some of the reasons for postponing childbearing? What effect will this have on population growth in the country, and why?

2. When a population is at carrying capacity, its *growth rate* is zero but its *birth rate* is not zero. Does the population stay exactly at carrying capacity, or can it vary a bit? Imagine a deer population that is at carrying capacity, and that it is mating season. The deer, of course, don't know anything about carrying capacities, but they do know about mating. What will happen to the deer's population size over the next few years?

3. Normally we think about a country undergoing demographic transition. But is it possible that the entire global population does this, too? Review the changes that happen to birth and death rates during the three phases of demographic transition. Also, review global population growth. Do you think global population growth can be explained by the way birth and death rates change during the demographic transition of a single country?

4. The age pyramids of Uganda and Italy both change *size* between 2012 and 2050, but they do not change in terms of basic *shape*. What types of changes must occur in order for an age pyramid to change shape? What are the potential consequences of this shape change?

The Informed Citizen

1. Ultimately there are only three options for reducing a population's growth rate: decrease the birth rate, increase the death rate, or both. Which of these options do you find most acceptable? What are your suggestions for implementing your option? Do your suggestions work equally well in developing and developed countries, or do you need a different plan for each?

2. War is sometimes thought to have an ecological basis: an effort to reduce population pressure by acquiring more resources like gold, oil, or territory. Do you think war could be a density-dependent factor that limits the size of the human population? How does mortality from wars compare to mortality from disease? Do some online research to find mortality estimates from both causes. For diseases, you may want to research the Spanish flu epidemic of 1918 and the current AIDS pandemic.

3. Go online to the Census Bureau's International Database (http://www.census.gov/population/international/data/idb/informationGateway.php). Select a country of your choice and view its 1950, 2012, and (projected) 2050 age pyramids. For some countries, data from 1950 may not be available, so try for as early a year as possible. Analyze the pyramid's general shape to determine whether the country's population is increasing, relatively stable, or shrinking. What sorts of population-related problems do you predict this country will have in the future? Are there "bulges" or "pits" in the pyramid that indicate specific events in the country's history? Do some background research on the country to find out. You may also want to look at some of the other data and tables available for your country on the Census Bureau's website.

Answers to Selected Questions

Chapter 1
2. e
3. a
5. c
6. b

Chapter 2
1. c
3. a
5. d
10. e
12. b

Chapter 3
1. b
6. e
9. d
11. a

Chapter 4
3. c
4. a
5. d
6. b

Chapter 5
2. b
6. a
9. d
15. e

Chapter 6
4. c
5. a
13. d
15. b

Chapter 7
1. e
3. a
8. d

Chapter 8
1. b
2. a
5. e
8. c

Chapter 9
1. b
5. c
10. a
12. e

Chapter 10
1. c
2. a
9. e
11. b

Chapter 11
3. b
4. a
9. d
15. c

Chapter 12
3. b
7. d
12. c
13. e

Chapter 13
2. b
4. a
7. c
9. e

Glossary

adaptation a change by a population to better align with specific environmental conditions

adult stem cell a cell from an adult that can renew itself, but that is limited in its flexibility

age pyramid a snapshot of a population at a given point in time; displays the number of individuals in age categories

allele different version of the same gene

allele frequency the number of alleles present in a population

amino acid a subunit of protein

amniocentesis a method for examining fetal cells that can be done between 14 and 17 weeks of pregnancy

amnion sac in which the embryo is suspended

anaphase I in meiosis, the stage in the first round of cell division when paired chromosomes separate and move to opposite poles, leaving a haploid set of chromosomes on each side of the cell

anaphase II in meiosis, the stage in the second round of cell division when the two chromatids of each chromosome detach from each other and move to opposite ends of the cell; each half of the cell has a complete haploid set of chromosomes

androgen-insensitivity syndrome (AIS) a disorder of sexual development in which an individual who is XY does not develop as a male because the individual's cells lack receptors to bind to testosterone

anemia a disease in which a person is deficient in hemoglobin

angiogenesis a process by which successful tumors secrete chemical signals that cause blood vessels to sprout from the circulatory system

angiogenesis inhibitor a drug that stops the growth of blood vessels, inhibiting the ability of tumors to attach to the circulatory system and grow

antibiotic drug used to kill a pathogen

antibody a molecule that binds specifically to an antigen and kills pathogens

anticodon a triplet nucleotide sequence at one end of tRNA

APGAR scale an assessment of a newborn baby conducted at one and five minutes after birth to assess a baby's activity, pulse, grimace response, appearance, and respiration

apoptosis cell death

artificial insemination sperm from the male partner are placed in the female partner's vagina

asexual reproduction reproduction that involves a single parent and no sex

assisted reproductive technology (ART) reproductive procedure that is used to bring healthy sperm and eggs together so that fertilization is possible

atom the building block of every physical thing in existence; there is one type of atom for each element

background extinction rate the extinction rate during "typical" times, not during a mass extinction

benign harmless

biodiversity the variety of species living in an ecosystem

biogeochemical cycle process that moves matter among the four reservoirs: the biosphere, atmosphere, hydrosphere, and lithosphere

biogeographic realm one of eight landmasses of the world with its own distinctive floras and faunas

biogeography the study of where different species live throughout the world

bioleaching the action of bacteria that have been genetically engineered to produce enzymes that dissolve metals from their ores, making the metal easier to extract

biological control an approach to controlling pests by introducing predators of a pest into an area

biological species concept it defines species as genetically distinct, independently evolving units

bioremediation the biological cleanup of pollution and contamination

blitzkrieg hypothesis a theory proposed by Paul Martin that humans hunted most of the large mammals on the North American continent to the point of extinction

bulb short, underground stem with fleshy leaves that store nutrients and food

Calvin cycle the second stage of photosynthesis in which the solar energy captured from light reactions is used to make sugar

Cambrian explosion the geological period of time from about 542–490 million years ago when every major group of animal appeared in the fossil record

cancer a disease of uncontrolled cell proliferation

carbon fixation the attachment of carbon dioxide from the atmosphere to another carbon-containing compound

carpel the female part of a flower

carrying capacity the maximum number of individuals the environment can support

cause and effect every event or outcome in nature has a source; if a scientist sets up the correct conditions, the results can be predicted in advance

cell the smallest and most basic unit of life

cell cycle four distinct phases of cell life: mitosis (the chromosomes separate into daughter cells), Gap 1 (the cell grows), DNA synthesis (the chromosomes replicate), and Gap 2 (the cell grows and prepares again for mitosis)

cell membrane outer boundary of the cell body, where the cell encounters its environment

cell theory the realization that all living organisms are made of cells; it was first proposed in 1839 by Matthias Schleiden and Theodor Schwann

cervical cap a barrier contraceptive that keeps sperm out of the uterus

cervix the lower part of the uterus and the opening to the vagina

character displacement when species evolve, a change in physical or behavioral characters that allows them to reduce competition

chemical bonding sharing of electrons in the outer shell of more than one atom

chemoprevention the action of natural or manufactured materials used to prevent or halt cancer development

chemotherapy a treatment method using drugs that impair the ability of cells to replicate; these medications may be taken orally or injected directly into the bloodstream

chlorophyll a green pigment used by plants in photosynthesis

chloroplast an organelle in a photosynthetic cell in which photosynthesis takes place

chorionic villus sampling (CVS) procedure used to isolate fetal cells that can be done between 10 and 12 weeks of pregnancy

chromatid one of two copies of the DNA and proteins that make up a replicated chromosome; replicated chromosomes are composed of two chromatids

chromosome structure that contains most of the genetic information of the cell; housed in the nucleus

classification the assignment of every species to a genus, family, order, class, phylum, and kingdom

cleavage the start of embryogenesis when the cells divide into new cells

cloning the process of isolating and copying a gene

codon a specific mRNA sequence of three nucleotides

coevolution the adaptation of one species to another

colonoscopy a procedure used to examine the large intestine, rectum, and anus to screen men and women for colon or rectal cancer

commensalism an ecological interaction in which one species benefits and the other is neither helped nor harmed

competition occurs when two or more species attempt to utilize the same resource; it has a negative effect on both species

competitive exclusion principle posits that two competitors cannot coexist

compounding a principle that underlies exponential growth; occurs if new entities are themselves capable of creating new entities

computerized tomography (CT) scan an advanced type of X-ray procedure that provides a more detailed look at internal structures by sending thin X-ray beams at different angles

condom a barrier contraceptive that covers the penis and prevents sperm from entering the female reproductive tract

conjoined twin identical twins whose bodies are physically attached at some location

conjugation two bacterial cells physically join together and transfer DNA from one cell to the other

contraceptive birth control

controlled experiment a test or manipulation in which a scientist keeps all variables (possible factors that could affect the outcome of the test) the same except for the one under investigation

cross mating

cyst a fluid-filled sac

cystic fibrosis a disease in which a thick, abnormal mucus accumulates in the lungs and the digestive system

cytokinesis the stage in cell division when the cells are pinched in two

cytoplasm in a cell, the cytosol and organelles together

cytoskeleton maintains the shape of the cell and the positions of the organelles

cytosol a viscous substance in a cell where small molecules are dissolved and organelles are embedded

Darwin's finches the finches on the Galapagos Islands that occupied different habitats and developed different behaviors

decomposer organism that causes dead bodies to rot

demographic transition the change in population growth patterns seen over time as developing countries become developed countries

demography the science that analyzes population growth

density dependent a factor responsive to population crowding

descent with modification a phrase used by Darwin to describe the fact that species changed and that evolution occurs

diagnostic test precise evaluation of risks to an embryo or fetus

diaphragm a barrier contraceptive placed in the vagina as close to the cervix as possible; it bars the way to the uterus

differentiation the change from cells that have unlimited potential to cells that are specialized

diploid cell cell with two complete sets of chromosomes

discrimination when one species recognizes another as not its own

disorder of sexual development (DSD) a condition sometimes noted at birth in which the genitalia are atypical

diversity the variety of life forms in the natural world

diversity index a mathematical formula that measures species diversity by taking into account richness, abundance, and evenness

DNA (deoxyribonucleic acid) a type of nucleic acid located in the cell nucleus; it is the physical material of which genes are made

dominant a trait that is seen even when one allele is present

doubling time the amount of time required for a population to double in size

ecological pyramid a diagram of the changes in trophic level

ecology the study of the interactions among species and their physical environments

ecosystem service naturally occurring ecological process that supports our agriculture, technology, culture, and more; can be divided into four groups: provisioning, regulating, cultural, and supporting

ectopic pregnancy a pregnancy in which an embryo develops in the fallopian tube rather than in the uterus; ectopic pregnancies cannot develop successfully to term, and if left untreated can lead to rupture of the fallopian tubes

electron negatively charged subatomic particle that orbits the nucleus of an atom

electrophoresis a widely used method for separating mixtures of molecules, including proteins, DNA, and RNA

element a substance composed of only one type of atom

embryo the earliest stage of development

embryogenesis the development of cells into embryos and bodies that are able to perform the tasks necessary for life

embryonic stem cell cell that can produce any type of cell in the body

empirical evidence information that one gets from direct observation, from experience, or from the results of experiments and other tests of hypotheses

endemic a species that is specific to a location and found nowhere else in the world

endomembrane system produces important molecules and delivers them to key locations within and outside the cell; consists of the endoplasmic reticulum, the Golgi apparatus, and lysosomes

endometriosis a painful condition in which uterine tissue grows elsewhere in the body, leading to inflammation and disruption of ovulation

endometrium the lining of the uterus produced during the menstrual cycle to support the development of an embryo if a pregnancy occurs

endoplasmic reticulum (ER) a series of membrane folds located near the nucleus; smooth endoplasmic reticulum handles the synthesis of fats and lipids, and rough endoplasmic reticulum produces proteins

endosperm cells that surround a plant embryo and form the bulk of the seed

enzyme protein that makes it easier for chemical reactions to occur

epidemic spread of a disease over a larger area and number of people than expected

epididymis coiled tube sitting atop each testis whose cells secrete chemical signals that help sperm to complete differentiation

equilibrium a state when the population's size does not change

estrogen a hormone secreted by the ovaries that promotes the development of the Müllerian ducts, oviducts, the uterus, and the upper end of the vagina

eugenics the notion that the way to improve human stock is to prevent biologically defective people from having children and encourage worthy people to have many

evolution theory that explains how all living organisms are related and how existing populations adapt to their environments and new species arise

evolutionary tree diagram that takes into account the degrees of similarity among species based on their physical characteristics and genes

exploitation an interaction between species in which one species benefits while the other is harmed

exponential growth growth that proceeds at an increasing rate over a given time frame

fact something that has actual existence

factor a gene that possesses the information for the production of a specific protein

falsifiable able to be proved wrong

fertilization the process of a sperm cell from the father being fused with an egg cell from the mother

fetal alcohol spectrum disorder (FASD) encompasses a range of health problems experienced by children who were exposed to alcohol prenatally

fetal alcohol syndrome (FAS) the most severe FASD; symptoms may include poor growth, abnormal facial features, hyperactivity, poor reasoning skills, vision problems, hearing problems, and intellectual disabilities

fetus the stage of development that follows the embryo

fitness the number of alleles, or forms of a gene, an organism passes on to the next generation

flower part of a plant that is used for sexual reproduction; it produces seeds and fruit

follicle a primary oocyte surrounded by a single layer of cells

food web a diagram showing a multilinked structure of "who eats whom (or what)"

founder's effect a special case of genetic drift in which a few individuals of a population leave or are forced to leave and begin a new population

fruit any structure that develops from a flower's ovary

gamete sex cell

gastrulation a stage in embryonic development in which cells and tissues move to new locations, where they will grow into organs

gene hereditary unit consisting of DNA

gene expression the process during which the information encoded in genes is used

gene therapy the correction of a faulty gene

generality how widely a scientific investigation applies to situations other than the specific ones scientists tested

genetic code the information encoded in DNA; it is the language of nucleotides

genetic counseling an approach used to inform couples about their relative risk of passing on certain specific genes associated with a disease

genetic determinism the idea that our genes determine, direct, or cause everything about us

genetic diversity the degree of genetic difference among individuals in a species

genetic drift random changes in the frequency of an allele

genetic engineering the process of taking genes from one species and inserting them into another species

genetic essentialism the notion that being human means having a human genome

genetically modified organism (GMO) an organism whose genetic material has been altered

genital external sex organ

genome the term for all genes in an organism

germ-line gene therapy the correction of genetic problems in the eggs and sperm so that harmful mutations can never be passed on

glyphosate a broad-spectrum herbicide; it is the active ingredient in Roundup

Golgi apparatus the place where many proteins acquire their final structure

growth factor a signal molecule that enables normal cells to divide

growth inhibitor a signal molecule that prevents cell division

growth rate how fast a population grows

habitat diversity the range of different types of natural homes or environments in which organisms live

haploid cell a cell with just one set of chromosomes

HbA normal hemoglobin that carries oxygen molecules throughout the body

HbS abnormal hemoglobin; although it can carry oxygen, it forms long fibers once it has given up its oxygen, causing red blood cells to sickle

hemoglobin the protein that carries oxygen in red blood cells

hemophilia a disease in which an individual cannot form blood clots in the event of an injury

Herceptin a drug that is used to treat a certain type of breast cancer by binding to the erbB-2 protein and stopping it from triggering cell division

homologue one in a pair of chromosomes

horizontal gene transfer the transference of genes between individuals of the same generation, and often between individuals of different species

hormone chemical signal responsible for physiological or developmental response, including secondary sex characteristics

human development the process by which an individual grows and matures from a single cell embryo inside a mother's womb to a baby that can survive on its own

human embryonic germ cell a cell that comes from a six- to nine-week-old embryo and fetus; these cells ultimately form the cells that produce eggs and sperm

human embryonic stem cell a cell that comes from an inner cell mass and can be isolated from an embryo that is five to seven days old and grown in a lab

human papillomavirus (**HPV**) a sexually transmitted virus that is responsible for most of the world's cancers of the genitals and anus

hypothalamus a part of the brain that secretes the gonadotropin-releasing hormone (GnRH)

hypothesis possible cause or mechanism that could explain observations and facts

immune response the body's ability to fight off infection

immunotherapy treatment that directly targets a cancer or boosts the immune system to make other cancer therapies more effective

***in vitro* fertilization** (**IVF**) immature oocytes are extracted and placed in a petri dish in a solution that permits them to mature, and they are combined with sperm from a male partner and then transferred into the uterus

indicator species a species whose presence indicates a healthy, functioning ecosystem

infertility failure to conceive and become pregnant after a year of regular, unprotected intercourse

inflorescence a plant structure containing many flowers

interphase the stage in cell division when cells prepare materials that will be necessary for meiosis or mitosis to occur

intrauterine device (**IUD**) a contraceptive inserted into a woman's uterus to slow the development of an endometrium

intrauterine insemination sperm are placed directly into the uterus

keystone species a single species that has a disproportionate effect on an ecosystem's biodiversity

kin selection selection for behaviors that help your relatives survive and reproduce

latitudinal gradient the decrease in the number of species as one moves away from the equator into the polar regions

leaf a broad or long, flat structure attached to the stem by a stalk; it is the place in the plant where photosynthesis occurs and food molecules are made

leukemia a type of blood cell cancer in which the tumor suppressor gene is mutated

light reaction the first stage of photosynthesis in which solar energy is captured

linear growth growth that proceeds at the same rate over a given time frame

logistic growth a pattern of population growth that takes density dependence into account; starts off exponential and stops when the population reaches carrying capacity

lymphocyte a type of white blood cell

lysosome digests waste materials and worn-out organelles and recycles the molecules so they can be reused

macromolecule large molecule made up of many atoms

magnetic resonance imagery (**MRI**) **scan** a medical imaging technique that uses a magnetic field and radio waves to visualize the inside of a body

malaria a disease caused by a parasite carried by the Anopheles mosquito; it is often fatal

malignant cell growth that continues in an uncontrolled manner, resulting in cancer

mammogram a form of X-ray used to detect possible breast cancer

mass extinction a period when over 75% of Earth's species disappeared

materialism the idea that effects in the natural world all have natural causes, rather than supernatural ones

meiosis a special type of cell division that produces eggs or sperm

melanoma an invasive and deadly skin cancer

menstruation the shedding of the lining of the uterus that supports fetal development

messenger RNA (**mRNA**) codons that identify precisely which amino acids should be joined together and in what order

metabolism all of the chemical reactions in cells that sustain life

metaphase I in meiosis, the stage in the first round of cell division when paired chromosomes line up in single file down the middle of the cell

metaphase II in meiosis, the stage in the second round of cell division when chromosomes line up in single file in the middle of the cell

metastasis a process by which individual cancer cells leave a tumor, enter the bloodstream, and produce secondary tumors at new sites

miasma harmful, toxic vapor supposedly exhaled by sick people or exuded by garbage or sewers into the air around them

microorganism a living thing that cannot be seen without the use of a microscope, such as bacteria

mimicry the close resemblance of an organism to another or to its surroundings

mitochondrion organelle responsible for extracting energy from food molecules

mitosis a type of cell division during which all of the chromosomes in a cell are copied or replicated and each new cell receives genetic information identical to the parent cell

modern synthesis the integration of genetics and evolutionary biology; one of its main architects was Sir Ronald Aylmer Fisher

molecular medicine a field of medicine that aims to understand the relationship between health and genes, proteins, and other molecules, in order to diagnose and treat disease

molecule formed through the chemical bonding of atoms

monohybrid a type of cross in which scientists pay attention to one trait at a time

mutation a physical change in DNA; five examples of mutations are point, deletion, duplication, inversion, and translocation

mutualism an interaction between species in which both species benefit

mutualistic an ecological interaction in which both species benefit

natural selection the mechanism whereby individuals with certain heritable traits have an increased chance of surviving and producing offspring

necrosis tissue death

neural tube the tissue that starts from a flat sheet of cells and rolls up, ultimately becoming the brain and spinal cord

neurulation formation of the earliest stages of the central nervous system

neutron neutral subatomic particle inside the nucleus of an atom

niche the particular set of resources a species uses to survive; includes biological factors like food sources and nesting and physical factors like temperature

nucleotide a subunit of a DNA molecule

nucleus central sphere of an atom made of protons and neutrons, with electrons orbiting; **Or** the most prominent organelle inside the cell; it houses the chromosomes that contain the genetic information

observation what you can see, hear, smell, taste, or feel physically

oncogene a gene that tells the cell to divide in the absence of normal instructions to do so

oogenesis the production of fertile eggs

orbital the outermost shell of an atom that can accommodate a specific number of electrons

organelle specialized structure in eukaryotic cells

ovary female sex organ where eggs are produced

oviduct a small tube down which an egg travels when it is propelled from the ovary

ovulation the release of an egg from the ovary

pangenesis an idea that persisted for more than 2000 years and that held that each part of the body produced a characteristic seed that traveled to reproductive organs; this idea was debunked by August Weismann in the 1880s

Pap test a method used to detect cervical cancer by collecting and examining cervical cells

pathogen disease-causing organism, such as a bacterium, fungus, or worm, that invades the body and does damage

pathologist a doctor who specializes in the study of disease

pedigree analysis a family history study; it shows a pattern of inheritance

pelvic inflammatory disease (PID) infertility caused by blocked oviducts that may be a consequence of a sexually transmitted disease or infection caused by an IUD

petal showiest part of the flower

pheromone chemical attractant used by females to attract males of their species

phloem part of the stem that carries food made in the leaves through the stems to the roots and other parts of the plant

photosynthesis a process that plants use to capture solar energy and then use it to make sugar

pituitary gland a part of the brain that secretes follicle-stimulating hormone (FSH) and luteinizing hormone (LH)

placenta an organ that connects a developing fetus to the mother's uterine wall and provides for an exchange of nutrients and waste elimination

pluripotent a cell that is capable of producing any of the cells in the body but none of the support structures needed for the development of a baby

polygynous a mating system in which one male mates with several females

prediction educated speculation about what an outcome will be

primary consumer herbivore

primary oocyte immature egg

primary producer photosynthetic organism that is the first link in the food web in terms of capturing solar energy

primary sex determination the step in sexual development that determines whether ovaries or testes form

productivity an indicator of how effectively plants convert sunlight into food; measured by the weight of plant material produced in the ecosystem

progesterone a hormone secreted by the corpus luteum that is used in the development of the endometrium or for preventing the development of a second follicle during one menstrual cycle

prophase I in meiosis, the stage in the first round of cell division when the chromosomes become shorter and thicker and line up with their homologues

prophase II in meiosis, the stage in the second round of cell division when chromosomes, still in their compact form, attach to the newly formed spindle

prostaglandin a chemical signal that causes muscle contractions in the female reproductive tract

prostate gland secretes a milky white fluid that neutralizes the acidity of the vagina

proton positively charged subatomic particle inside the nucleus of an atom

proto-oncogene a gene that encodes for a protein that regulates cell division

protozoan a simple, single-celled organism

pseudogene a gene carrying a mutation that makes its products nonfunctional

pseudohermaphroditism a disorder of sexual development in which an individual who is XY develops testes but is missing an enzyme that lets testosterone send the right signals until puberty; having started life as a girl, such an individual develops into a man

pseudoscience fake science

quackery promoting the use and/or purchase of remedies even when there is no scientific evidence or plausible rationale for their effectiveness

quorum a population of bacteria large enough to overcome the body's defenses

radiation energy transmitted as waves or subatomic particles;

radiation therapy radiation delivered either externally or internally to kill cancer cells

recessive a trait that is not seen when only one allele is present

red blood cell a cell that carries oxygen through the blood

reinforcement natural selection that acts to reduce hybrid formation

replication the process by which the double helix of DNA unzips and new strands form

reproductive cloning creating an embryo with the intention of producing a baby

resistance the ability of an organism to survive exposure to a drug that previously had been able to kill or disable it

resistance gene gene that has mutated in such a way that the proteins it makes take on a different function or perform old functions in a new way

ribosomal RNA (rRNA) forms ribosomes, clamp-like structures that hold mRNA and tRNA in the correct orientation so that the amino acids can be connected together

ribosome studs the surface of the rough endoplasmic reticulum (RER) and threads the protein into the interior of the RER as it is being made

rivet hypothesis the idea that ecosystems are like airplanes with rivets; if too many species (or rivets) are lost, the ecosystem will crash quickly (just like an airplane)

root an organ that anchors a plant to its surface and enables it to absorb water and nutrients from the soil

root hair tiny lateral extension of a root's outer cells that absorbs water and minerals from the soil

scientific method set of procedures scientists use in their investigations; includes four steps: observations and facts, hypotheses and predictions, testing, and evaluation and interpretation of results

scientific name the unique classification of each species from kingdom to species

screening test evaluation to see whether an embryo, fetus, baby, child, or adult is at risk for a particular problem

scrotum a pouch that holds the testes

secondary compound chemical made by plants that is usually poisonous and is not needed by the plant for its metabolism

secondary consumer carnivore

secondary sex determination the step in sexual development that governs the development of sex-related body characteristics such as external genitals, structure of pelvic bone, voice tone, and locations of body fat and hair

seed a plant embryo surrounded by a protective coating

seed coat a protective cover over the seed that keeps it from drying out

semen the final combination of sperm and secretions by the prostate gland, seminal vesicles, and bulbourethral glands

sepal the outer protective cover of a flower bud

sexual reproduction reproduction that occurs when an egg cell from a female fuses with a sperm cell from a male to form a new individual

sickle cell a type of abnormal red blood cell that is shaped like a sickle

sickle cell disease an inherited genetic disorder in which red blood cells sickle easily, leading to potentially serious physical consequences

signal transducer a molecule inside the cell that relays the information once a signal molecule binds to a receptor

somatic cell nuclear transfer (SCNT) therapeutic cloning in which a nucleus is taken from a healthy cell and placed into an egg that has had its own nucleus removed; the transplanted nucleus then directs the development of an embryo from which embryonic stem cells can be removed

somatic gene therapy the insertion of a normally functioning gene into a patient to cure a genetic disorder

speciation a process by which species branch off from existing species to form new species

species abundance the number of individuals of each species present in an ecological community

species evenness the pattern of species abundance

species richness the diversity of species alive in an area

species-area relationship the relationship between the area of an island, or any isolated area, and the number of species it has

spermicide drug that kills sperm

spindle formed from the microtubules of the cytoskeleton; this is a structure that helps to distribute chromosomes to new cells

stability the ability of an ecosystem to resist environmental disturbances and return to its original configuration after them

stamen the male part of a flower

stem connects a plant's roots to its leaves and flowers; it provides the structural support for the plant

stem cell a cell that can renew itself and give rise to other cells

surrogacy an embryo is placed in the uterus of a woman who is not the biological mother in order to establish a healthy pregnancy that results in the birth of a baby

telophase I and cytokinesis in meiosis, the stage in the first round of cell division when the cell pinches and divides in two; each new cell has a complete set of organelles and chromosomes

telophase II in meiosis, the stage in the second round of cell division when the nuclear membrane reforms, enclosing the chromosomes

test a procedure that sets up the conditions the predictions require

testability a procedure for determining the evidence in support of a hypothesis

testis male sex organ where sperm are produced

testosterone a hormone that promotes the development of the Wolffian ducts, the prostate gland, and the penis and regulates the descent of the testes into the scrotum

theory an idea, supported by evidence, that provides a bigger picture than a hypothesis of how some aspect of nature works; it may weave together supporting evidence from several scientific fields

theory of island biogeography the proposition that holds that the number of species found on an island is determined by the rates of immigration and colonization by new species and extinction of established species

therapeutic cloning producing embryos as sources of healthy embryonic stem cells for the medical treatment of individuals

total fertility the number of children a woman could have, on average, if she lived at least until the end of her reproductive years

totipotent a cell that is capable of producing any of the cells in a developing embryo

trait characteristic

transcription the process of copying or transcribing genetic information from DNA to RNA

transcription factor a molecule that regulates which genes are active in a cell

transduction the transfer of bacterial DNA from one bacterium to another by a virus

transfer RNA (tRNA) a carrier molecule; at one end is the anticodon, and at the other is a binding site to which a specific amino acid is attached

transformation the process by which a cell takes in and uses DNA from a foreign source

translation the process by which the information in RNA is converted to a new language: amino acid sequences

transmission the passage of a pathogen from one host to another

trophic level level that organisms occupy on the food web

tubal ligation a surgical procedure in which a woman's oviducts are cut and the ends are sealed so that the eggs cannot get into the uterus

tumor a lump that results from the production of extra cells

tumor suppressor a gene that encodes for a protein that tells a cell not to divide

ultrasound a device that relies on high-frequency sound waves to send back images of internal organs

uterus the womb; the hollow, muscular organ located between the bladder and the rectum of females

vaccination the administration of material to provoke the development of immunity to a specific pathogen

variation traits that differ, often by geographic location

vasectomy a surgical procedure in which the vas deferens is snipped and tied off to prevent sperm from entering semen

vertical gene transfer the transference of genes from parents to offspring

virulence a measure of a pathogen's ability to cause damage or death

Xeroderma pigmentosum (XP) a rare disorder in which individuals are extremely sensitive to ultraviolet (UV) light and more easily develop skin cancers if they are exposed to sunlight or other sources of UV light

X-ray a type of relatively high-energy radiation

xylem part of a stem that transports water and minerals from the roots to the leaves and flowers

zygote a one-celled embryo produced as a result of fertilization

Credits

CHAPTER 1

Chapter Opening Photo PHILIPPE HUGUEN/AFP/Getty Images; **1.1** ollyy / Shutterstock .com; **Technology Connection, p. 6** Tom McHugh / Science Source; Robert J. Erwin / Science Source; **1.2** AP Photo/Harry Cabluck; **1.3** Nastya Pirieva / Shutterstock.com; **1.4** Svetlana Valoueva / Shutterstock.com; cybervelvet / Shutterstock.com; **1.8** Artpose Adam Borkowski / Shutterstock.com; aodaodaodaod / Shutterstock.com; Elena Schweitzer / Shutterstock.com; **1.9** Eric V. Grave / Science Source; **Scientist Spotlight, p. 13** Chip Somodevilla/Getty Images; **1.11** Copyright © 2011, National Science Teachers Association (NSTA). Reprinted with permission from *The Science Teacher*, Vol. 78, No. 9, December 2011.; **Life Application, p. 18** Alfred Pasieka / Science Source; **1.15** Andrew Astbury / Shutterstock.com; **1.20** Elena Schweitzer / Shutterstock.com

CHAPTER 2

Chapter Opening Photo emin kulyev / Shutterstock.com; **2.1** Josef & Peter Schafer, {{PD-1996}}; **2.2** Universal History Archive/Getty Images; **2.3** University of Michigan Museum of Art, Collection of the University of Michigan Health System, Gift of Pfizer Inc., UMHS.26; **Life Application, p. 40** Public Health Image Library, Centers for Disease Control and Prevention; **Scientist Spotlight, p. 41** Science Source; PAUL J. RICHARDS/AFP/Getty Images; **2.7** Stefan Redel / Shutterstock.com; Sergey Panychev / Shutterstock.com; chinahbzyg / Shutterstock .com; Fotokostic / Shutterstock.com; Jan Martin Will / Shutterstock.com; VannPhotography / Shutterstock.com; Vladislav Gurfinkel / Shutterstock.com; **2.8** oksankash / Shutterstock.com; James Steidl / Shutterstock.com; **Technology Connection, p. 50 A and B** Public Health Image Library, Centers for Disease Control and Prevention; **Technology Connection, p. 50** Jarrod Erbe / Shutterstock.com; **2.9** Photosani / Shutterstock.com; Marco Brivio; Mike Tan C.T. / Shutterstock.com; Biophoto Associates / Science Source; Vlododymyr Goinyk / Shutterstock.com; Paul Maguire / Shutterstock.com; **2.10** Michael Richardson; **2.11** Art Lien; **2.12** Crepesoles / Shutterstock.com; iStockphoto/Juanmonino

CHAPTER 3

Chapter Opening Photo Denis Kukareko / Shutterstock.com; **3.1** Image courtesy of the Hensel Family and Figure 8 Films; **3.3** Science Source; **Scientist Spotlight, p. 63** Science Source; **3.5** Lebendkulturen.de / Shutterstock. com; Juan Gaertner / Shutterstock.com; **3.6** Michelangelus / Shutterstock.com; **3.7** CNRI / Science Source; **3.12** David M. Phillips / Science Source; **3.13** Michael Abbey / Science Source; Michael Abbey / Science Source; Michael Abbey / Science Source; Michael Abbey / Science Source; Michael Abbey / Science Source; **3.14** Claude Edelmann / Science Source; **Technology Connection, p. 79** Dmitry Kalinovsky / Shutterstock.com; **Life Application, p. 80** Rick's Photography / Shutterstock.com; **3.28** Blank Archives/Getty Images; **3.29** ASSOCIATED PRESS

CHAPTER 4

Chapter Opening Photo Eric Isselee / Shutterstock.com; **4.1** Henrik Larsson / Shutterstock .com; **4.2** Sebastian Kaulitzki / Shutterstock.com; **4.9** Tom Hollyman / Science Source; **4.12** © Bettmann/CORBIS; **4.13** image courtesy of the Mendel Museum and Abbot Lukáš Evžen Mastinec; **4.18** africa924 / Shutterstock.com; **4.19** Rachelle Burnside / Shutterstock .com; John Copland / Shutterstock.com; Kletr / Shutterstock.com; Migel / Shutterstock.com; **Scientist Spotlight, p. 110, A and B** Science Source; **4.31** Sebastian Kaulitzki / Shutterstock .com; **Life Application, p. 120** iStockphoto/ralcro

CHAPTER 5

Chapter Opening Photo Photo Researchers; **5.1** Science Source; Science Source; Scott Camazine / Science Source; Science Source; **5.4** Sakuoka / Shutterstock.com; CNRI / Science Source; **5.14** Nightman1965 / Shutterstock.com; **5.16** Dmitry Matrosov / Shutterstock.com; Michael Drager / Shutterstock.com; WM_idea / Shutterstock.com; Hemanta Kumar Raval / Shutterstock.com; OZaiachin / Shutterstock.com; Birgit Reitz-Hoffman / Shutterstock.com; 3DDock / Shutterstock.com; wonderisland / Shutterstock.com; **Scientist Spotlight, p. 142** Science Source; **5.20** Monkey Business Images / Shutterstock.com; **5.21 A** Levent Konuck / Shutterstock.com; **5.21 B** Simon Fraser / Science Source; **Technology Connection, p. 147 A** iStockphoto/yumiyum; **B** Nightman1965 / Shutterstock.com; **5.23** Science Sourcel; **5.24** Véronique Burger / Science Source; **Life Application, p. 150** Serg64 / Shutterstock.com

CHAPTER 6

Chapter Opening Photo Sergey Uryadnikov; **6.1** FRANCK FIFE/AFP/GettyImages; **6.3** Kletr / Shutterstock.com; **6.7** Carlo Allegri/Getty Images; **Life Application, p. 162** Bettmann/ Corbis / AP Images; **6.15** DUSAN ZIDAR / Shutterstock.com; Ikeskinen / Shutterstock.com; PHANIE / Photo Researchers, Inc.; Gary Parker / Science Source; Ray Ellis / Science Source; Scott Camazine / Science Source; GARO / PHANIE / Science Source; Phanie / Science Source; **How Do We Know?, p. 171** David M. Phillips / Science Source; **6.16** SPL / Science Source; Scimat / Science Source; SCIMAT / Science Source; Steve Gschmeissner / Science Source; **Technology Connection, p. 176** Tinydevil / Shutterstock.com; **6.19** Will & Deni McIntyre / Science Source; **6.20** Saturn Stills / Science Source; **6.21** ZouZou / Shutterstock.com; **Scientist Spotlight, p. 178** Science Source

CHAPTER 7

Chapter Opening Photo Monika Wisniewska / Shutterstock.com; **7.1** Eyerounds.org, University of Iowa; **7.2** iofoto / Shutterstock.com; Max Sudakov / Shutterstock.com; Drozdowski / Shutterstock.com; Lev Kropotov / Shutterstock.com; CoolR / Shutterstock.com; Sari ONeal / Shutterstock.com; **7.4** Dr. Jeremy Burgess / Science Source; **7.6** Martin Shields / Science Source; **7.7 A** Curioso / Shutterstock.com; **B** Mary Lane / Shutterstock.com; **C** Tamara Kulikova / Shutterstock.com; **7.11** Biehler Mitchell / Shutterstock.com; Alex James Bramwell / Shutterstock.com; Josef F. Steufer / Shutterstock.com; Paul Henjum; **7.15** Scott Camazine / Science Source; **7.16** Victor M. Vicente Selvas; **7.17** International Rice Research Institute; **Scientist Spotlight, p. 201** Vince Bucci/AFP/Getty Images; **Technology Connection, p. 203** Steve Jurvetson; **Life Application, p. 205** *Nicolle Rager Fuller, National Science Foundation* **7.21 A** LIU JIN/ AFP/Getty Images; **B** PO3 Patrick Kelley; **7.22** National Human Genome Research Institute; **7.23 A** Ted Kinsman / Science Source; **B** Blue Rose Man; **C** Nathan Shaner; **7.24** Associated Press

CHAPTER 8

Chapter Opening Photo Paul Cowan / Shutterstock.com; **8.1** M.E. Grenander Department of Special Collections & Archives, University at Albany, State University of New York; **8.2** M.E. Grenander Department of Special Collections & Archives, University at Albany, State University of New York; **8.3** American Philosophical Society; **8.4** University Hospitals of Cleveland, Case Western Reserve University, Dr. Guyuron; **8.5** sam100 / Shutterstock.com; **Scientist Spotlight, p. 223** Acey Harper//Time Life Pictures/Getty Images; **8.6** NYPL/Science Source; **8.8** Getty Images; **8.11** SARAH LEEN/National Geographic Creative; **8.12** Kara Jade Quan-Montgomery / Shutterstock.com; **8.14** Alex Wong/Getty Images; Chip Somodevilla/ Getty Images; **8.16** Simon Fraser / Science Source; Department of Health and Human Services; **Technology Connection, p. 238** Martin Shields / Science Source; **8.17** Jaren Jai Wicklund / Shutterstock.com; Vitalinka / Shutterstock.com; PUNIT PARANJPE/AFP/Getty Images

CHAPTER 9

Chapter Opening Photo Rita Kochmarjova / Shutterstock.com; **9.5** BIOPHOTO ASSOCIATES; **9.7** Ludmila Yilmaz / Shutterstock.com; Kerry L. Werry / Shutterstock.com; **9.9** Murray, Patti / Animals Animals; Pal Teravagimov / Shutterstock.com; worldswildlifewonders / Shutterstock .com; Christian Musat / Shutterstock.com; **9.10** H. Zell; **Scientist Spotlight, p. 257** A. Barrington Brown / Science Source; **9.17** K. Thalia Grant; **9.18** Stubblefield Photography / Shutterstock.com; **9.24** Rubberball/Mike Kemp; **9.27** Three Lions; **9.28** SPL / Science Source; **9.29** Rita Kochmarjova / Shutterstock.com; IrinaK / Shutterstock.com; Eric Isselee / Shutterstock

Index